SOLUTIONS MANUAL FOR ORGANIC CHEMISTRY

T. W. Graham Solomons
Jack E. Fernandez
University of South Florida

JOHN WILEY & SONS, INC.

New York London Sydney Toronto

ISBN 0-471-81218-8

Printed in the United States of America

10 9 8 7 6 5 4 3 2 1

TO THE STUDENT

This manual presents solutions to all the problems in the text including the end-of-chapter problems. In many instances we have given not only the answer, but also an explanation of the reasoning that leads to the solution. Although many problems have more than one solution, we have generally given only one. Thus, you should not necessarily assume that your answer is incorrect if it differs from the one given here.

The heart of organic chemistry lies in problem-solving. This is as true for the practicing organic chemist as for the beginning student. But, a problem in organic chemistry is like a riddle; once you have seen the answer, it is impossible for you to go through the process of solving it. The essential value of a problem lies in the mental exercise of the problem-solving process, and you cannot get this exercise if you already know the answer. The best way to use this manual, therefore, is to check the problems you have worked, or to find explanations for unsolved problems *only* after you have made serious attempts at working them.

Problems involving syntheses are always intriguing to chemists and students alike. These problems are probably unlike any that you have seen in other courses. In them you are asked to put together a series of reactions that will convert one compound into another. To do this, it is not enough to start your reasoning with the starting materials because you must also keep in mind the desired product. In some instances you can work from both ends simultaneously, and this is best done by trying to find an intermediate that will link the starting materials and products in a sequence of reactions. In many cases, however, a synthesis problem is best solved by reasoning backward, by starting your thinking with the product and by trying to discover a series of reactions that will lead back to the starting compounds. Sherlock Holmes in *A Study in Scarlet* said:

Most people if you describe a train of events to them, will tell you what the result would be. They can put these events together in their minds, and argue from them that something will come to pass. There are a few people, however, who, if you told them a result, would be able to evolve from their own inner conciousness what the steps were which led up to that result. This power is what I mean when I talk of reasoning backward, or analytically.

This power is what the organic chemistry student must develop.

1
A STUDY OF THE COMPOUNDS OF CARBON

1.1

Using the elemental symbol to denote the nucleus and inner shell electrons:

(a)  (Na·) + (:Cl:) ⟶ Na⁺ + (:Cl:)⁻

(b)  (Mg:) + 2 (:F:) ⟶ Mg⁺² + 2 (:F:)⁻

(c)  (K·) + (·Br:) ⟶ K⁺ + (:Br:)⁻

1.2

(a) 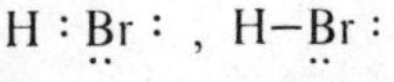 H : Br̈ : , H–Br̈ :

(b) 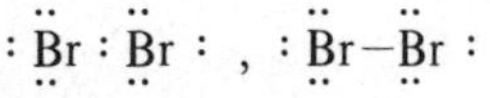 : Br̈ : Br̈ : , : Br̈–Br̈ :

(c) 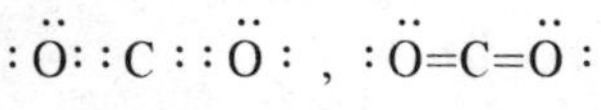 : Ö : : C : : Ö : , : Ö=C=Ö :

(d) 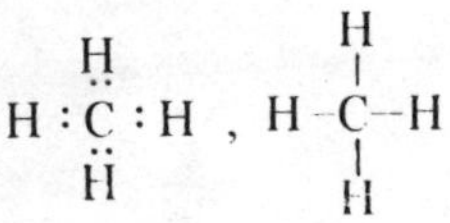
$$\text{H : C : H} \quad , \quad \text{H–C–H}$$
with H above and H below each C

(e) 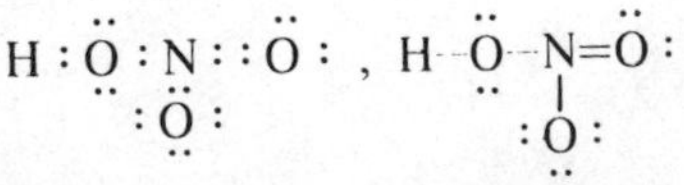 H : Ö : N : : Ö : , H–Ö–N=Ö
  : Ö :   : Ö :

(f) 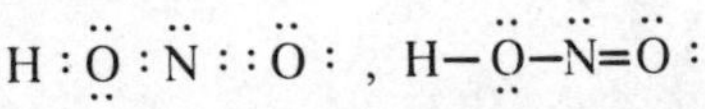 H : Ö : N : : Ö : , H–Ö–N=Ö

(g) H : Ö : Ö : H , H–Ö–Ö–H

(h)
$$\text{H : Si : H} \quad , \quad \text{H–Si–H}$$
with H above and H below the Si

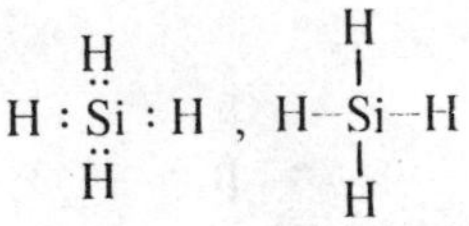

(i) 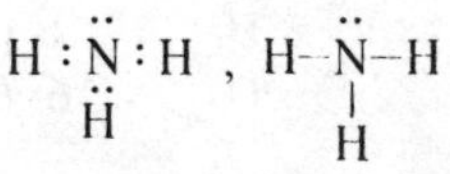
$$\text{H : N̈ : H} \quad , \quad \text{H–N̈–H}$$
with H below each N

(j) : Cl̈ : P : Cl̈ : , : Cl̈–P–Cl̈ :
      : Cl̈ :   : Cl̈ :

(k) : F̈ : N̈ : F̈ : , : F̈–N̈–F̈ :
      : F̈ :   : F̈ :

(l)
$$\text{H : C : Cl̈ :} \quad , \quad \text{H–C–Cl̈ :}$$
with H above and H below each C

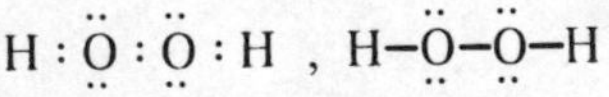

(m) H : Ö : , H–Ö :
      H   H

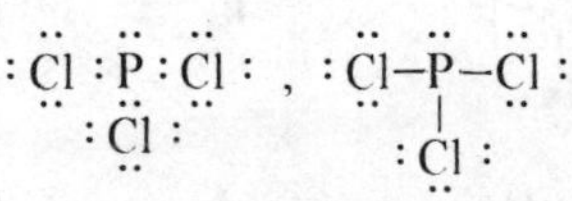

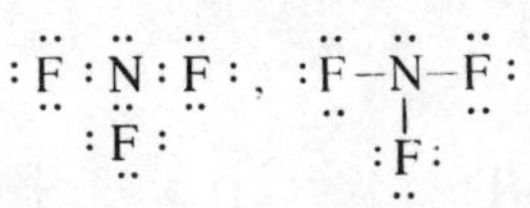

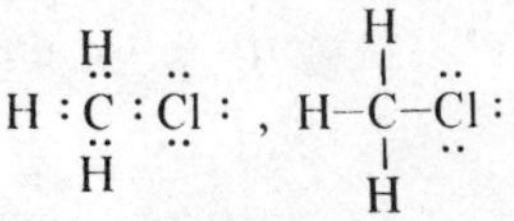

(n) : Ö : H⁻ , : Ö–H⁻

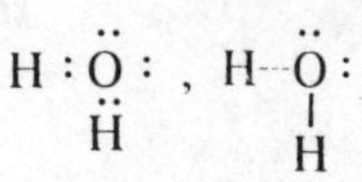

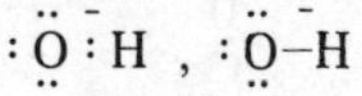

(o) $\left[:\ddot{O}:N::\ddot{O}: \atop \quad\ :\ddot{O}: \right]^{-}$, $\left[:\ddot{O}-N=\ddot{O}: \atop \qquad\ :\ddot{O}: \right]^{-}$ (q) $\left[\begin{matrix} H \\ H:N:H \\ H \end{matrix} \right]^{+}$ $:\ddot{\underset{..}{Cl}}:^{-}$, $\left[\begin{matrix} H \\ | \\ H-N-H \\ | \\ H \end{matrix} \right]^{+}$ $:\ddot{\underset{..}{Cl}}:^{-}$

(p) $\left[:\ddot{O}:N::\ddot{O}: \right]^{-}$, $\left[:\ddot{O}-N=\ddot{O}: \right]^{-}$ (r) Mg^{+2} $\left[\begin{matrix} :\ddot{O} \\ :\ddot{O}:S:\ddot{O}: \\ :\ddot{O} \end{matrix} \right]^{-2}$, Mg^{2+} $\left[\begin{matrix} :\ddot{O} \\ \| \\ :\ddot{O}-S-O: \\ \| \\ :\ddot{O} \end{matrix} \right]^{-2}$

1.3

Formal charge = group number – [½ (number of shared electrons) + (number of unshared electrons)]

Charge on the ion = number of protons + number of electrons
               = sum of all formal charges;

		Formal Charge	Total Charge
(a) $H-\underset{\underset{H}{\|}}{\overset{\overset{H}{\|}}{B}}-H$	H B	$1-(1+0)=\ \ 0$ $3-(4+0)=-1$	-1
(b) $H-\ddot{O}-\underset{:\ddot{O}}{\overset{\|}{S}}-\ddot{O}-H$	H O (OH) O (=O) S	$1-(1+0)=\ \ 0$ $6-(2+4)=\ \ 0$ $6-(2+4)=\ \ 0$ $6-(4+2)=\ \ 0$	0
(c) $:\ddot{F}-\underset{\underset{:\ddot{F}:}{\|}}{\overset{\overset{:\ddot{F}:}{\|}}{B}}-\ddot{F}:$	F B	$7-(1+6)=\ \ 0$ $3-(4+0)=-1$	-1
(d) $H-\underset{\underset{H}{\|}}{\ddot{O}}-H$	H O	$1-(1+0)=\ \ 0$ $6-(3+2)=+1$	$+1$
(e) $:\ddot{F}-\underset{\underset{:\ddot{F}:}{\|}}{N}-\ddot{F}:$	F N	$7-(1+6)=\ \ 0$ $5-(3+2)=\ \ 0$	0
(f) $H-\underset{\underset{H}{\|}}{\ddot{C}}-H$	H C	$1-(1+0)=\ \ 0$ $4-(3+2)=-1$	-1
(g) $H-\underset{\underset{H}{\|}}{C}-H$	H C	$1-(1+0)=\ \ 0$ $4-(3+0)=+1$	$+1$
(h) $H-\underset{\underset{H}{\|}}{\overset{.}{C}}-H$	H C	$1-(1+0)=\ \ 0$ $4-(3+1)=\ \ 0$	0
(i) $H-\ddot{C}-H$	H C	$1-(1+0)=\ \ 0$ $4-(2+2)=\ \ 0$	0

1.4

(a) No formal charges   (d) No formal charges

(b) No formal charges   (e) $CH_3-\overset{\overset{\displaystyle CH_3}{|}}{\underset{\underset{\displaystyle :\ddot{O}:^-}{|}}{\overset{+}{N}}}-CH_3$

(c) No formal charges   (f) $CH_3-\overset{+}{\underset{\underset{:\ddot{O}:^-}{|}}{N}}=\ddot{O}:$

1.5

(a) $\overset{\longrightarrow}{\overset{+}{H}-Br}$ (c) H–H (dipole moment = 0)

(b) $\overset{\longrightarrow}{\overset{+}{I}-Cl}$ (d) Cl–Cl (dipole moment = 0)

1.6

(a) Tetrahedral   (c) Trigonal planar   (e) Tetrahedral

(b) Tetrahedral   (d) Tetrahedral   (f) Linear

1.7

The two C=O bond moments are opposed and cancel each other:

$$\overset{\longleftarrow\ \ \overset{+}{}\ \ \overset{+}{}\longrightarrow}{O = C = O}$$

If the bond angle were other than 180°, then the individual bond moments would not cancel. There would be a resultant dipole moment.

1.8

The direction of polarity of the N–H bond is opposite to that of the N–F bond.

In NH_3, the resultant N–H bond polarities and the polarity of the unshared electron pair are in the same direction.

In NF_3 the resultant N–F bond polarities partially cancel the polarity of the unshared electron pair.

1.9

BF_3 is trigonal planar. Its B–F bonds are all necessarily equal in polarity, and the F–B–F bond angles are all equal (120°).

Therefore, the resultant of the three bond moments is zero.

1.10

Assuming a 100-g sample, the amounts of the elements are:

	Weight	Moles (A)			B		
C	57.45	$\dfrac{57.45}{12.01}$	=	4.78	$\dfrac{4.78}{0.300}$ = 15.9	=	16
H	5.40	$\dfrac{5.40}{1.008}$	=	5.36	$\dfrac{5.36}{0.300}$ = 17.9	=	18
N	8.45	$\dfrac{8.45}{14.01}$	=	0.603	$\dfrac{0.603}{0.300}$ = 2.01	=	2
S	9.61	$\dfrac{9.61}{32.06}$	=	0.300	$\dfrac{0.300}{0.300}$ = 1.00	=	1
O*	19.09	$\dfrac{19.09}{16.00}$	=	1.19	$\dfrac{1.19}{0.300}$ = 3.97	=	4
	100.00						

(* by difference from 100)

The empirical formula is thus $C_{16}H_{18}N_2SO_4$. The empirical formula weight (334.4) is within the range given for the molecular weight (330 ± 10), thus the molecular formula for Penicillin G is the same as the empirical formula.

1.11

H H H   H H H   H   H H
| | | | | | | | |
H–C–C–C–OH   H–C–C—C–H   H–C–O–C–C–H
| | | | | | | | |
H H H   H OH H   H   H H

1.12

The equal length O–O bonds in ozone can be accounted for by the two equivalent resonance structures,

The hybrid structure thus has two equivalent (1½) bonds joining the middle oxygen to the other two. The unshared electron pair on the middle oxygen accounts for the fact that the molecule is bent.

1.13

(c)
$$
\begin{array}{c}
\quad\;\; H\!-\!\overset{H}{\underset{}{C}}\!-\!H \\[2pt]
H - \overset{H}{\underset{H}{C}} - \overset{\overset{\displaystyle}{}}{\underset{\underset{\displaystyle}{}}{C}} - \overset{H}{\underset{H-\overset{H}{\underset{H}{C}}-H}{C}} - \overset{H}{\underset{H}{C}} - H
\end{array}
$$

(f)
$$
H - \overset{H}{\underset{H}{C}} - \overset{H}{\underset{H}{C}} - \overset{H}{\underset{H}{C}} - \overset{H}{\underset{H}{C}} - \overset{..}{\underset{..}{O}}H
$$

(d)
$$
H - \overset{H}{\underset{H}{C}} - \overset{:\overset{..}{C}l:}{\underset{H}{C}} - \overset{:\overset{..}{C}l:}{\underset{H}{C}} - \overset{H}{\underset{H}{C}} - H
$$

(g)
$$
H - \overset{H}{\underset{H}{C}} - \overset{:\overset{..}{O}}{\underset{}{\overset{\|}{C}}} - \overset{H}{\underset{H}{C}} - \overset{H-\overset{H}{\underset{H}{C}}-H}{\underset{H}{C}} - \overset{H}{\underset{H}{C}} - H
$$

(e)
$$
H - \overset{H}{\underset{H}{C}} - \overset{:\overset{..}{O}H}{\underset{H}{C}} - \overset{H}{\underset{H}{C}} - \overset{H}{\underset{H}{C}} - H
$$
OH

(h)
$$
H - \overset{H}{\underset{H}{C}} - \overset{H}{\underset{H}{C}} - \overset{:\overset{..}{O}H}{\underset{H}{C}} - \overset{H-\overset{H}{\underset{H}{C}}-H}{\underset{H}{C}} - \overset{H}{\underset{H}{C}} - H
$$
OH

1.14

(a) and (d) are structural isomers.

(e) and (f) are structural isomers.

1.15

(a) To calculate the percentage composition from the molecular formula, first determine the weight of each element in one mole of the compound. For $C_6H_{12}O_6$,

$$C_6 \;=\; 6 \times 12.01 = 72.06 \qquad \frac{72.06}{180.2} = 0.400 = 40.0\%$$

$$H_{12} \;=\; 12 \times 1.008 = 12.10 \qquad \frac{12.10}{180.2} = 0.0671 = 6.7\%$$

$$O_6 \;=\; 6 \times 16.00 = 96.00 \qquad \frac{96.00}{180.2} = 0.533 = 53.3\%$$

Molecular Wt.   180.16

Then determine the percentage of each element using the formula.

$$\text{Percentage of A} \;=\; \frac{\text{Weight of A}}{\text{Molecular Weight}} \times 100$$

(b) $C_2 = 2 \times 12.01 = 24.02$ $\dfrac{24.02}{75.07} = 0.320 = 32.0\%$

$H_5 = 5 \times 1.008 = 5.04$ $\dfrac{5.04}{75.07} = 0.067 = 6.7\%$

$N = 1 \times 14.01 = 14.01$ $\dfrac{14.01}{75.07} = 0.187 = 18.7\%$

$O_2 = 2 \times 16.00 = 32.00$ $\dfrac{32.00}{75.07} = 0.426 = 42.6\%$

(c) $C_3 = 3 \times 12.01 = 36.03$ $\dfrac{36.03}{280.77} = 0.128 = 12.8\%$

$H_5 = 5 \times 1.008 = 5.04$ $\dfrac{5.04}{280.77} = 0.018 = 1.8\%$

$Br_3 = 3 \times 79.90 = 239.70$ $\dfrac{239.70}{280.77} = 0.854 = 85.4\%$

1.16

(a) 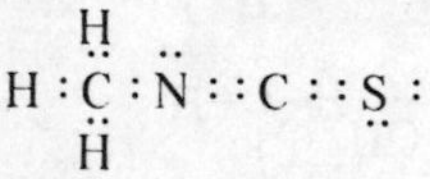 $H : \overset{H}{\underset{H}{C}} : \ddot{N} : : C : : \ddot{\underset{\cdot}{S}} :$

(d) $H : \overset{H}{\underset{H}{C}} : \ddot{N} : : C : : \ddot{O} :$

(b) $H : \overset{\cdot\cdot}{\underset{H}{C}} : C : : : \overset{+}{N} : \ddot{\underset{\cdot\cdot}{O}} :$

(e) $H : C : : C : : \ddot{O} :$

(c) $H : \overset{H}{\underset{H}{C}} : \ddot{O} : \overset{:\ddot{O}}{\underset{+}{N}} : \ddot{\underset{\cdot\cdot}{O}} :$

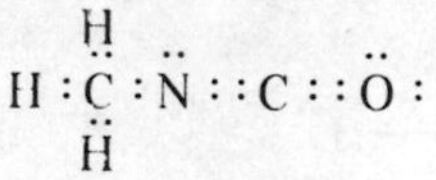

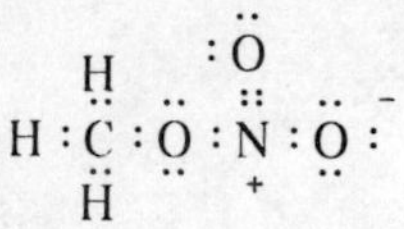

(f) 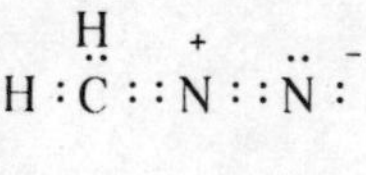 $H : C : : \overset{+}{N} : : \ddot{N} :$

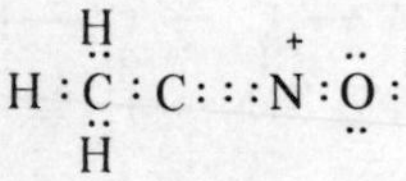

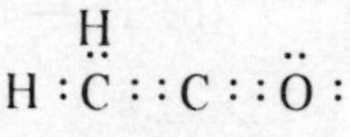

1.17

(a)

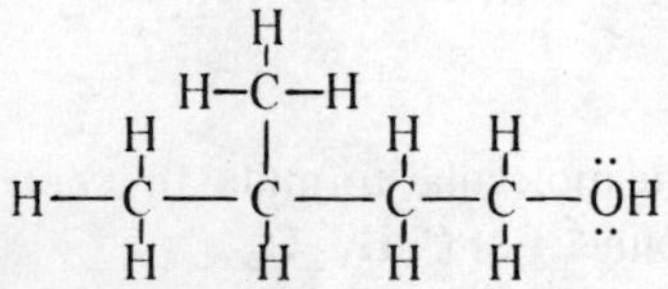

(c)

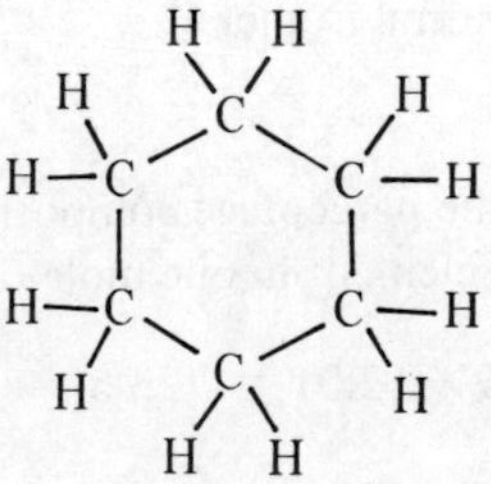

(b)

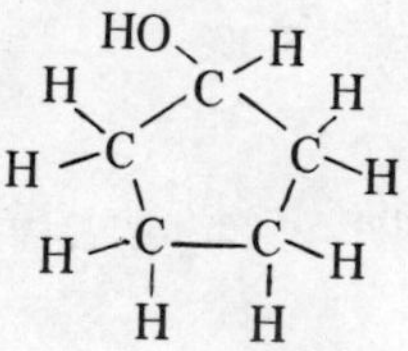

(d)

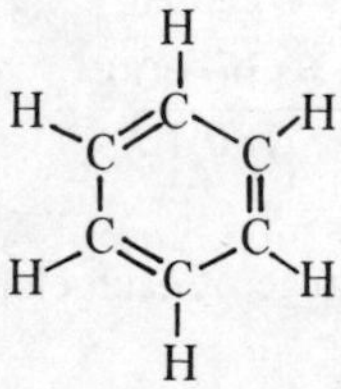

1.18

Electron Configuration   Orbital Arrangement

(a) Be $1s^2 2s^2$

(b) B $1s^2 2s^2 2p_x^1$

(c) C $1s^2 2s^2 2p_x^1 2p_y^1$

(d) N $1s^2 2s^2 2p_x^1 2p_y^1 2p_z^1$

(e) O $1s^2 2s^2 2p_x^2 2p_y^1 2p_z^1$

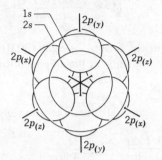

1.19

(a) NH_2 (b) CH   (c) C_2H_4O (d) C_5H_7N (e) CH_2 (f) CH

1.20

Empirical Formula	Empirical Formula Weight	$\left(\dfrac{\text{Molecular Wt.}}{\text{Emp. Form. Wt.}}\right)$	Molecular Formula
(a) CH_2O	30	$\dfrac{179}{30} \cong 6$	$C_6H_{12}O_6$
(b) CHN	27	$\dfrac{80}{27} \cong 3$	$C_3H_3N_3$
(c) CCl_2	83	$\dfrac{410}{83} \cong 5$	C_5Cl_{10}

1.21

First we must determine the empirical formula. Assuming that the difference between the percentages given and 100 percent is due to oxygen,

C: 40.04 $\dfrac{40.04}{12.01} = 3.33$ $\dfrac{3.33}{3.33} = 1$

H: 6.69 $\dfrac{6.69}{1.008} = 6.64$ $\dfrac{6.64}{3.33} \cong 2$

O: $\dfrac{53.27}{100.00}$ $\dfrac{53.27}{16.00} = 3.33$ $\dfrac{3.33}{3.33} = 1$

The empirical formula is thus CH_2O.

     To determine the molecular formula we must first determine the molecular weight. At standard temperature and pressure, the volume of one mole of an ideal gas is 22.4 liters. Assuming ideal behavior,

$$\frac{1.00\ g}{0.746\ liter} = \frac{M}{22.4\ liters} \quad \text{Where M = Molecular weight.}$$

$$M = \frac{(1.00)\ (22.4)}{0.746} = 30.0\ g$$

The empirical formula weight (30.0) equals the molecular weight, thus the molecular formula is the same as the empirical formula.

1.22

As in problem 1.21, the molecular weight is found by the equation

$$\frac{1.251\ g}{1.00\ liter} = \frac{M}{22.4\ liter}$$

$$M = (1.251)\ (22.4)$$

$$M = 28.02$$

To determine the empirical formula, we must determine the amount of carbon in 3.926 g of carbon dioxide, and the amount of hydrogen in 1.608 g of water.

C: $\left(3.926\ g\ CO_2\right)\left(\frac{12.01\ g\ C}{44.01\ g\ CO_2}\right) = 1.071\ g\ carbon$

H: $\left(1.608\ g\ H_2O\right)\left(\frac{2.016\ g\ H}{18.016\ g\ H_2O}\right) = \underline{0.179\ g\ hydrogen}$
$\hspace{8cm} 1.250\ g\ sample$

The weight of C and H in a 1.250 g sample is 1.250 g. Therefore there are no other elements present in ethene.

To determine the empirical formula we proceed as in problem 1.10 except that the sample size is 1.250 instead of 100 g.

C: $\frac{1.071}{12.01} = 0.0892 \quad \frac{0.0892}{0.0892} = 1$

H: $\frac{0.179}{1.008} = 0.178 \quad \frac{0.178}{0.0892} = 2$

The empirical formula is thus CH_2. The empirical formula weight (14) is one-half the molecular weight. Thus the molecular formula is C_2H_4.

1.23

Using the procedure of problem 1.10,

C: $\quad 59.10 \quad \frac{59.10}{12.01} = 4.92 \quad \frac{4.92}{0.817} = 6.02 \cong 6$

H: $\quad 4.92 \quad \frac{4.92}{1.008} = 4.88 \quad \frac{4.88}{0.817} = 5.97 \cong 6$

N: $\quad 22.91 \quad \frac{22.91}{14.01} = 1.64 \quad \frac{1.64}{0.817} = 2$

O: $\quad \frac{13.07}{100.00} \quad \frac{13.07}{16.00} = 0.817 \quad \frac{0.817}{0.817} = 1$

The empirical formula is thus $C_6H_6N_2O$. The empirical formula weight is 123.13 which is equal to the molecular weight within experimental error. The molecular formula is thus the same as the empirical formula.

1.24

C:	40.88	$\dfrac{40.88}{12.01} = 3.40$	$\dfrac{3.40}{0.619} = 5.5$	$5.5 \times 2 = 11$
H:	3.74	$\dfrac{3.74}{1.008} = 3.71$	$\dfrac{3.71}{0.619} = 6$	$6 \times 2 = 12$
Cl:	21.95	$\dfrac{21.95}{35.45} = 0.619$	$\dfrac{0.619}{0.619} = 1$	$1 \times 2 = 2$
N:	8.67	$\dfrac{8.67}{14.01} = 0.619$	$\dfrac{0.619}{0.619} = 1$	$1 \times 2 = 2$
O:	$\dfrac{24.76}{100.00}$	$\dfrac{24.76}{16.00} = 1.55$	$\dfrac{1.55}{0.619} = 2.5$	$2.5 \times 2 = 5$

The empirical formula is thus $C_{11}H_{12}Cl_2N_2O_5$. The empirical formula weight (323) is equal to the molecular weight, therefore the molecular formula is the same as the empirical formula.

1.25
(a) Tetramethyllead, because it is a relatively low-boiling liquid, must be a molecular compound. Its bonds are covalent. Lead fluoride, a high-melting solid, is ionic. A higher temperature (greater energy) is required to separate ions of opposite charges than to separate neutral molecules.

(b) The C—Pb bonds are covalent because lead and carbon have similar electronegativities. Lead and fluorine, on the other hand, have widely differing electronegativities, and Pb—F bonds are therefore ionic.

1.26

$$H : \ddot{C}l : \quad \longleftrightarrow \quad H \bar{:} \ \ddot{C}l : ^+ \quad \longleftrightarrow \quad H^+ \ : \ddot{C}l : ^-$$

(I)   (II)   (III)

(a) III is more reasonable than II because in III the negative charge resides on chlorine, the more electronegative atom.

(b) III would contribute more than II.

(c) The dipole moment, $\overset{+\longrightarrow}{H-Cl}$, can be accounted for by a hybrid of structures I, II, and III, but in which III contributes more than II.

1.27
The three equivalent resonance structures,

when combined result in equivalent N—O bonds that are intermediate in length between single and double bonds.

1.28

```
     ··                  ··                 ··                 ··
    : O              : O               : O              : O :⁻
  ··  ‖  ··  ⁻    ··⁻ ‖  ··           ‖  ··           ··  |  ··
  : O–S–O :      : O–S=O :        : O=S–O :         : O=S=O :
      ‖             |                 |                |
    : O            : O :             : O :⁻           : O :⁻
      ··             ··                ··               ··


     ··               ··  ⁻              ··                ··
   : O :            : O              : O :             O :
  ··  |  ··  ⁻    ··⁻ |  ··         ··  |⁺ ··  ⁻      ·· ‖⁺ ··  ⁻
  : O=S–O :      : O–S=O :        : O=S–O :        : O–S–O :
      ‖               ‖                |                |
    : O            : O :             : O :⁻           : O :⁻
      ··             ··                ··               ··


     ··  ⁻           ··  ⁻             ··  ⁻
   : O :            : O :            : O :
  ··⁻ |⁺ ··      ··⁻ |⁺ ··  ⁻      ··⁻ |⁺² ··  ⁻
  : O–S=O :      : O–S–O :        : O–S–O :
      |               ‖               |
    : O :            : O            : O :
      ··             ⁻                ··
```

Yes. Note that *each* oxygen atom in the first six structures appears three times doubly bonded to S and three times singly bonded to S. In the next four structures *each* oxygen appears doubly bonded once and singly bonded three times. Thus each oxygen has the same average type of bonding, hence the same bond length.

Unlike second period elements, sulfur has five vacant $3d$ orbitals that can accept electrons. Its valence shell can therefore expand, in principle, to as many as 18 valence electrons. Actually, structures with more than two double bonds and two single bonds to sulfur can be considered insignificant.

2

FUNCTIONAL GROUPS AND FAMILIES OF ORGANIC COMPOUNDS: THE MAJOR REACTION TYPES

2.1

(a) 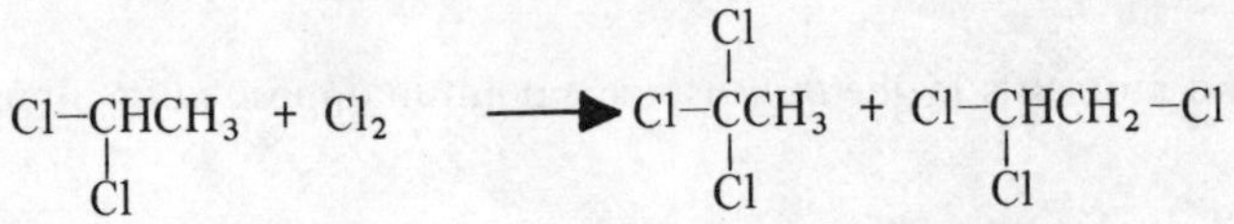 $Cl-CHCH_3 + Cl_2 \longrightarrow Cl-CCH_3 + Cl-CHCH_2-Cl$
            $|$ $|$ $|$
           Cl Cl Cl
(with Cl shown above the center carbon)

1,1-dichloroethane $\longrightarrow$ two trichloroethanes

(b) 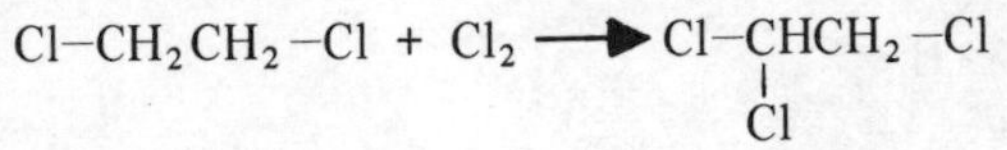 $Cl-CH_2CH_2-Cl + Cl_2 \longrightarrow Cl-CHCH_2-Cl$
                                                          $|$
                                                        Cl

1,2-dichloroethane $\longrightarrow$ one trichloroethane (the same as the second compound above)

(c)   See above   (d) Yes, the trichloroethanes are isomers.

(e) 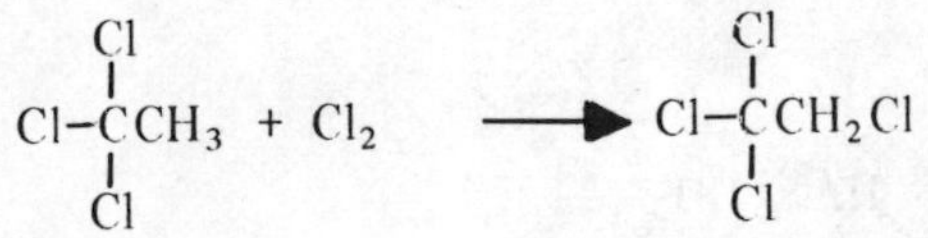 $Cl-CCH_3 + Cl_2 \longrightarrow Cl-CCH_2Cl$
           $|$ $|$
          Cl Cl
(with Cl above the center carbon in both)

1,1,1–trichloroethane $\longrightarrow$ one tetrachloroethane

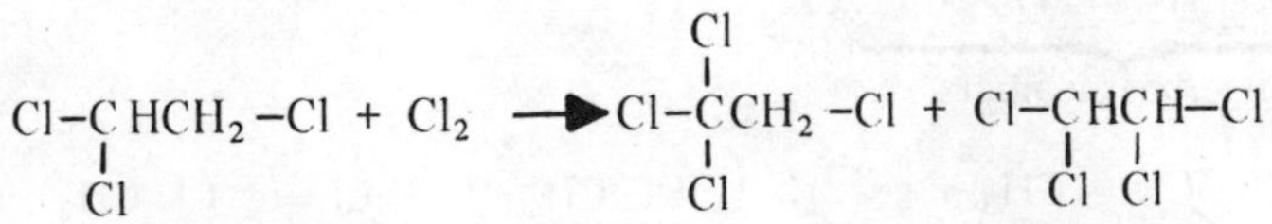

$Cl-CHCH_2-Cl + Cl_2 \longrightarrow Cl-CCH_2-Cl + Cl-CHCH-Cl$
      $|$ $|$ $|$ $|$
     Cl Cl Cl Cl
(with Cl above the center carbon on the left and product)

1,1,2 trichloroethane $\longrightarrow$ two tetrachloroethanes, one of which is the same as the one formed from 1,1,1-trichloroethane.

Thus there are only two tetrachloroethanes

(f)   See (e) above.

(g)   Only one pentachloroethane is possible:

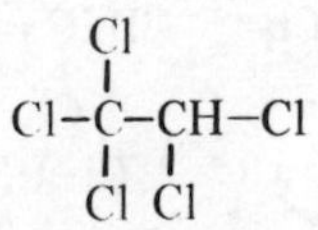

$$Cl-\underset{\underset{Cl}{|}}{\overset{\overset{Cl}{|}}{C}}-\underset{\underset{Cl}{|}}{CH}-Cl$$

2.2

(a) Counting the double bond as two electron pairs located in the region of space between the two carbon atoms, each carbon has three atoms attached to it:

11

$$\underset{H}{\overset{H}{\diagdown}} C = C \underset{H}{\overset{H}{\diagup}}$$

The maximum separation of the electrons bonding the three atoms about each carbon atom occurs when they are equally spaced about that atom; i.e., when the bond angles are $\sim 120°$ and the molecule is planar.

(b) The H—C—H bond angle is less than $120°$ because the repulsive force between the electron pairs of the C—H bonds is less than that between the electron pairs of the C—H bonds and the four electrons of the carbon—carbon double bond.

2.3

(a) The C—X bond moments in the *trans*-isomers point in opposite directions and therefore cancel:

$$\text{trans-} \qquad \text{cis-} \qquad X = \text{Cl or Br}$$

In the *cis*- isomers the bond moments are additive

(b) The C—Cl bond moment is larger than the C—Br bond moment because Cl is more electronegative than Br.

(c) Yes   (d) Yes

2.4

(a)

cis - trans isomers

(b)

cis - trans isomers   *cis - trans* isomers

$CH_3CH=CCl_2$ $Cl-CHCH=CH_2$ $Cl-CH_2C=CH_2$
                          | |
                          Cl Cl

(c) $Cl_3CCH=CH_2$ $Cl-CHC=CH_2$ $Cl-CH_2CH=CCl_2$ $CH_3C=CCl_2$
                          | | |
                          Cl Cl Cl

cis - trans isomers   *cis - trans* isomers

(d)　$Cl_3CC=CH_2$

　　　　$\underset{Cl}{|}$

$$\underset{H}{\overset{Cl_3C}{\diagdown}}C=C\underset{H}{\overset{Cl}{\diagup}} \qquad \underset{H}{\overset{Cl_3C}{\diagdown}}C=C\underset{Cl}{\overset{H}{\diagup}}$$

cis - trans isomers

$Cl_2CHCH=CCl_2$

$$\underset{Cl}{\overset{Cl_2CH}{\diagdown}}C=C\underset{Cl}{\overset{H}{\diagup}} \qquad \underset{Cl}{\overset{Cl_2CH}{\diagdown}}C=C\underset{H}{\overset{Cl}{\diagup}}$$

cis - trans isomers

$ClCH_2C=CCl_2$

　　　$\underset{Cl}{|}$

(e)

$$\underset{Cl}{\overset{Cl_3C}{\diagdown}}C=C\underset{H}{\overset{Cl}{\diagup}} \qquad \underset{Cl}{\overset{Cl_3C}{\diagdown}}C=C\underset{Cl}{\overset{H}{\diagup}}$$

cis - trans isomers

$Cl_3CCH=CCl_2$ 　 $Cl_2CHC=CCl_2$

　　　　　　　　　　　　$\underset{Cl}{|}$

(f)　See (a-e) above.

2.5

(a)　$\underset{\underset{CH_3}{|}}{\overset{\overset{Cl}{|}}{CH_3CCH_3}}$ 　(b)　$\underset{\underset{I}{|}}{CH_3CHCH_2CH_3}$

2.6

(a) Yes, counting the electrons in the triple bond together and as occupying the region between the two carbon atoms, the remaining electrons (in the C–H bonds) are at a maximum separation from the electrons in the triple bond when the bond angle is $180°$.

(b)　$H-C\equiv\!\equiv\!\equiv C-\overset{|}{\underset{|}{C}}-$

(c)　$H-C\equiv\!\equiv\!\equiv C-\overset{H}{\underset{H}{C\diagdown\!\!\!_{H}}}$　or　$H-\bigcirc\!\!\equiv\!\!\bigcirc-\overset{H}{\bigcirc\diagdown\!\!_{H}}$

2.7

(a) Yes, the halogen adds to the less hydrogenated carbon, i.e., to the one that already bears a chlorine atom.

(b)　$CH_3C\equiv CH + HCl \longrightarrow \underset{\underset{Cl}{|}}{CH_3C=CH_2} \overset{HCl}{\longrightarrow} \underset{\underset{Cl}{|}}{\overset{\overset{Cl}{|}}{CH_3CCH_3}}$

2.8

(a)　$HC\equiv CH + 2HCl \longrightarrow CH_3CHCl_2$

(b) $CH_2{=}CH_2 + Cl_2 \xrightarrow[\text{dark}]{CCl_4} CH_2ClCH_2Cl$

(c) $CH_2{=}CH_2 + Br_2 \xrightarrow[\text{dark}]{CCl_4} CH_2BrCH_2Br$

(d) $HC{\equiv}CH + 2Br_2 \xrightarrow[\text{dark}]{CCl_4} CHBr_2CHBr_2$

(e) $HC{\equiv}CH + HCl \longrightarrow CH_2{=}CHCl \xrightarrow{HBr} CH_3CHClBr$

(f) $CH_3C{\equiv}CH + HCl \longrightarrow CH_3\overset{\overset{\displaystyle Cl}{|}}{C}{=}CH_2 \xrightarrow{HBr} CH_3\overset{\overset{\displaystyle Cl}{|}}{\underset{\underset{\displaystyle Br}{|}}{C}}CH_3$

(g) $CH_3C{\equiv}CH + 2HBr \longrightarrow CH_3CBr_2CH_3$

(h) $CH_3CH_3 + Br_2 \xrightarrow[\substack{\text{excess} \\ CH_3CH_3}]{\text{light}} CH_3CH_2Br + HBr$

$CH_2{=}CH_2 + HBr \longrightarrow CH_3CH_2Br$

2.9

(a) $K_a = \dfrac{[H_3O^+][CF_3COO^-]}{[CF_3COOH]} = 1$

let $[H_3O^+] = [CF_3COO^-] = X$

then $[CF_3COOH] = 0.1 - X$

$\therefore \quad \dfrac{(X)(X)}{0.1 - X} = 1 \quad$ or $\quad X^2 = 0.1 - X$

$X^2 + X - 0.1 = 0$

Using the quadratic formula, $X = \dfrac{-b \pm \sqrt{b^2 - 4ac}}{2a}$,

$X = \dfrac{-1 \pm \sqrt{1 + 0.4}}{2} = \dfrac{-1 \pm \sqrt{1.4}}{2} = \dfrac{-1 \pm 1.183}{2} = \dfrac{+0.183}{2}$

$X = 0.0915$ (We can exclude negative values of X.)

$[H_3O^+] = [CF_3COO^-] = 0.0915 \underline{M}$

(b) Percentage ionized $= \dfrac{[H_3O^+]}{0.1} \times 100 = \dfrac{(0.0915)(100)}{0.1}$

Percentage ionized $= 91.5\%$

2.10

Molecules of N-propylamine can form hydrogen bonds to each other,

$$CH_3CH_2CH_2N\overset{\displaystyle H}{\underset{\displaystyle H}{\Big\langle}} \cdots :NCH_2CH_2CH_3 \,,$$

whereas molecules of trimethylamine, because they have no hydrogens attached to nitrogen, cannot form hydrogen bonds to each other.

2.11

(a) Alkyne   (b) Carboxylic acid   (c) Alcohol

(d) Aldehyde   (e) Alkane   (f) Ketone

2.12

(a) Carbon-carbon double bonds, hydroxyl group

(b) Ketone group, hydroxyl group, carbon-carbon double bond

(c) Carbon-carbon double bond, ester group

(d) Amide groups

(e) Aldehyde group, hydroxyl groups

(f) Carbon-carbon double bond, ether linkage

(g) Carbon-carbon double bond, hydroxyl group

2.13

(a) Elimination   (b) Substitution   (c) Substitution

(d) Oxidation   (e) Condensation   (f) Condensation

2.14

(a) $CH_3CH_2CCl_3$ $CH_3CHCHCl_2$ CH_3CCH_2Cl $ClCH_2CH_2CHCl_2$
              (with Cl below on 2nd C)   (with Cl above and below on 3rd C)

$ClCH_2CHCH_2Cl$ (five trichloropropanes)
(with Cl below)

(b) CH_3CHCCl_3 $ClCH_2CH_2CCl_3$ CH_3CCHCl_2 $Cl_2CHCH_2CHCl_2$
(with Cl below)   (with Cl above and below)

$ClCH_2CHCHCl_2$ $ClCH_2CCH_2Cl$ (six tetrachloropropanes)
(with Cl below)   (with Cl above and below)

2.15

(a) CH_3CHCH_2Br (with CH_3 above) $+ KOH \xrightarrow{\text{ethanol}} CH_3C=CH_2$ (with CH_3 above) $+ KBr + H_2O$

(b) By dehydrobromination as shown in (a) above, and then by adding HBr to the resulting alkene, isobutyl bromide can be converted into tert-butyl bromide:

$$CH_3C\!=\!CH_2 + HBr \xrightarrow{\ CCl_4\ } CH_3\overset{\overset{\displaystyle CH_3}{|}}{\underset{\underset{\displaystyle Br}{|}}{C}}CH_3 \quad \text{(Markovnikov's rule)}$$

(with CH_3 substituent on the first carbon)

2.16
(a) Primary   (b) Secondary   (c) Tertiary   (d) Secondary

(e) Secondary   (f) Tertiary

2.17
(a) Secondary   (b) Primary   (c) Tertiary   (d) Primary

(e) Secondary

2.18
(a) $2H_3O^+ + CO_3^{-2} \longrightarrow [H_2CO_3] + 2H_2O \longrightarrow 3H_2O + CO_2$

(b) $H_3O^+ + CH_3\overset{\overset{\displaystyle O}{\|}}{C}O^- \longrightarrow H_2O + CH_3\overset{\overset{\displaystyle O}{\|}}{C}OH$

(c) $CO_3^{-2} + H_2O \longrightarrow HCO_3^- + OH^-$

(d) $:H^- + H_2O \longrightarrow H_2 + OH^-$

(e) $:CH_3^- + H_2O \longrightarrow CH_4 + OH^-$

(f) $:CH_3^- + HC\!\equiv\!CH \longrightarrow HC\!\equiv\!C:^- + CH_4$

(g) $H_3O^+ + NH_3 \longrightarrow NH_4^+ + H_2O$

(h) $NH_4^+ + NH_2^- \longrightarrow 2NH_3$

(i) $CH_3CH_2O^- + H_2O \longrightarrow CH_3CH_2OH + OH^-$

2.19
Oxygen-containing compounds contain either $=\overset{..}{O}:$ or $-\overset{..}{O}-$. Both of these are Brønsted-Lowry bases in the presence of the strong proton donor, sulfuric acid. The equation for the reaction using an ether as an example is

$$R\!-\!\overset{..}{\underset{..}{O}}\!-\!R + H_2SO_4 \rightleftharpoons \underbrace{R\!-\!\overset{\overset{\displaystyle H}{|}}{\underset{..}{O}}{}^+\!-R}_{\text{Salt}} + HSO_4^-$$

The salt is soluble in the highly polar H_2SO_4.

2.20
(a) Ethyl alcohol because its molecules can form hydrogen bonds to each other. Methyl ether molecules have no hydrogens attached to oxygen.

(b) Ethylene glycol because its molecules have more OH groups and will therefore participate in more extensive hydrogen bonding.

(c) Heptane because it has a higher molecular weight. (Neither compound can form hydrogen bonds.)

(d) 1-Propanol because its molecules can form hydrogen bonds to each other. Acetone molecules have no hydrogens attached to oxygen.

(e) *Cis*-1,2-dichloroethane because its molecules have a higher dipole moment.

(f) Propanoic acid because its molecules can form hydrogen bonds to each other.

2.21

(a) Reduction with lithium aluminum hydride ($LiAlH_4$).

(b) Oxidation with chromic acid (H_2CrO_4).

(c) Reduction with $LiAlH_4$, or preferably with the milder sodium borohydride ($NaBH_4$).

(d) Oxidation with H_2CrO_4.

(e) Condensation with methyl alcohol in the presence of an acid catalyst.

(f) Self-condensation of methyl alcohol at $150°$ (under pressure) in the presence of a strong acid catalyst.

2.22

(a) $CH_3OCH_2CH_3$ (b) $CH_3CH_2CH_2OH$ (c) $CH_3\overset{\displaystyle OH}{\underset{|}{C}}HCH_3$

(d) $CH_3\overset{\displaystyle O}{\overset{||}{C}}OCH_2CH_3$ $CH_3CH_2\overset{\displaystyle O}{\overset{||}{C}}OCH_3$ (e) $CH_3CH_2CH_2CH_2X$

(f) $CH_3CH_2CHXCH_3$ (g) $CH_3\overset{\displaystyle CH_3}{\underset{\underset{\displaystyle X}{|}}{\overset{|}{C}}}CH_3$ (h) $CH_3\overset{\displaystyle CH_3}{\overset{|}{C}}H\overset{\displaystyle C}{\underset{\underset{\displaystyle O}{||}}{}}H$ or $CH_3CH_2CH_2\overset{C}{\underset{\underset{\displaystyle O}{||}}{}}H$

(i) $CH_3\overset{\displaystyle O}{\overset{||}{C}}CH_2CH_3$ (j) $CH_3CH_2\overset{\displaystyle CH_3}{\overset{|}{C}}HNH_2$ (k) $CH_3CH_2CH_2NHCH_3$

(l) $CH_3CH_2N(CH_3)_2$ (m) $CH_3CH_2CH_2\overset{\displaystyle O}{\overset{||}{C}}NH_2$ (n) $CH_3\overset{\displaystyle O}{\overset{||}{C}}NHCH_2CH_3$

2.23

Staggered   Eclipsed

2.24

Basic strength depends upon ability to accept a proton. In $(CF_3)_3N$:, the high electronegativity of fluorine reduces the availability of the lone electron pair on nitrogen. In $(CH_3)_3N:$, the lone pair is more available to bond with a proton. (An alternate view is that the conjugate acid is rendered less stable by the presence of the electronegative fluorines.)

$$CF_3-\underset{\underset{CF_3}{|}}{\overset{\overset{CF_3}{|}}{N}}: \; + \; HA \; \rightleftharpoons \; CF_3-\underset{\underset{CF_3}{|}}{\overset{\overset{CF_3}{|}}{\overset{+}{N}}}-H \; + \; A^-$$

2.25
An ester group.

2.26

(a) $CH_3CH_2\overset{\overset{O}{\|}}{C}-OCH_2CH_3 \xrightarrow[\text{(a)}]{HO^-} CH_3CH_2\overset{\overset{O}{\|}}{C}-O^- + CH_3CH_2OH$

$\xrightarrow[\text{(b)}]{H_3O^+} CH_3CH_2\overset{\overset{O}{\|}}{C}-OH + H_2O$

(b) $CH_3\overset{\overset{O}{\|}}{C}-O-\!\!\bigcirc \xrightarrow[\text{(a)}]{OH^-} CH_3\overset{\overset{O}{\|}}{C}-O^- + HO-\!\!\bigcirc$

$\xrightarrow[\text{(b)}]{H_3O^+} CH_3\overset{\overset{O}{\|}}{C}-OH$

(c) $CH_3CH_2CH_2\overset{\overset{O}{\|}}{C}-NH_2 \xrightarrow[\text{(a)}]{OH^-} CH_3CH_2CH_2\overset{\overset{O}{\|}}{C}-O^- + NH_3 \xrightarrow[\text{(b)}]{2H_3O^+}$

$CH_3CH_2CH_2\overset{\overset{O}{\|}}{C}-OH + NH_4^+ + 2H_2O$

(d) $CH_3\overset{\overset{O}{\|}}{C}-\underset{\underset{CH_3}{|}}{N}-CH_3 \xrightarrow[\text{(a)}]{OH^-} CH_3\overset{\overset{O}{\|}}{C}-O^- + (CH_3)_2NH$

$\xrightarrow[\text{(b)}]{2H_3O^+} CH_3\overset{\overset{O}{\|}}{C}-OH + (CH_3)_2NH_2^+ + 2H_2O$

(e) $CH_3CH_2\overset{\overset{O}{\|}}{C}-NHCH_2CH_3 \xrightarrow[\text{(a)}]{OH^-} CH_3CH_2\overset{\overset{O}{\|}}{C}-O^- + CH_3CH_2NH_2$

$\xrightarrow[\text{(b)}]{2H_3O^+} CH_3CH_2\overset{\overset{O}{\|}}{C}-OH + CH_3CH_2NH_3^+ + 2H_2O$

3

ALKANES AND CYCLOALKANES: THEIR STRUCTURES, PROPERTIES, AND SYNTHESES

3.1

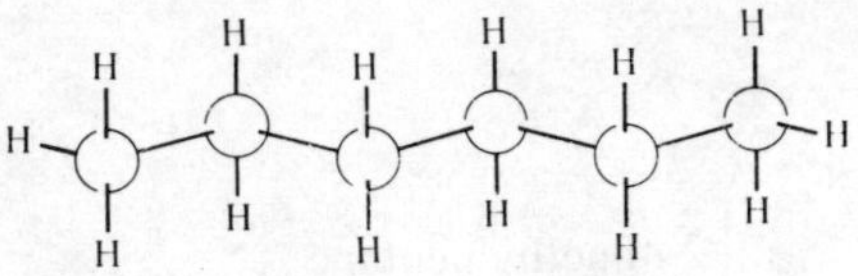

(1) $CH_3CH_2CH_2CH_2CH_2CH_3$

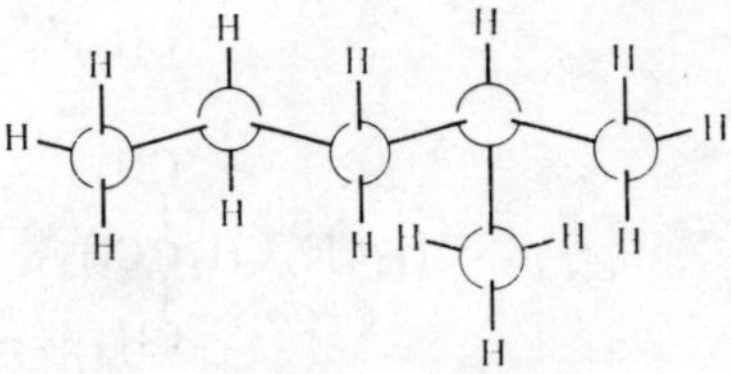

(2) $CH_3CH_2CH_2CHCH_3$
$\quad\quad\quad\quad\quad\quad |$
$\quad\quad\quad\quad\quad\quad CH_3$

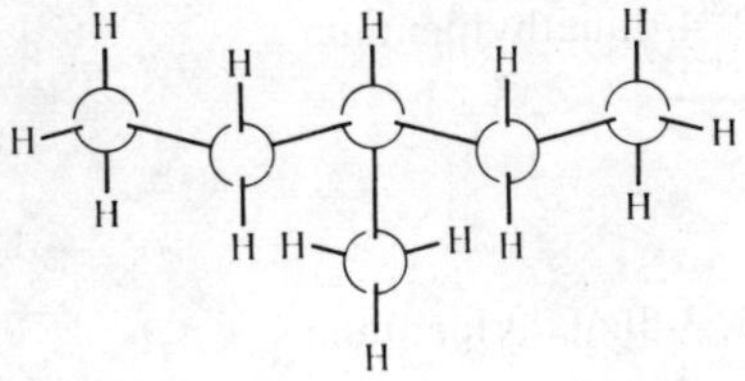

(3) $CH_3CH_2CHCH_2CH_3$
$\quad\quad\quad\quad |$
$\quad\quad\quad\quad CH_3$

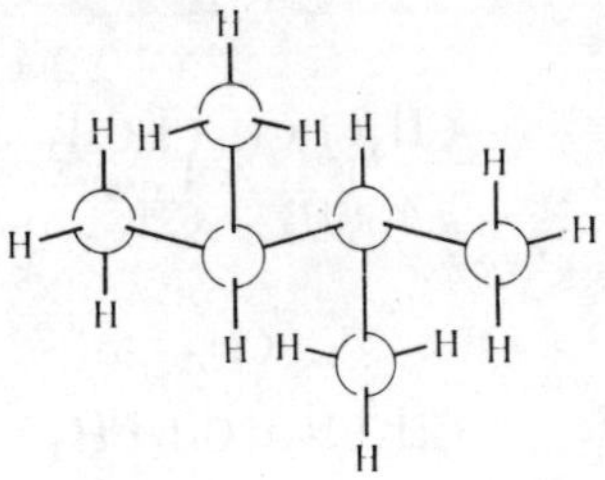

$\quad\quad\quad\quad\quad\quad CH_3$
$\quad\quad\quad\quad\quad\quad |$
(4) $CH_3CHCHCH_3$
$\quad\quad\quad\quad\quad\quad |$
$\quad\quad\quad\quad\quad\quad CH_3$

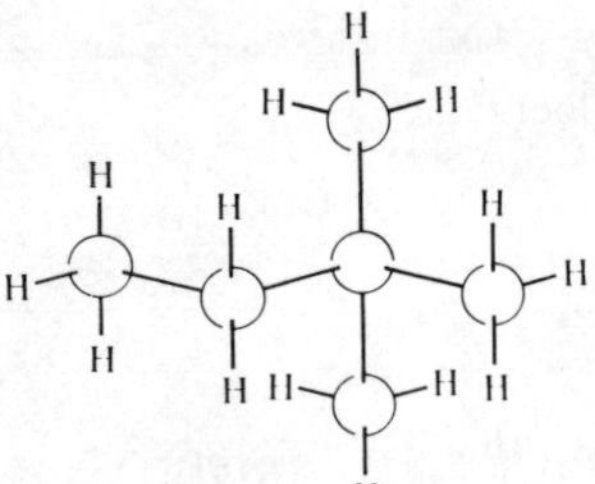

$\quad\quad\quad\quad CH_3$
$\quad\quad\quad\quad |$
(5) $CH_3CH_2CCH_3$
$\quad\quad\quad\quad |$
$\quad\quad\quad\quad CH_3$

3.2

(a)   Refer to problem 3.1 above:

(1) Hexane, (2) 2-methylpentane, (3) 3-methylpentane, (4) 2,3-dimethylbutane,
(5) 2,2-dimethylbutane.

(b) $CH_3CH_2CH_2CH_2CH_2CH_2CH_3$ heptane

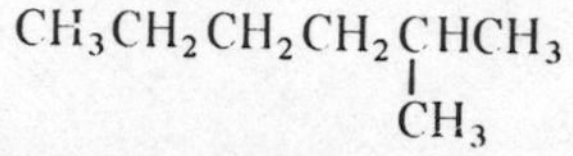

$$CH_3CH_2CH_2CH_2\underset{\underset{\displaystyle CH_3}{|}}{C}HCH_3$$
2-methylhexane

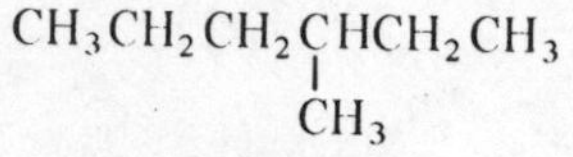

$$CH_3CH_2CH_2\underset{\underset{\displaystyle CH_3}{|}}{C}HCH_2CH_3$$
3-methylhexane

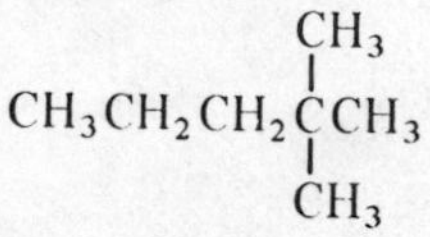

$$CH_3CH_2CH_2\overset{\overset{\displaystyle CH_3}{|}}{\underset{\underset{\displaystyle CH_3}{|}}{C}}CH_3$$
2, 2-dimethylpentane

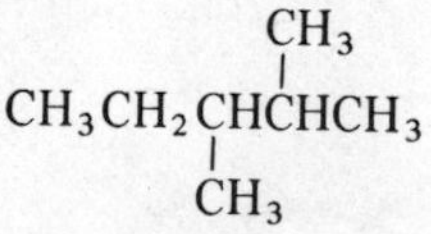

$$CH_3CH_2\overset{\overset{\displaystyle CH_3}{|}}{C}H\underset{\underset{\displaystyle CH_3}{|}}{C}HCH_3$$
2, 3-dimethylpentane

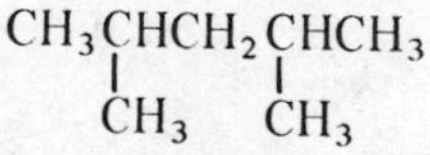

$$CH_3\underset{\underset{\displaystyle CH_3}{|}}{C}HCH_2\underset{\underset{\displaystyle CH_3}{|}}{C}HCH_3$$
2, 4-dimethylpentane

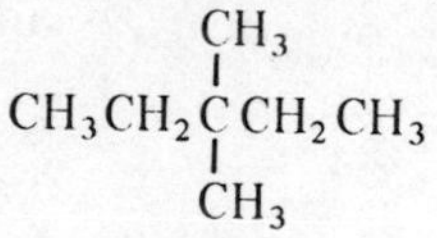

$$CH_3CH_2\overset{\overset{\displaystyle CH_3}{|}}{\underset{\underset{\displaystyle CH_3}{|}}{C}}CH_2CH_3$$
3, 3-dimethylpentane

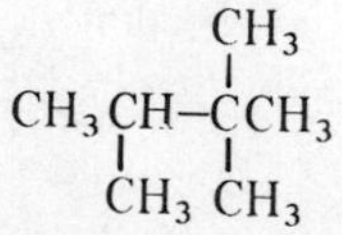

$$CH_3\underset{\underset{\displaystyle CH_3}{|}}{C}H-\overset{\overset{\displaystyle CH_3}{|}}{\underset{\underset{\displaystyle CH_3}{|}}{C}}CH_3$$
2, 2, 3-trimethylbutane

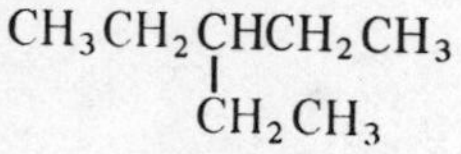

$$CH_3CH_2\underset{\underset{\displaystyle CH_2CH_3}{|}}{C}HCH_2CH_3$$
3-ethylpentane

3.3

(a)

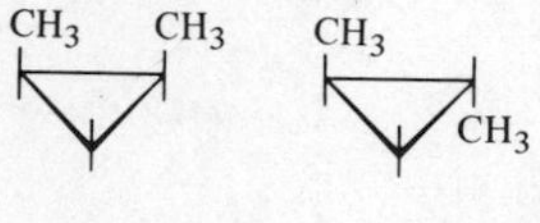

(cis)   *(trans)*

(b)

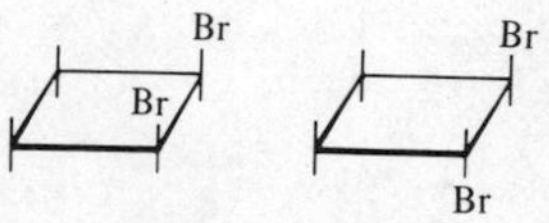

(cis)   *(trans)*

3.4

(a)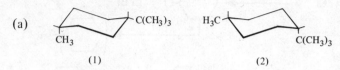

(1)   (2)

(b) No. In (1), the methyl group is axial and the *tert*-butyl group is equatorial; in (2) the situation is reversed.

(c) The *tert*-butyl group is larger than the methyl; conformation (1) is more stable because the *tert*-butyl group is equatorial.

(d) The preferred conformation at equilibrium is (1).

3.5

(a) Conformations of *cis*-isomer are equivalent, (e, a) and (a, e).

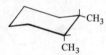

(a, e)   (e, a)

(b) Conformations of *trans*-isomer are not equivalent, (e, e) and (a, a).

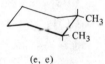

(e, e)   (a, a)

(c) The *trans*-(e, e) conformation is more stable than the *trans*-(a,a).

(d) The *trans*-(e, e) would be more highly populated at equilibrium.

3.6

(a) $CH_3I + 2 Li \xrightarrow{\text{ether}} CH_3Li + LiI$

$2 CH_3Li + CuI \longrightarrow (CH_3)_2CuLi + LiI$

$(CH_3)_2CuLi + CH_3CH_2CH_2CH_2{-}Cl \longrightarrow CH_3CH_2CH_2CH_2CH_3 + CH_3Cu$

                                                         + LiCl

(b) $(CH_3)_2CH{-}Br + 2Li \xrightarrow{\text{ether}} (CH_3)_2CHLi + LiBr$

$\xrightarrow{\text{CuI}} [(CH_3)_2CH]_2CuLi + LiI$

$[(CH_3)_2CH]_2CuLi + CH_3CH_2{-}Br \longrightarrow CH_3\overset{\underset{\displaystyle CH_3}{|}}{C}HCH_2CH_3 + (CH_3)_2CHCu$

                                                         + LiBr

(c) —Br +2Li $\xrightarrow{\text{ether}}$ —Li + Li Br

$\xrightarrow{\text{CuI}}$ $\left(\text{}\right)_2$—Cu Li + LiI

$\left(\text{}\right)_2$—CuLi + CH$_3$−I $\longrightarrow$ —CH$_3$ + —Cu + LiI

(d) —Br + 2Li $\xrightarrow{\text{ether}}$ —Li + Li Br

$\xrightarrow{\text{CuI}}$ $\left(\text{}\right)_2$—CuLi + LiI

$\left(\text{}\right)_2$—CuLi + CH$_3$CH$_2$CH$_2$−Br $\longrightarrow$ —CH$_2$CH$_2$CH$_3$ + —Cu

+ LiBr

3.7

(a) CH$_3$CHCHCH$_2$CH$_3$
        | |
        Cl Cl

(b) CH$_3$CCH$_3$ with CH$_3$ above and I below the central C

(c) CH$_3$CH$_2$CHCH$_2$CH$_3$
              |
              CH$_2$
              CH$_3$

(d) CH$_3$CH−CH−CHCH$_2$CH$_2$CH$_2$CH$_2$CH$_2$CH$_3$
         | | |
        CH$_3$ CH$_3$ CH$_3$

(e) CH$_3$CH$_2$CH$_2$CHCH$_2$CH$_2$CH$_2$CH$_2$CH$_3$
                 |
                 CH
             CH$_3$ CH$_3$

(f) with CH$_3$ and CH$_3$

(g) with CH$_3$, CH$_3$

(h) H$_3$C, CH$_3$

(i) H—CH(CH$_3$)$_2$

(j) with CH$_3$, H, H, CH(CH$_3$)$_2$

(k) CH$_3$CHCH$_2$CH$_2$CH$_2$−Cl
        |
        CH$_3$

(l) CH$_3$CCH$_2$CCH$_2$CH$_2$CH$_2$CH$_3$
with CH$_3$, CH$_3$ above and CH$_3$, CH$_3$ below

(m) CH$_3$CCH$_2$−Cl
with CH$_3$ above and CH$_3$ below

(n) CH$_3$CHCH$_2$CH$_3$
with CH$_3$ above

3.8

(a) 3,4-dimethylhexane   (e) ethylcyclohexane

(b) 2-methylbutane   (f) cyclopentylcyclopentane

(c) 2,4-dimethylpentane   (g) 6-isobutyl-2-methyldecane

(d) 3-methylpentane

3.9

	Common	IUPAC
(a) $CH_3CH_2CH_2—Cl$	n-propyl chloride	1-chloropropane
$CH_3\overset{\mid}{\underset{\underset{Cl}{\mid}}{C}HCH_3}$	isopropyl chloride	2-chloropropane

(b) $CH_3CH_2CH_2CH_2—Br$ n-butyl bromide   1-bromobutane

$CH_3CH_2\overset{\overset{Br}{\mid}}{C}HCH_3$ sec-butyl bromide   2-bromobutane

$CH_3\overset{\underset{\underset{CH_3}{\mid}}{}}{C}HCH_2—Br$ isobutyl bromide   1-bromo-2-methylpropane

$CH_3\overset{\overset{CH_3}{\mid}}{\underset{\underset{CH_3}{\mid}}{C}}—Br$ tert-butyl bromide   2-bromo-2-methylpropane

3.10

(a) $CH_3\overset{\overset{CH_3}{\mid}}{\underset{\underset{CH_3}{\mid}}{C}}CH_3$ 2,2-dimethylpropane (neopentane)

(b) $CH_3\overset{\overset{CH_3}{\mid}}{C}HCH_2CH_3$ 2-methylbutane (isopentane)

(c) $CH_3CH_2CH_2CH_2CH_3$ pentane

(d) cyclopentane structure   cyclopentane

(e) $CH_3\overset{\overset{CH_3}{\mid}}{C}H-\overset{\overset{CH_3}{\mid}}{C}HCH_3$ 2,3-dimethylbutane

3.11

(a) $CH_3\overset{\overset{CH_3}{\mid}}{C}H\underset{\underset{CH_3CH\,CH_3}{\mid}}{C}H\overset{\overset{CH_3}{\mid}}{C}HCH_3$ 2,4-dimethyl-3-isopropylpentane

(b) 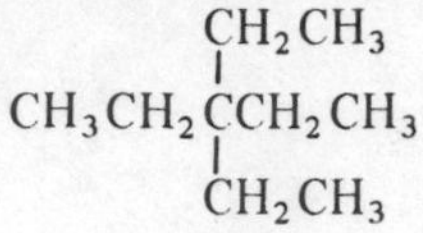 3,3-diethylpentane

(c) 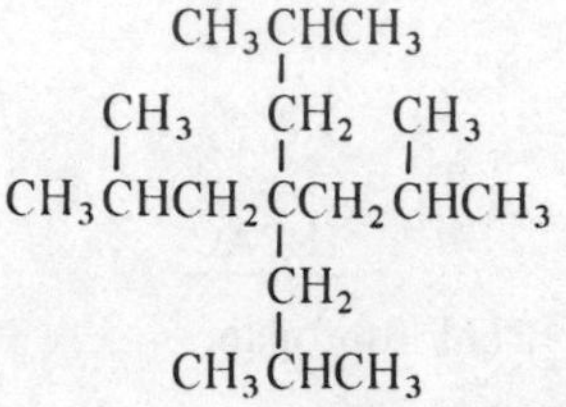 4,4-diisobutyl-2,6-dimethylheptane

(d) 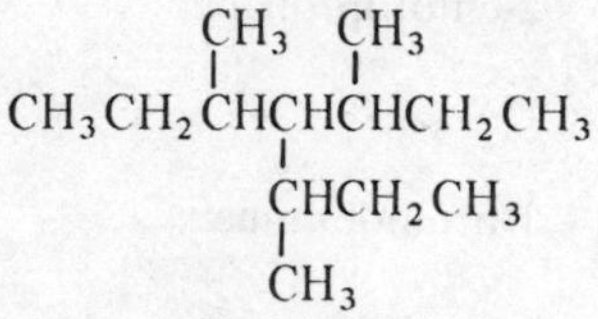 4-*sec*-butyl-3,5-dimethylheptane

(e) 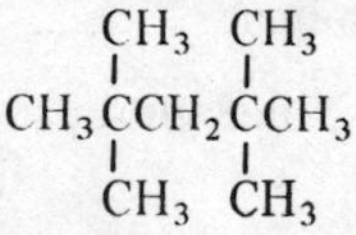 2,2,4,4-tetramethylpentane

(f) 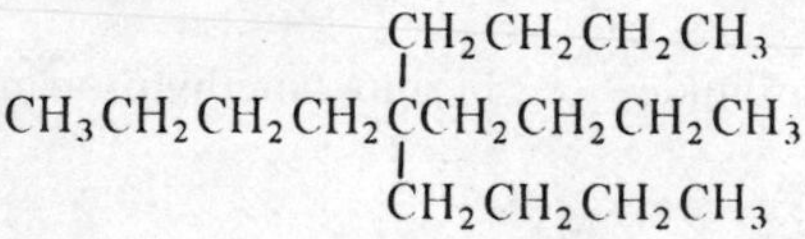 5,5-dibutylnonane

3.12

(a) neopentane 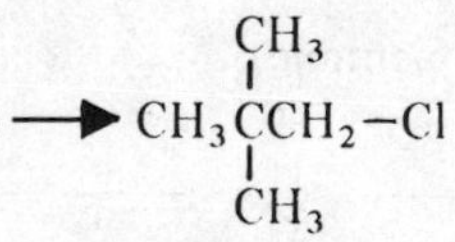

(b) pentane 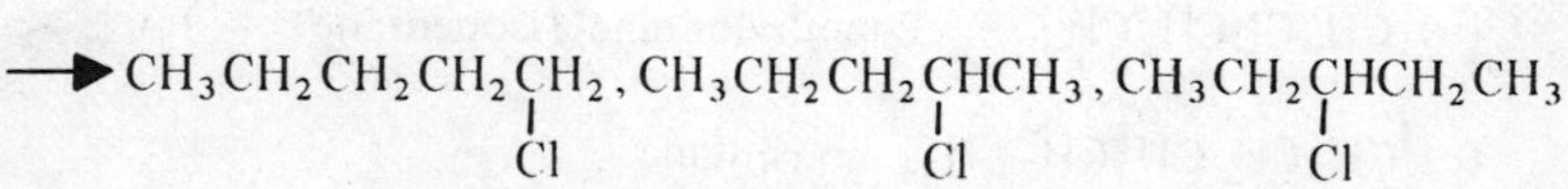

(c) isopentane 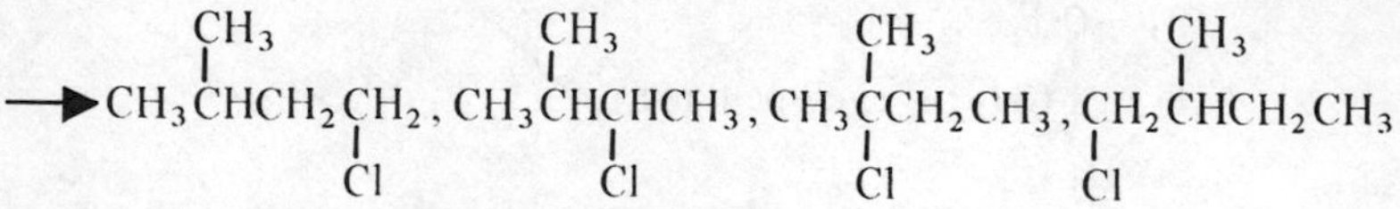

(d) neopentane 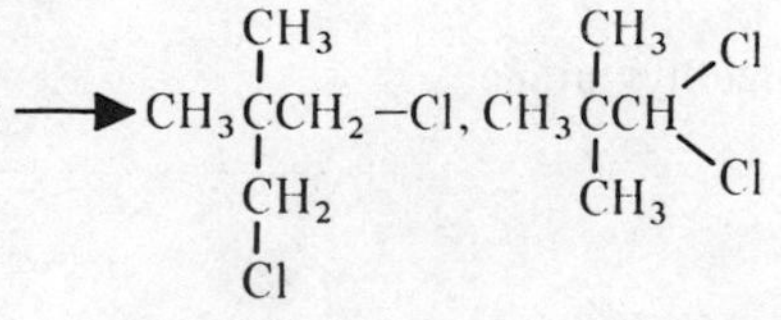

3.13

The methyl groups are larger than the hydrogen atom. The resulting mutual repulsions among the methyl groups cause a larger than tetrahedral bond angle.

3.14

(a) PE

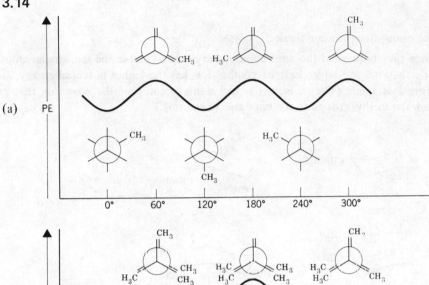

(b) PE

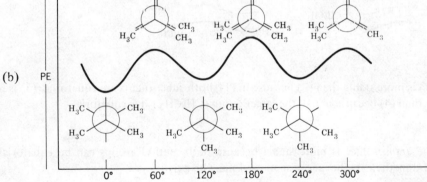

(c) PE

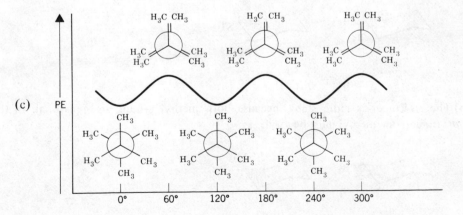

3.15

(a) Hexane. Branched chain hydrocarbons have lower boiling points than their normal isomers.

(b) Hexane. Boiling point increases with molecular weight.

(c) Pentane. (See (a) above).

(d) Chloroethane, because it has a higher molecular weight, and is more polar.

3.16

(a) The *trans*-isomer is more stable.

(b) Since they both yield the same combustion products and in the same molar amounts, the one that has the larger heat of combustion has the higher potential energy, and is therefore less stable (The *cis*-isomer is less stable because of the crowding that exists between the methyl groups on the same side of the ring.)

3.17

(a)

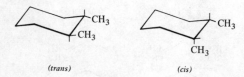

```
(1) (e, e)                    (2) (a, a)
```

(b)

```
(3) (a, e)                    (4) (e, a)
```

(c) (1) is more stable than (2) because in (1), both substituents are equatorial. (3) is more stable than (4) because in (3), the larger group ($CH(CH_3)_2$) is equatorial.

3.18

(a) The *trans*-isomer is more stable because both methyl groups can be equatorial. In *cis*-1,1-dimethylcyclohexane, one methyl must be axial.

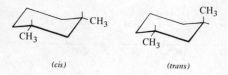

```
(trans)              (cis)
```

(b) The *cis*-isomer is more stable because both methyl groups are equatorial. In the *trans*-isomer, one methyl must be axial.

```
(cis)                (trans)
```

(c) The *trans*-isomer is more stable for the same reason as in (a).

(trans)   *(cis)*

3.19

In *cis*-1, 3-*di-tert*-butylcyclohexane, the two substituents are both equatorial (see problem 1.38 b above), whereas in the *trans*-isomer, one of the *tert*-butyl groups must be axial. The instability of a chair conformation with such a large group in an axial position forces the molecule into a less strained twist conformation:

trans (chair conformation)

3.20

3.21

(a) Wurtz reaction:

$$2\ CH_3CH_2CH_2-Cl\ +\ 2Na\ \longrightarrow\ CH_3CH_2CH_2CH_2CH_2CH_3\ +\ 2\ NaCl$$

Corey-House Synthesis:

$$CH_3CH_2CH_2-Cl\ +\ 2Li\ \longrightarrow\ CH_3CH_2CH_2Li\ +\ LiCl$$

$$\downarrow CuI$$

$$(CH_3CH_2CH_2)_2CuLi$$

$$CH_3CH_2CH_2CH_2CH_2CH_3 \xleftarrow{\ CH_3CH_2CH_2Cl\ }$$

(b) $CH_3\underset{\underset{Br}{|}}{C}HCH_2CH_2CH_2CH_3 + Zn \xrightarrow{H^+} CH_3CH_2CH_2CH_2CH_2CH_3 + ZnBr_2$

(c) $CH_3CH_2CH_2CH_2CH=CH_2 + H_2 \xrightarrow{Ni} CH_3CH_2CH_2CH_2CH_2CH_3$

4

CHEMICAL REACTIVITY: REACTIONS OF ALKANES AND CYCLOALKANES

4.1

(a) $CH_3-H + F-F \longrightarrow CH_3-F + H-F$
(D=104) (D=38)   (D=108) (D=136)

+ 142 kcal/mole is required for bond cleavage   − 244 kcal/mole is evolved in bond formation $\left.\begin{array}{}\\ \\ \\\end{array}\right\}$ ΔH = + 142 − 244 = − 102 kcal/mole (exothermic)

(b) $CH_3-H + Cl-Cl \longrightarrow CH_3-Cl + H-Cl$
(D=104) (D=58)   (D=83.5) (D=103)

+ 162 kcal/mole is required for bond cleavage   − 186.5 kcal/mole is evolved in bond formation $\left.\begin{array}{}\\ \\ \\\end{array}\right\}$ ΔH = + 162 − 186.5 = − 24.5 kcal/mole (exothermic)

(c) $CH_3-H + Br-Br \longrightarrow CH_3-Br + H-Br$
(D=104) (D=46)   (D=70) (D=87.5)

+ 150 kcal/mole   − 157.5 kcal/mole   = −7.5 kcal/mole = ΔH (exothermic)

(d) $CH_3-H + I-I \longrightarrow CH_3-I + H-I$
(D-104) (D-36)   (D=56) (D=71)

+ 140 kcal/mole   − 127 kcal/mole   = + 13 kcal/mole = ΔH (endothermic)

4.2

(a)

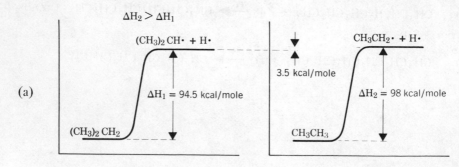

28

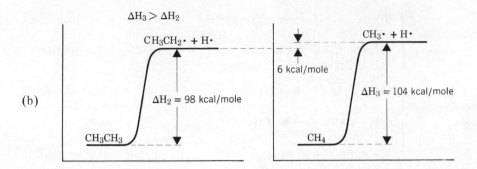

(b)

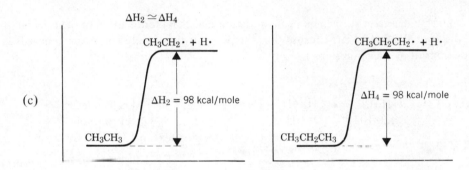

(c)

(d) The radicals produced are both primary radicals, and they are otherwise structurally similar, therefore they are of essentially equal stability.

4.3

Bond dissociation energies of the following C—Cl bonds are:

$$CH_3-Cl \longrightarrow CH_3 \cdot + Cl \cdot \qquad \Delta H = 83.5 \text{ kcal/mole}$$

$$CH_3CH_2-Cl \longrightarrow CH_3CH_2 \cdot + Cl \cdot \qquad \Delta H = 81.5 \text{ kcal/mole}$$

$$(CH_3)_2CH-Cl \longrightarrow (CH_3)_2CH \cdot + Cl \cdot \qquad \Delta H = 81 \text{ kcal/mole}$$

$$(CH_3)_3C-Cl \longrightarrow (CH_3)_3C \cdot + Cl \cdot \qquad \Delta H = 78.5 \text{ kcal/mole}$$

Since in each case the same kind of compound (an alkyl chloride) is decomposed into the same kinds of products (an alkyl free radical and a chlorine atom), it follows that the energy required (ΔH) is a measure of the instability of the radical relative to the alkyl halide. In other words, the less stable the free radical, the more energy will be required to break the bond between it and the chlorine atom. Bond dissociation energies for these alkyl chlorides are, respectively, 83.5, 81.5, 81, 78.5. They are in the same order as the stabilities of the free radicals produced: $CH_3 \cdot < CH_3CH_2 \cdot < (CH_2)_2CH \cdot < (CH_3)_3C \cdot$

4.4

Chain-
initiating
step

$$Br-Br \longrightarrow 2 Br \cdot \qquad \Delta H = +46 \text{ kcal/mole}$$

$(D=46)$

Chain-
propagating
steps
$$\begin{cases} \text{Br} \cdot + \text{CH}_3\text{--H} \longrightarrow \text{CH}_3 \cdot + \text{HBr} & \Delta H = +16.5 \text{ kcal/mole} \\ \qquad (D=104) \qquad\qquad\quad (D=87.5) \\ \text{CH}_3 \cdot + \text{Br--Br} \longrightarrow \text{CH}_3\text{--Br} + \text{Br} \cdot & \Delta H = -24 \text{ kcal/mole} \\ \qquad (D=46) \qquad\quad (D=70) \end{cases}$$

Chain-
terminating
steps
$$\begin{cases} \text{CH}_3 \cdot + \text{Br} \cdot \longrightarrow \text{CH}_3\text{--Br} & \Delta H = -70 \text{ kcal/mole} \\ \qquad\qquad\qquad (D=70) \\ \text{CH}_3 \cdot + \text{CH}_3 \cdot \longrightarrow \text{CH}_3\text{--CH}_3 & \Delta H = -88 \text{ kcal/mole} \\ \qquad\qquad\qquad\quad (D=88) \\ \text{Br} \cdot + \text{Br} \cdot \longrightarrow \text{Br--Br} & \Delta H = -46 \text{ kcal/mole} \\ \qquad\qquad\quad (D=46) \end{cases}$$

4.5
It would be incorrect to include chain-initiation and chain-termination in the calculation of the overall value of ΔH because those steps occur only rarely (once for hundreds or thousands of propagation steps.)

4.6
(a) $\text{CH}_3 \cdot + \text{H--Cl} \longrightarrow \text{CH}_3\text{--H} + \text{Cl} \cdot$ $\Delta H = -1 \text{ kcal/mole}$
$\quad\quad (D=103) \qquad (D=104)$ $E_{\text{act}} = +2.8 \text{ kcal/mole}$
$\qquad\qquad\qquad\qquad\qquad\qquad$ (See text, p 136; E_{act} for the re-
$\qquad\qquad\qquad\qquad\qquad\qquad$ verse reaction is 3.8 kcal/mole)

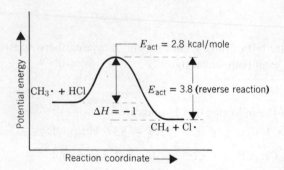

(b) $\text{CH}_3 \cdot + \text{H--Br} \longrightarrow \text{CH}_3\text{--H} + \text{Br} \cdot$ $\Delta H = -16.5 \text{ kcal/mole}$
$\quad\quad (D=87.5) \qquad (D=104)$ $E_{\text{act}} = +2.1 \text{ kcal/mole}$

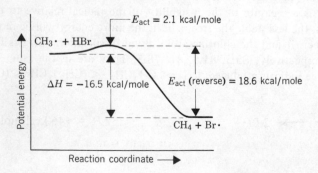

(c) $CH_3—CH_3 \longrightarrow 2\,CH_3\cdot$ ΔH $= +88$ kcal/mole
       $(D=88)$ E_{act} $= +88$ kcal/mole

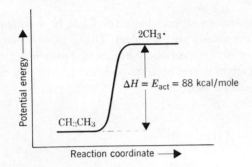

$\Delta H = E_{act}$ for any reaction in which bonds are broken but no bonds are formed.

(d) $Br—Br \longrightarrow 2\,Br\cdot$ ΔH $= +46$ kcal/mole
       $(D=46)$ E_{act} $= 46$ kcal/mole

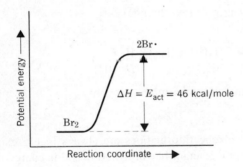

(e) $2\,Cl\cdot \longrightarrow Cl—Cl$ ΔH $= -58$ kcal/mole
       $(D=58)$ E_{act} $= 0$ kcal/mole

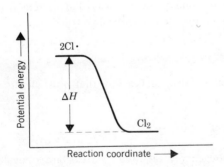

4.7

If all 10 hydrogen atoms of isobutane were equally reactive, the relative amounts of reaction at primary hydrogens and at tertiary hydrogens would be 9/1, i.e., the ratio of isobutyl chloride to *tert*-butyl chloride would be 9/1. Since the ratio is instead 62.5/37.5

(1.67), the tertiary hydrogen atom must be more reactive than the primary hydrogen atoms.

4.8

Laboratory preparation of alkyl halides by direct chlorination can be accomplished in good yield when all hydrogens in the alkane are equivalent. This is true of neopentane, and cyclopentane. (In these cases, the preparation would be practical only for monochlorination, where an excess of hydrocarbon would be employed, or for complete chlorination where an excess of chlorine would be used.)

4.9

(a) $Cl \cdot + CH_3CH_2-H \longrightarrow CH_3CH_2 \cdot + H-Cl$
    $(D=98)$ $(D=103)$

$\Delta H = 98 - 103 = -5 \text{ kcal/mole (exothermic)}$

(b) $Cl \cdot + (CH_3)_2CH-H \longrightarrow (CH_3)_2CH \cdot + H-Cl$
    $(D=94.5)$ $(D=103)$

$\Delta H = 94.5 - 103 = -8.5 \text{ kcal/mole (exothermic)}$

(c) $Cl \cdot + CH_3CH_2CH_2-H \longrightarrow CH_3CH_2CH_2 \cdot + H-Cl$
    $(D=98)$ $(D=103)$

$\Delta H = 98 - 103 = -5 \text{ kcal/mole (exothermic)}$

4.10

The hydrogen abstraction steps in alkane fluorinations are always highly exothermic. Thus the transition states are even more reactant-like in structure and in energy than they are in alkane chlorinations. The type of C—H bond being broken ($1°$, $2°$, or $3°$) has practically no effect on the relative rates of the reactions.

4.11

(a) Homolysis is cleavage of a covalent bond in such a way that the electrons of the ruptured bond are divided equally between the atoms involved:

$$:\ddot{C}l-\ddot{C}l: \longrightarrow :\ddot{C}l \cdot + \cdot \ddot{C}l:$$

(b) Heterolysis is cleavage of a covalent bond in such a way that both electrons of the ruptured bond remain with one atom. Ions are formed:

$$H-\ddot{C}l: \longrightarrow H^+ + :\ddot{C}l:^-$$

(c) Bond dissociation energy (D) is the energy required to dissociate a covalent bond homolytically:

$$H-H \longrightarrow 2H \cdot \quad D = 104 \text{ kcal/mole}$$

(d) A free radical is an atom or group that has an unpaired electron.

$$: \overset{\cdot\cdot}{\underset{\cdot\cdot}{Br}} \cdot \qquad \text{or} \qquad CH_3 \cdot$$

(e) A carbocation is an ion that has a trivalent carbon atom that bears a positive charge:

$$CH_3 - \overset{\overset{\displaystyle CH_3}{|}}{\underset{\underset{\displaystyle CH_3}{\diagdown}}{\overset{\diagup +}{C}}}$$

(f) A carbanion is an ion that has a trivalent carbon atom that bears an unshared electron pair and a negative charge.

$$H - \overset{\overset{\displaystyle H}{|}}{\underset{\underset{\displaystyle H}{|}}{C}} :^{-}$$

4.12

$$CH_3\overset{\overset{\displaystyle CH_3}{|}}{\underset{\underset{\displaystyle \cdot}{}}{C}}CH_2CH_3 \;>\; CH_3\overset{\overset{\displaystyle CH_3}{|}}{\underset{\underset{\displaystyle \cdot}{}}{CH}}CHCH_3 \;>\; \cdot CH_2\overset{\overset{\displaystyle CH_3}{|}}{CH}CH_2CH_3 \;\cong\; CH_3\overset{\overset{\displaystyle CH_3}{|}}{CH}CH_2CH_2 \cdot$$

4.13

(a) $\quad CH_3\overset{\overset{\displaystyle Br}{|}}{\underset{\underset{\displaystyle CH_3}{|}}{C}}CH_2CH_3$, $\quad$ because the tertiary hydrogen atom is much more reactive than either the primary or secondary hydrogen atoms.

(b) $\;CH_3\overset{\overset{\displaystyle CH_3}{|}}{CH}CH_2CH_3 + Cl\cdot \longrightarrow \cdot CH_2\overset{\overset{\displaystyle CH_3}{|}}{CH}CH_2CH_3$

$$CH_3 - \overset{\overset{\displaystyle CH_3}{|}}{\underset{\underset{\displaystyle \cdot}{}}{C}} - CH_2CH_3$$

$$CH_3 - \overset{\overset{\displaystyle CH_3}{|}}{\underset{\underset{\displaystyle \circ}{}}{CH}} - CHCH_3$$

$$CH_3 - \overset{\overset{\displaystyle CH_3}{|}}{CH} - CH_2CH_2 \cdot$$

$\xrightarrow{\;Cl_2\;}$

$$Cl - CH_2\overset{\overset{\displaystyle CH_3}{|}}{CH}CH_2CH_3 + Cl\cdot$$
$+$
$$CH_3\overset{\overset{\displaystyle CH_3}{|}}{\underset{\underset{\displaystyle Cl}{|}}{C}} - CH_2CH_3 + Cl\cdot$$
$+$
$$CH_3\overset{\overset{\displaystyle CH_3}{|}}{CH} - \overset{\underset{\underset{\displaystyle Cl}{|}}{}}{CH}CH_3 + Cl\cdot$$
$+$
$$CH_3\overset{\overset{\displaystyle CH_3}{|}}{CH}CH_2CH_2 - Cl + Cl\cdot$$

(c) Chlorine is more reactive than bromine, and is therefore less selective. See pages 146-149.

4.14

(a) $Cl_2 \longrightarrow 2\,Cl\cdot$ (b) $2\,Cl\cdot \longrightarrow Cl_2$

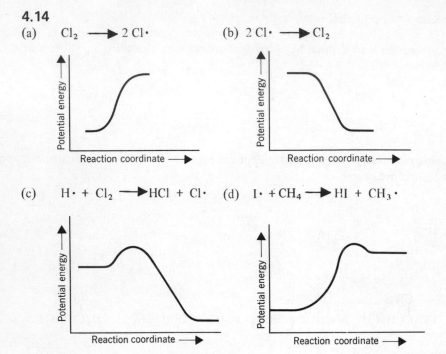

(c) $H\cdot + Cl_2 \longrightarrow HCl + Cl\cdot$ (d) $I\cdot + CH_4 \longrightarrow HI + CH_3\cdot$

4.15

(a) $Cl_2 \xrightarrow[\text{light}]{\text{heat or}} 2\,Cl\cdot$ Chain-initiating step

(b)

$$\left. \begin{array}{l} \begin{array}{l} \Delta H = -5 \\ \text{kcal/mole} \end{array} \quad \begin{array}{l} Cl\cdot + CH_3CH_2{-}H \longrightarrow H{-}Cl + CH_3CH_2\cdot \\ \qquad\qquad (D{=}98) \qquad\quad (D{=}103) \end{array} \\[14pt] \begin{array}{l} \Delta H = -23.5 \\ \text{kcal/mole} \end{array} \quad \begin{array}{l} CH_3CH_2\cdot + Cl_2 \longrightarrow CH_3CH_2{-}Cl + Cl\cdot \\ \qquad\quad (D{=}58) \qquad\qquad (D{=}81.5) \end{array} \end{array} \right\} \begin{array}{l} \text{Chain-} \\ \text{propagating} \\ \text{steps} \end{array}$$

$$\left. \begin{array}{l} CH_3CH_2\cdot + Cl\cdot \longrightarrow CH_3CH_2Cl \\ 2CH_3CH_2\cdot \longrightarrow CH_3CH_2CH_2CH_3 \\ 2Cl\cdot \longrightarrow Cl_2 \end{array} \right\} \text{Chain-terminating steps}$$

(c) The bond dissociation energy of the $CH_3CH_2{-}H$ bond (98 kcal/mole) is smaller than that of the $CH_3{-}H$ bond (104 kcal/mole), therefore ethane reacts with $Cl\cdot$ faster than methane does.

4.16

(a) (2) $Cl\cdot + CH_3{-}CH_3 \longrightarrow CH_3{-}Cl + CH_3\cdot$ $\Delta H = +4.5$ kcal/mole
    $\qquad\qquad (D{=}88) \qquad\qquad (D{=}83.5)$ $E_{act} > +4.5$ kcal/mole

(3) $CH_3\cdot + Cl_2 \longrightarrow CH_3{-}Cl + Cl\cdot$ $\Delta H = -25.5$ kcal/mole
    $\qquad\quad (D{=}58) \quad (D{=}83.5)$ E_{act} is small

In this reaction, step 2 is endothermic ($\Delta H = +4.5$ kcal/mole) and thus E_{act} must be greater than $+4.5$ kcal/mole. Although we do not know the exact E_{act} of the reaction that yields ethyl chloride (problem 4.15), we can assume that it is less than 3.8 kcal/mole (E_{act} for the corresponding step in the chlorination of methane). Therefore we conclude that the reaction here with an E_{act} greater than $+ 4.5$ kcal/mole, will not compete with the reaction of problem 4.15.

(b) (1) $\quad$ F–F $\longrightarrow$ 2 F· $\qquad\qquad\qquad\qquad$ $\Delta H = + 38$ kcal/mole
$\qquad$ (D=38)

(2) F· + CH$_3$–CH$_3$ $\longrightarrow$ $\quad$ CH$_3$–F + CH$_3$· $\qquad$ $\Delta H = - 20$ kcal/mole
$\qquad$ (D=88) $\qquad\qquad\qquad$ (D=108) $\qquad\qquad$ $E_{act} > 0$

(3) CH$_3$· + F–F $\longrightarrow$ $\quad$ CH$_3$–F + F· $\qquad\qquad$ $\Delta H = - 70$ kcal/mole
$\qquad\quad$ (D=38) $\qquad\qquad$ (D=108) $\qquad\qquad\qquad$ $E_{act} > 0$

Since the propagation steps are both highly exothermic it is possible for each E_{act} to be quite small, and therefore for the reaction to take place.

4.17

(a) CH$_3$–H + F–F $\longrightarrow$ CH$_3$· + H–F + F· $\quad$ $\Delta H = + 6$ kcal/mole
$\quad$ (D=104) $\quad$ (D=38) $\qquad\qquad\qquad$ (D=136) $\qquad$ $E_{act} > 6$ kcal/mole

$\quad$ CH$_3$· + F· $\longrightarrow$ CH$_3$–F $\qquad\qquad\qquad\qquad\quad$ $\Delta H = -108$ kcal/mole
$\qquad\qquad\qquad\quad$ (D=108) $\qquad\qquad\qquad\qquad\qquad$ $E_{act} = 0$

If E_{act} for the first step is not much greater than 6 kcal/mole, this mechanism is likely.

(b) CH$_3$–H + Cl–Cl $\longrightarrow$ CH$_3$· + H–Cl + Cl· $\quad$ $\Delta H = + 59$ kcal/mole
$\quad$ (D=104) $\quad$ (D=58) $\qquad\qquad\qquad$ (D=103) $\qquad$ $E_{act} \geq 59$ kcal/mole

$\quad$ CH$_3$· + Cl· $\longrightarrow$ CH$_3$–Cl $\qquad\qquad\qquad\qquad$ $\Delta H = -83.5$ kcal/mole
$\qquad\qquad\qquad\quad$ (D=83.5) $\qquad\qquad\qquad\qquad$ $E_{act} = 0$

This mechanism is highly unlikely because the E_{act} for the first step must be ≥ 59 kcal/mole.

4.18

4.19

(b) CH$_3$CH$_2$CH$_2^+$ + I$^-$ $\longrightarrow$ CH$_3$CH$_2$CH$_2$–I

4.20

(a) $\quad$ CH$_3$–H $\qquad$ D = 104, $\qquad$ CH$_3$CH$_2$–H $\qquad$ D = 98 kcal/mole. (Recall that here, $E_{act} = D$.)

CH_3CH_2-H bond rupture requires less energy, therefore spontaneous homolysis (cracking) occurs at a lower temperature.

(b) CH_3-CH_3 D = 88 kcal/mole = E_{act}

C—C bond rupture requires less energy than C—H bond rupture, therefore C—C bond rupture occurs at a lower temperature than CH_3CH_2-H bond rupture.

(c) $CH_3CH_2-CH_2CH_3$ D = 82 kcal/mole = E_{act}

$CH_3CH_2CH_2-CH_3$ D = 85 kcal/mole = E_{act}

Here again the bond with the lower bond dissociation energy will undergo spontaneous homolysis (cracking) at the lower temperature.

4.21

(1) $CH_3CH_2CH_3 \longrightarrow CH_3CH_2 \cdot + CH_3 \cdot$ D = 85 kcal/mole
(2) $CH_3CH_2CH_3 \longrightarrow CH_3CH_2CH_2 \cdot + H \cdot$ D = 98 kcal/mole
(3) $CH_3CH_2CH_3 \longrightarrow CH_3CHCH_3 + H \cdot$ D = 94.5 kcal/mole

(a) Since E_{act} is equal to D, we can assume that (1) is the most likely chain-initiating step.

(b) $CH_3 \cdot + CH_3CH_2CH_3 \longrightarrow CH_3-H + \cdot CH_2CH_2CH_3$ ΔH = -6 kcal/mole
        $(D=98)$ $(D=104)$

Since ΔH is negative, E_{act} need not be large.

(c) $CH_3 \cdot + CH_3CH_2CH_3 \longrightarrow CH_4 + CH_3CHCH_3$ ΔH = -9.5 kcal/mole
        $(D=94.5)$ $(D=104)$

On the basis of energy requirements, this is a likely alternative to step 1. On the basis of the probability factor, it is less likely because there are only two secondary hydrogen atoms compared with six primary hydrogen atoms.

5
ALKENES:
STRUCTURE AND
SYNTHESIS

5.1
Absorption of a photon of the correct frequency can excite a π electron into an anti-bonding orbital. Such an orbital has a nodal plane between the carbon atoms, thus rotation about the C—C bond can occur (see pp. 382-384).

5.2
(a) $CH_2{=}CCH_3$ + H_2 $\longrightarrow$ CH_3CHCH_3 (isobutane)
$\qquad$ | $\qquad\qquad\qquad\qquad\qquad\qquad$ |
$\qquad$ CH_3 $\qquad\qquad\qquad\qquad\qquad\qquad$ CH_3

(b) C_4H_8 + $6O_2$ $\longrightarrow$ $4CO_2$ + $4H_2O$

(c) Yes, because the same molar amounts of the same combustion products are formed from all of the C_4H_8 isomers.

(d) $CH_2{-}CH_2$ (cyclobutane) and $CH_2 \underset{}{\overset{CH_2}{\diagdown \diagup}} CHCH_3$ (methylcyclopropane)
$\quad\ \ |\qquad |$
$\quad\ \ CH_2{-}CH_2$

(e) Yes. (See answer to (c)).

5.3
(a) No.

(b) Protonation followed by loss of water gives a secondary carbocation which can lose a proton two different ways:

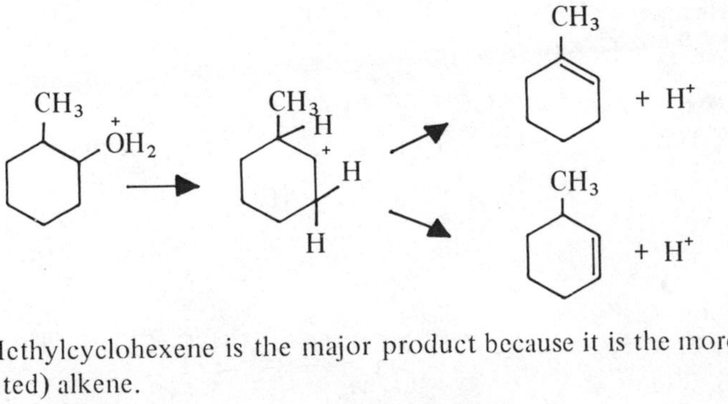

(c) 1-Methylcyclohexene is the major product because it is the more stable (more highly substituted) alkene.

5.4
(a) 2-Methyl-2-butene

(b) *Cis* - 4-octene

(c) 1-Bromo-2-methylpropene

(d) 4-Methylcyclohexene

5.5

(a)   (b)   (c) $CH_3C=CCH_2CH_3$
 $\overset{|}{C}H_3\ \overset{|}{C}H_3$

(d)   (e)

(f) $CH_2=CHCCl_3$ (g) $CH_2=CCH_3$ (h) $CH_3CH=CH_2$
$\overset{|}{C}H_3$

(i) $CH_2=CHCH_2CHCH_3$ (j) $CH_2=CH-\!\triangleleft$

5.6

(a) $CH_2=CHCH_2CH_2CH_3$

 1-Pentene *Cis* -2-pentene *Trans* -2-pentene

$CH_2=CCH_2CH_3$ $CH_2=CH-CHCH_3$
$\overset{|}{C}H_3 \overset{|}{C}H_3$

2-Methyl-2-butene 2-Methyl-1-butene 3-Methyl-1-butene

(b) $CH_2=CHCH_2CH_2CH_2CH_3$

 1-Hexene *Cis* -2-hexene *Trans* -2-hexene

     $CH_2=CCH_2CH_2CH_3$
$\overset{|}{C}H_3$

 *Cis* -3-hexene *Trans* -3-hexene 2-Methyl-1-pentene

$CH_2=CHCHCH_2CH_3$ $CH_2=CHCH_2CHCH_3$ $CH_3C=CHCH_2CH_3$
$\overset{|}{C}H_3 \overset{|}{C}H_3 \overset{|}{C}H_3$

 3-Methyl- 4-Methyl- 2-Methyl-
 1-pentene 1-pentene 2-pentene

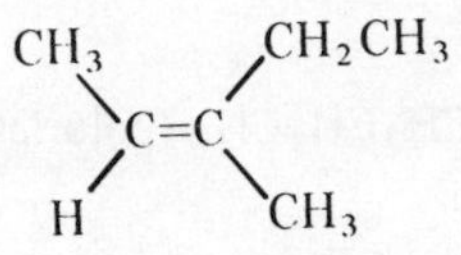
CH3
CH2CH3
C=C
H
CH3

Cis-3-methyl-
2-pentene*

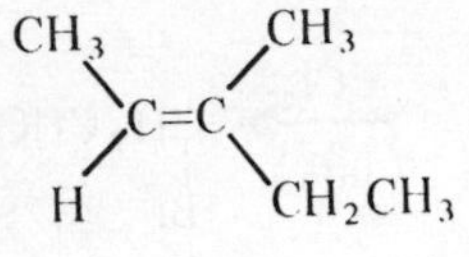
CH3
CH3
C=C
H
CH2CH3

Trans-3-methyl-
2-pentene*

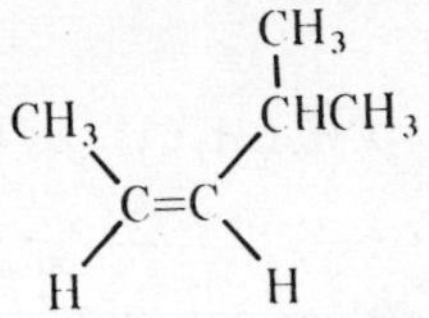
CH3
CH3
CHCH3
C=C
H
H

Cis-4-methyl-
2-pentene

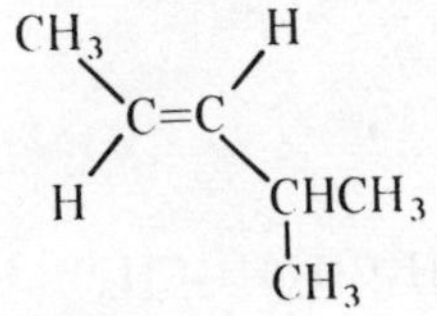
CH3
H
C=C
H
CHCH3
CH3

Trans-4-methyl-
2-pentene

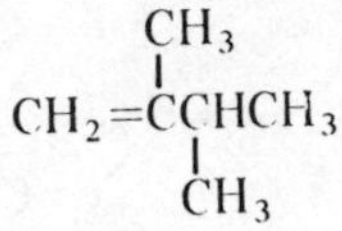
CH3
CH2=CCHCH3
CH3

2, 3-dimethyl-
1-Butene

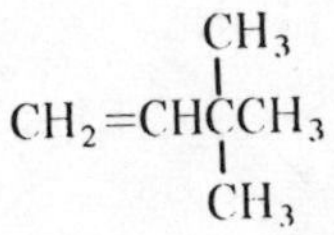
CH3
CH2=CHCCH3
CH3

3, 3-dimethyl-
1-Butene

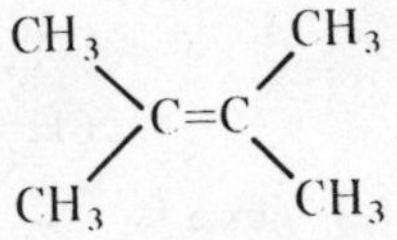
CH3
CH3
C=C
CH3
CH3

2, 3-Dimethyl-
2-butene

(c)

C_5H_{10} :

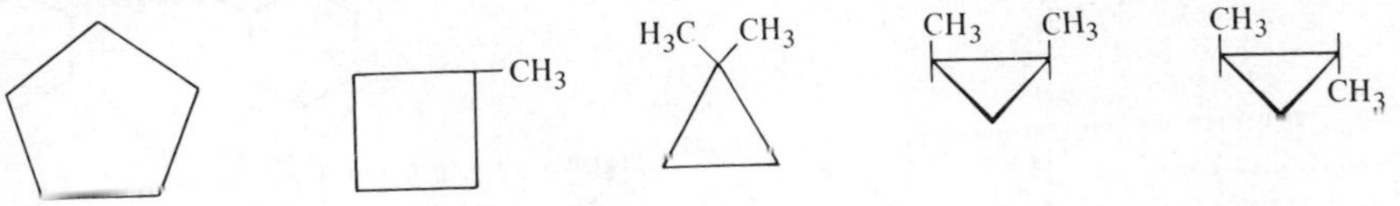
CH3
H3C
CH3
CH3
CH3
CH3
CH3

C_6H_{12} :

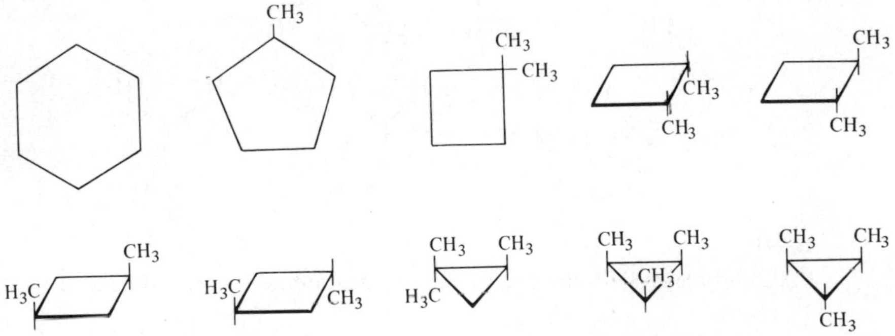
CH3
CH3
CH3
CH3
CH3
CH3
CH3
CH3
H3C
H3C
CH3
CH3
CH3
H3C
CH3
CH3
CH3
CH3
CH3
CH3

5.7

(a) 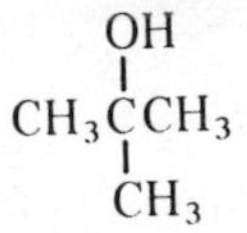 or 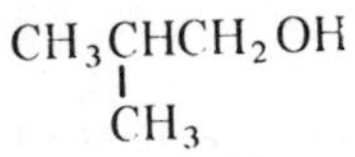
OH
CH3CCH3
CH3
CH3CHCH2OH
CH3

(c) 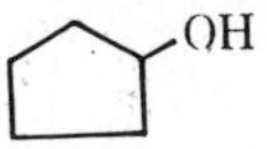
OH

(b) 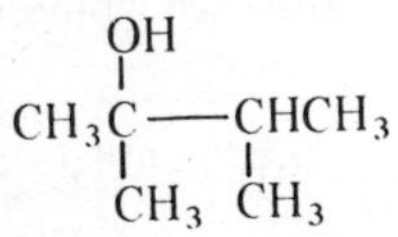 or 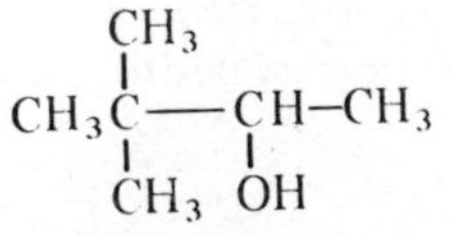
OH
CH3C—CHCH3
CH3 CH3
CH3
CH3C—CH—CH3
CH3 OH

(d) 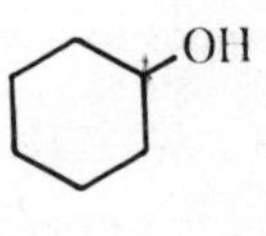
OH

* The *cis-trans* designation here is ambiguous. See pp. 274 - 276.

5.8

$$CH_2=CHCH_2CH_2CH_2CH_3 + Br_2 \xrightarrow[\substack{(dark)\\ R.T.}]{CCl_4} CH_2CHCH_2CH_2CH_2CH_3 \text{ (colorless)}$$

with Br and Br groups below.

Cyclohexane does not react with Br_2 in the dark at room temperature, thus the red-brown color of the bromine will persist in the solution.

5.9

(a) No   (b) No

(c) Yes

(d) No   (e) No   (f) No   (g) Yes

(h) Yes

(i) No

(j) No   (k) No   (l) Yes

5.10

(a) 2,3-dimethyl-2-butene > 2-methyl-2-pentene > *trans*-3-hexene > *cis*-2-hexene > 1-hexene.

(b) The only alkenes whose relative stabilities could be measured by comparative heats of hydrogenation are those that yield the same hydrogenation product; i.e., *trans*-3-hexene, 1-hexene, *cis*-2-hexene all yield hexane on hydrogenation.

5.11

Although *trans* molecules are usually more stable than their *cis* isomers, in the case of cyclooctene, the *trans* isomer is probably more strained than the *cis* isomer because the ring is too small to allow a strain-free *trans* configuration. Therefore we would expect the *trans* isomer to have the higher heat of hydrogenation.

5.12

(a) *Cis-trans* isomerization caused by rupture of the π-bond.

(b) Equilibrium should favor the *trans* isomer because it is more stable than the *cis* isomer.

5.13

(a) An sp^2 hybridized carbon atom has more s-character than one that is sp^3 hybridized, thus electrons in these sp^2 orbitals are closer to the nucleus than those in sp^3 orbitals. A bond between an sp^2 hybridized carbon and an sp^3 hybridized carbon is polarized toward the sp^2 hybridized carbon.

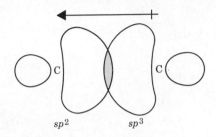

(b) The greater acidity of ethene can be explained by the greater stability of its conjugate base, $CH_2=\overset{..}{C}H^-$. The greater stability of $CH_2=\overset{..}{C}H^-$ compared with $CH_3CH_2^-$ can be explained by the greater s-character of the sp^2 orbital that holds the unbonded electron pair in $CH_2=\overset{..}{C}H^-$. In $CH_3CH_2^-$, the electrons are in an sp^3 orbital. Being farther from the nucleus, the sp^3 electrons have a higher potential energy.

5.14

(a) $\underset{\displaystyle \overset{|}{CH_3}}{CH_3CH=CCH_3}$ (major) + $\underset{\displaystyle \overset{|}{CH_3}}{CH_2=CHCHCH_3}$

(b) $\underset{\displaystyle \overset{|}{CH_3}}{CH_3CH_2C=CH_2}$ (c) $CH_3CH=CHCH_2CH_3$ (*trans*-predominates)

(d) $\underset{H}{\overset{CH_3}{>}}C=C\underset{CH_2CH_3}{\overset{H}{<}}$ (major) + $\underset{H}{\overset{CH_3}{>}}C=C\underset{H}{\overset{CH_2CH_3}{<}}$ + $CH_2=CHCH_2CH_2CH_3$

(e) ⬡=CH₂   (f) ⬡—CH₃ (major product) + ⬡=CH₂

5.15

$$\underset{\displaystyle \overset{|}{CH_3}}{\overset{\displaystyle \overset{OH}{|}}{CH_3CCH_2CH_3}} > \underset{\displaystyle \overset{|}{CH_3}}{\overset{\displaystyle \overset{OH}{|}}{CH_3CHCHCH_3}} > \underset{\displaystyle \overset{|}{CH_3}}{CH_3CHCH_2CH_2OH}$$

The order of reactivity is dictated by the order of stability of the intermediate carbocations: tertiary > secondary > primary

5.16

(a) *Cis*-1, 2-dimethylcyclopentane

(b) *Cis*-1, 2-dimethylcyclohexane:

(cyclohexane chair structure with CH₃ and CH₃ substituents)

(c) *Cis*-1, 2-dideuteriocyclohexane:

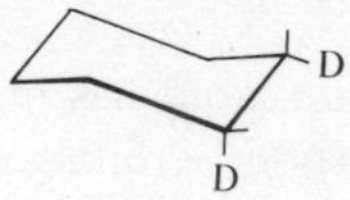

5.17

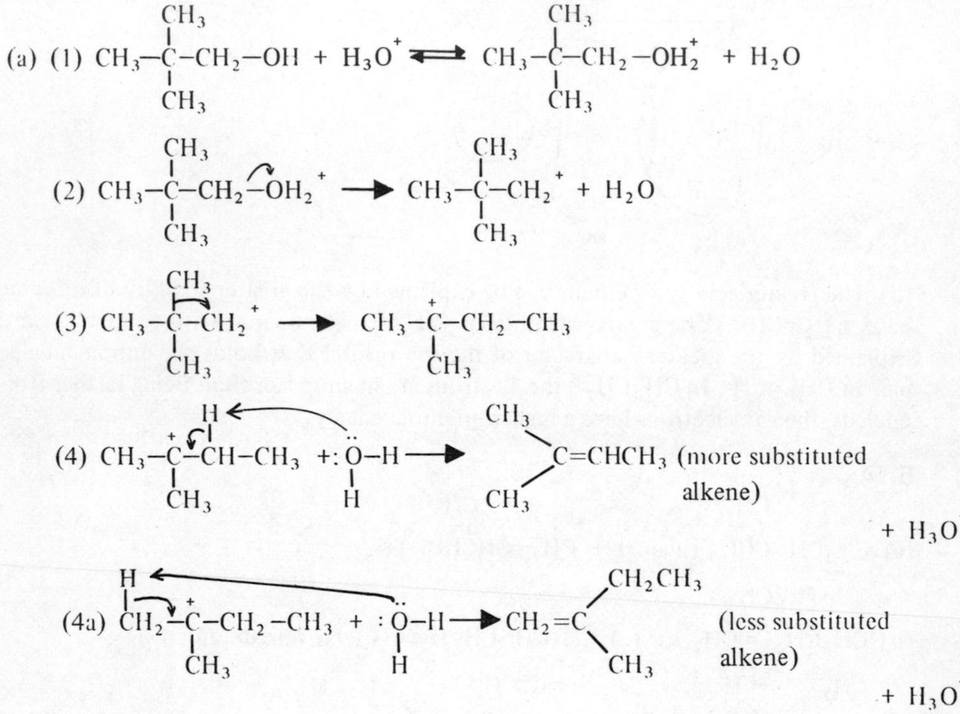

(a) (1) $CH_3-\underset{\underset{CH_3}{|}}{\overset{\overset{CH_3}{|}}{C}}-CH_2-OH + H_3O^+ \rightleftharpoons CH_3-\underset{\underset{CH_3}{|}}{\overset{\overset{CH_3}{|}}{C}}-CH_2-\overset{+}{O}H_2 + H_2O$

(2) $CH_3-\underset{\underset{CH_3}{|}}{\overset{\overset{CH_3}{|}}{C}}-CH_2-\overset{+}{O}H_2 \longrightarrow CH_3-\underset{\underset{CH_3}{|}}{\overset{\overset{CH_3}{|}}{C}}-CH_2^+ + H_2O$

(3) $CH_3-\underset{\underset{CH_3}{|}}{\overset{\overset{CH_3}{|}}{C}}-CH_2^+ \longrightarrow CH_3-\underset{\underset{CH_3}{|}}{\overset{+}{C}}-CH_2-CH_3$

(4) $CH_3-\underset{\underset{CH_3}{|}}{\overset{+}{C}}-\overset{\overset{H}{|}}{C}H-CH_3 + :\overset{\overset{..}{}}{\underset{\underset{H}{|}}{O}}-H \longrightarrow \underset{CH_3}{\overset{CH_3}{}}C=CHCH_3$ (more substituted alkene)

$+ H_3O^+$

(4a) $\overset{\overset{H}{|}}{C}H_2-\underset{\underset{CH_3}{|}}{\overset{+}{C}}-CH_2-CH_3 + :\overset{..}{\underset{\underset{H}{|}}{O}}-H \longrightarrow CH_2=C\overset{CH_2CH_3}{\underset{CH_3}{}}$ (less substituted alkene)

$+ H_3O^+$

(Steps 2 and 3 may occur at the same time.)

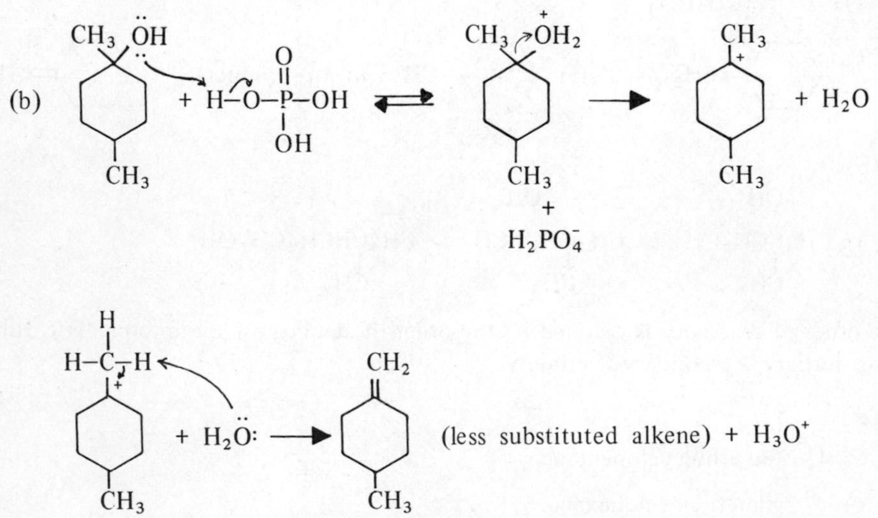

(b)

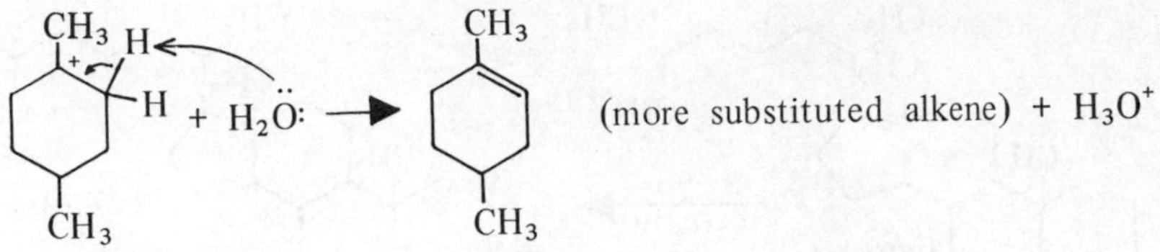

(more substituted alkene) + H_3O^+

(c)

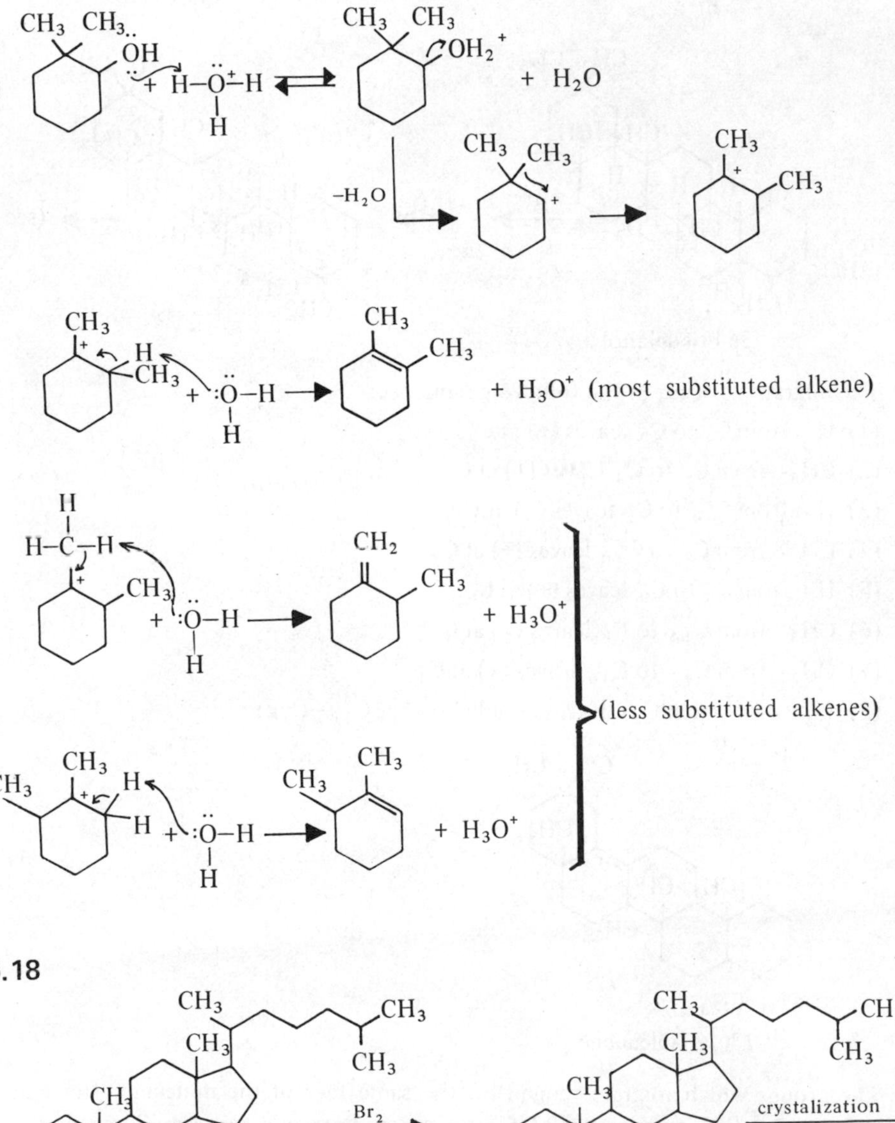

$+ H_2O$

$+ H_3O^+$ (most substituted alkene)

$+ H_3O^+$

$+ H_3O^+$

(less substituted alkenes)

5.18

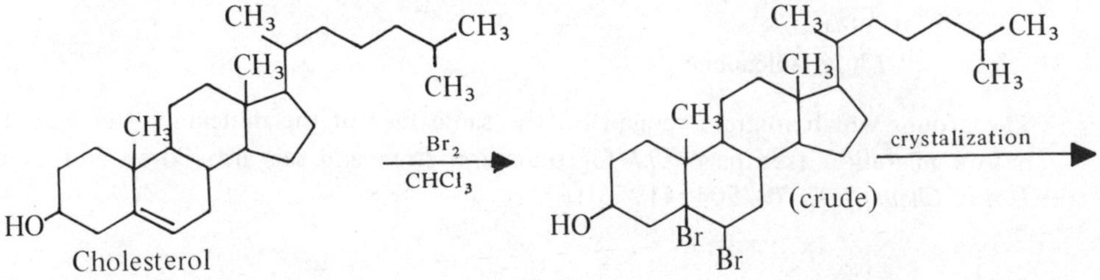

Cholesterol

$\xrightarrow[\text{CHCl}_3]{\text{Br}_2}$

(crude)

$\xrightarrow{\text{crystalization}}$

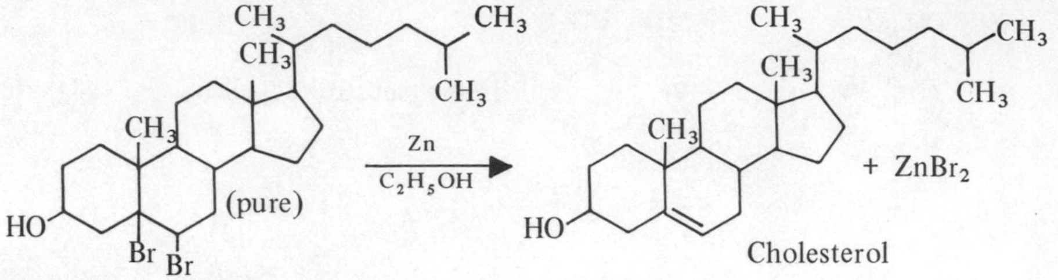

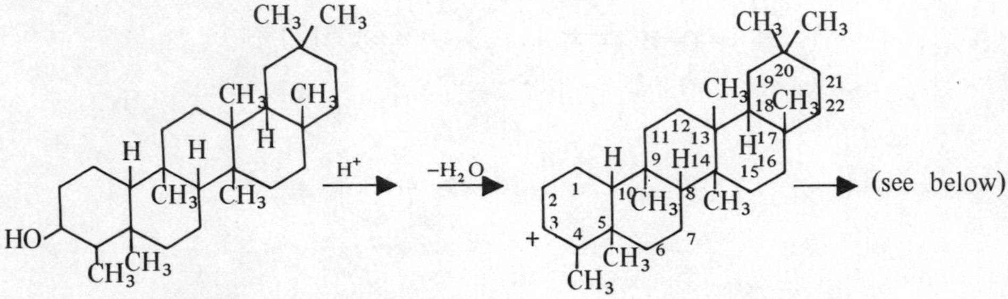

Cholesterol

3β-Friedelanol

The migrations occur in the following sequence:

(1) $H:^-$ from C_4 to C_3 leaves (+) at C_4.

(2) $CH_3:^-$ from C_5 to C_4 leaves (+) at C_5.

(3) $H:^-$ from C_{10} to C_5 leaves (+) at C_{10}.

(4) $CH_3:^-$ from C_9 to C_{10} leaves (+) at C_9.

(5) $H:^-$ from C_8 to C_9 leaves (+) at C_8.

(6) $CH_3:^-$ from C_{14} to C_8 leaves (+) at C_{14}.

(7) $CH_3:^-$ from C_{13} to C_{14} leaves (+) at C_{13}.

(8) Loss of H^+ from C_{18} leaves double bond at $C_{13}-C_{18}$:

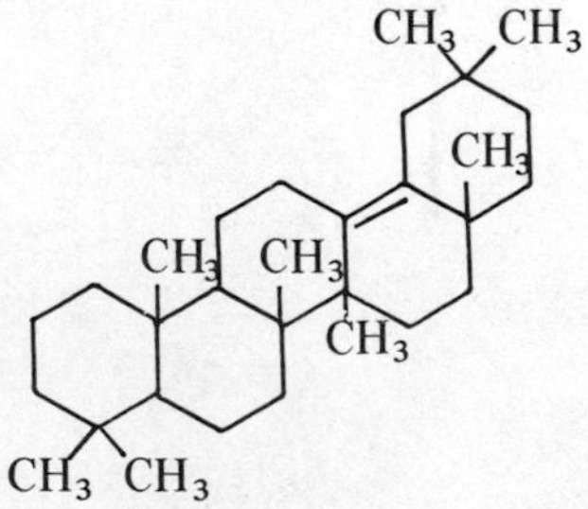

13(18)-Oleanene

The groups which migrate remain on the same face of the molecule after migration as before migration (see page 177 for transition state and see also Corey and Ursprung, *J. Am. Chem. Soc.*, **78**, 5041 (1956)).

6
REACTIONS OF ALKENES: ADDITION REACTIONS OF THE CARBON-CARBON DOUBLE BOND

6.1

$$CH_2-CH-CH_3 \quad \text{2-chloro-1-iodopentane}$$
$$\overset{|}{I} \quad \overset{|}{Cl}$$

6.2

(a) $CH_3CH_2CH{=}CH_2 + H{-}\ddot{I}: \rightleftharpoons CH_3CH_2\overset{+}{C}HCH_3 + :\ddot{I}:^- \rightarrow CH_3CH_2CHCH_3$
$$\overset{|}{I}$$

(b)
$$\begin{array}{cc} CH_3 & CH_3 \\ \diagdown \diagup \\ C{=}C \\ \diagup \diagdown \\ CH_3 & H \end{array} + \overset{\delta+}{:\ddot{I}}-\overset{\delta-}{\ddot{B}r}: \rightleftharpoons \begin{array}{cc} CH_3 & CH_3 \\ \diagdown \diagup \\ C-C-H \\ \diagup \overset{+}{|} \\ CH_3 & I \end{array} + :\ddot{B}r:^-$$

$$\rightarrow \begin{array}{c} CH_3 \\ | \\ CH_3C-CHCH_3 \\ | | \\ Br I \end{array}$$

(c) cyclohexene derivative $+ H{-}\ddot{C}l: \rightleftharpoons$ (cyclohexyl cation) $+ :\ddot{C}l:^- \rightarrow$ (product)

6.3

$$\begin{array}{c} CH_3 \\ | \\ CH_3-C-CH{=}CH_2 \\ | \\ CH_3 \end{array} + H{-}\ddot{C}l: \rightleftharpoons \begin{array}{c} CH_3 \\ | \\ CH_3-C-\overset{+}{C}H-CH_3 \\ | \\ CH_3 \end{array} + :\ddot{C}l:^-$$

$$\begin{array}{c} CH_3 \\ | \\ CH_3-C-\overset{+}{C}H-CH_3 \\ | \\ CH_3 \end{array} + :\ddot{C}l:^- \rightarrow \begin{array}{c} CH_3 \\ | \\ CH_3-C-CH{-}CH_3 \\ | | \\ CH_3 Cl \end{array}$$

or

$$\begin{array}{c} CH_3 \\ | \\ CH_3-\overset{+}{C}{-}CH{-}CH_3 \\ | \\ CH_3 \end{array} \rightarrow \begin{array}{c} CH_3 \\ | \\ CH_3-C-CH-CH_3 \\ | \\ CH_3 \end{array} \xrightarrow{:\ddot{C}l:^-} \begin{array}{c} Cl CH_3 \\ | | \\ CH_3-C-CH-CH_3 \\ | \\ CH_3 \end{array}$$

6.4

(a) $CH_3-CH=CH_2 + H-\overset{\cdot\cdot+}{\underset{H}{O}}-H \rightleftharpoons CH_3-\overset{+}{C}H-CH_3 + H_2O$

$CH_3-\overset{+}{C}H-CH_2 + :\overset{\cdot\cdot}{\underset{H}{O}}-H \rightleftharpoons CH_3-\underset{\underset{H}{\overset{+}{O}-H}}{CH}-CH_3$

$CH_3-\underset{\overset{H}{\underset{H}{:\overset{+}{O}-H}}}{CH}-CH_3 + :\overset{\cdot\cdot}{\underset{H}{O}}-H \rightleftharpoons CH_3-\underset{\overset{OH}{}}{CH}-CH_3 + H_3O^+$

(b) The product is isopropyl alcohol because the more stable isopropyl carbocation is produced in the first step. The formation of *n*-propyl alcohol would require the production of the less stable *n*-propyl carbocation.

6.5

$CH_3CH_2CH_2CH_2CH=CH_2 + Hg^+OAc \rightarrow CH_3CH_2CH_2CH_2-\overset{+}{C}H-CH_2-HgOAc$

$CH_3CH_2CH_2CH_2-\underset{\overset{HgOAc}{}}{\overset{+}{C}H}-CH_2 + :\overset{\cdot\cdot}{\underset{H}{O}}-CH_2CH_3 \rightarrow CH_3CH_2CH_2CH_2-\underset{\overset{H}{:\overset{+}{O}-CH_2CH_3}}{CH}-CH_2-HgOAc$

$\downarrow -H^+$

$CH_3CH_2CH_2CH_2-\underset{\overset{O-CH_2CH_3}{}}{CH}-CH_3 \xleftarrow[OH^-]{NaBH_4} CH_3CH_2CH_2CH_2-\underset{\overset{OCH_2CH_3}{}}{CH}-CH_2-HgOAc$

6.6

(a) $CH_3-\underset{\overset{CH_3}{}}{C}=CH-CH_3 + Hg(OAc)_2 + H_2O \xrightarrow{THF} CH_3-\underset{\overset{CH_3}{}}{\overset{OH}{C}}-\underset{}{\overset{CH_3}{C}H}-HgOAc$

$\xrightarrow[OH^-]{NaBH_4} CH_3-\underset{\overset{CH_3}{}}{\overset{OH}{C}}-CH_2-CH_3$

(b) $+ Hg(OAc)_2 + H_2O \xrightarrow{THF}$ $\xrightarrow[OH^-]{NaBH_4}$

(c) $+ Hg(OAc)_2 + CH_3OH \xrightarrow{THF}$ $\xrightarrow[OH^-]{NaBH_4}$

(d) $CH_3-\underset{\underset{CH_3}{|}}{\overset{\overset{CH_3}{|}}{C}}-CH=CH_2$ + $Hg(OAc)_2$ + CH_3OH $\xrightarrow{THF}$ $CH_3-\underset{\underset{CH_3}{|}}{\overset{\overset{CH_3}{|}}{C}}-\underset{\underset{OCH_3}{|}}{CH}-\overset{\overset{HgOAc}{|}}{CH_2}$

$\xrightarrow[OH^-]{NaBH_4}$ $CH_3-\underset{\underset{CH_3}{|}}{\overset{\overset{CH_3}{|}}{C}}-\underset{\underset{OCH_3}{|}}{CH}-CH_3$

6.7

(a) $3CH_3CH_2CH=CH_2$ + $\frac{1}{2}(BH_3)_2$ $\longrightarrow$ $(CH_3CH_2CH_2CH_2)_3B$ $\xrightarrow[OH^-]{H_2O_2}$

$3CH_3CH_2CH_2CH_2OH$ + H_3BO_3

(b) $3CH_3-\underset{\underset{CH_3}{|}}{C}=CH-CH_3$ + $\frac{1}{2}(BH_3)_2$ $\longrightarrow$ $(CH_3-\overset{\overset{CH_3}{|}}{CH}-\overset{\overset{CH_3}{|}}{CH})_3-B$ $\xrightarrow[OH^-]{H_2O_2}$

$3CH_3-\underset{\underset{CH_3}{|}}{CH}-\underset{\underset{OH}{|}}{CH}-CH_3$ + H_3BO_3

(c)

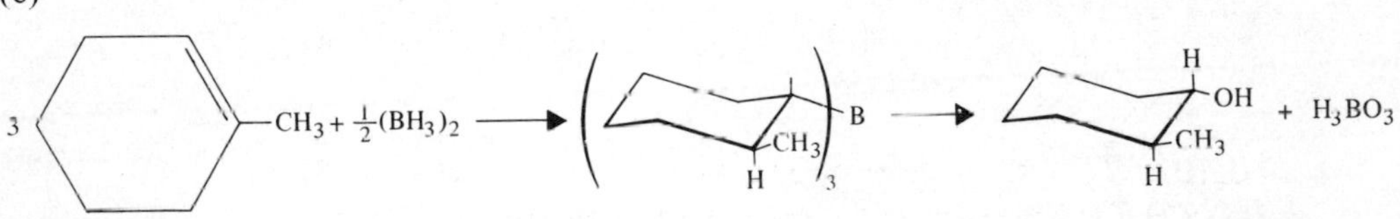

6.8

(a) $3CH_3-\overset{\overset{CH_3}{|}}{C}=CH_2$ + $\frac{1}{2}(BH_3)_2$ $\longrightarrow$ $(CH_3-\overset{\overset{CH_3}{|}}{CH}-CH_2)_3-B$

$\xrightarrow[heat]{CH_3COOD}$ $3CH_3-\overset{\overset{CH_3}{|}}{CH}-CH_2D$ + $(CH_3COO)_3B$

(b)

3 $-CH_2D$ + $(CH_3COO)_3B$

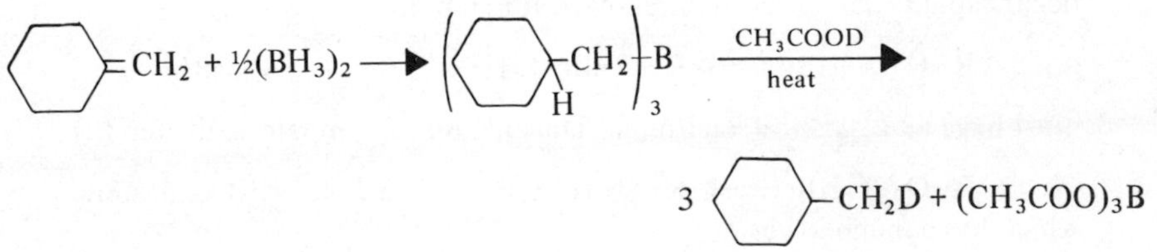

(c)

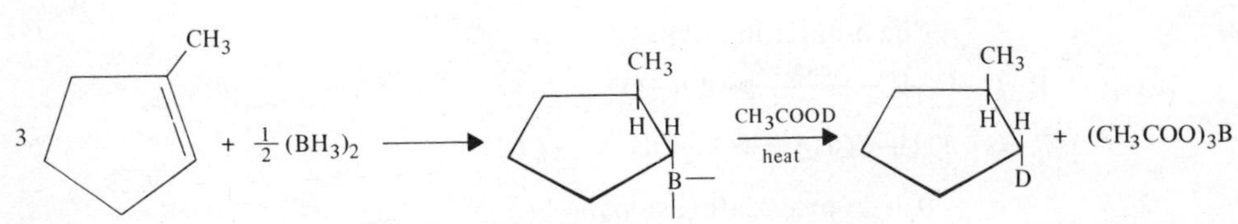

(d)

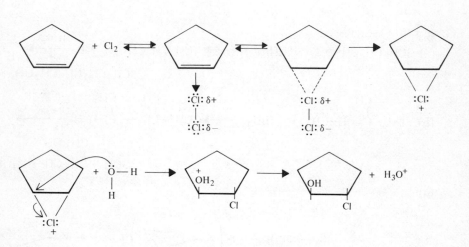

6.9

6.10

(a) $CH_3CH_2CH=CHCH_3$

(c) $CH_3CH_2CHCH=CH_2$
        |
        CH_3

(b) $CH_3C{=}CCH_3$
    (with CH_3 above each central carbon)

(d) [cyclohexene structure]

6.11

The initiation reaction (step 1) has no competitor and all of the subsequent steps must occur rapidly. The reaction disallowed for step 2,

$$R-\ddot{O}\cdot + HBr \xrightarrow{\text{//}} R-\ddot{O}-Br + H\cdot \qquad \Delta H \cong 39 \text{ kcal/mole}$$

must have an $E_{act} > 39$ kcal/mole. Thus it cannot compete with step (2),

$$R-\ddot{O}\cdot + HBr \longrightarrow R-\ddot{O}-H + Br\cdot \qquad \Delta H \cong -13 \text{ kcal/mole}$$

which has a much lower E_{act}.

6.12

Chain-initiating steps

(a) (1) $R-\ddot{O}-\ddot{O}-R \xrightarrow{\text{heat}} 2 R-\ddot{O}\cdot$

(2) $R-\ddot{O}\cdot + H-CCl_3 \longrightarrow R-\ddot{O}H + \cdot CCl_3$

Chain-propagating steps

(3) $CH_3CH_2CH_2CH=CH_2 + \cdot CCl_3 \longrightarrow CH_3CH_2CH_2CH-CH_2CCl_3$

(4) $CH_3CH_2CH_2\overset{\cdot}{C}HCH_2CCl_3 + H\text{–}CCl_3 \longrightarrow CH_3CH_2CH_2CH_2CH_2CCl_3$

$$+ \;\cdot CCl_3$$

then (3), (4), (3), (4), etc.

(b) (1) $R\text{–}\overset{\cdot\cdot}{\underset{\cdot\cdot}{O}}\text{–}\overset{\cdot\cdot}{\underset{\cdot\cdot}{O}}\text{–}R \xrightarrow{\text{heat}} 2\,R\text{–}\overset{\cdot\cdot}{\underset{\cdot\cdot}{O}}\cdot$

   (2) $R\text{–}\overset{\cdot\cdot}{\underset{\cdot\cdot}{O}}\cdot + CH_3CH_2\text{–}\overset{\cdot\cdot}{\underset{\cdot\cdot}{S}}\text{–}H \longrightarrow R\text{–}\overset{\cdot\cdot}{\underset{\cdot\cdot}{O}}H + CH_3CH_2\text{–}\overset{\cdot\cdot}{\underset{\cdot\cdot}{S}}\cdot$

       Chain-propagating steps

         CH_3 CH_3

(3) $CH_3\overset{|}{C}=CH_2 + \cdot\overset{\cdot\cdot}{\underset{\cdot\cdot}{S}}CH_2CH_3 \longrightarrow CH_3\overset{|}{\underset{\cdot}{C}}\text{–}CH_2\text{–}\overset{\cdot\cdot}{\underset{\cdot\cdot}{S}}\text{–}CH_2CH_3$

         CH_3 CH_3

(4) $CH_3\overset{|}{\underset{\cdot}{C}}CH_2SCH_2CH_3 + HSCH_2CH_3 \longrightarrow CH_3\overset{|}{C}HCH_2SCH_2CH_3 + \cdot\overset{\cdot\cdot}{S}CH_2CH_3$

then (3), (4), (3), (4), etc.

(c) (1) $R\text{–}\overset{\cdot\cdot}{\underset{\cdot\cdot}{O}}\text{–}\overset{\cdot\cdot}{\underset{\cdot\cdot}{O}}\text{–}R \xrightarrow{\text{heat}} 2\,R\text{–}\overset{\cdot\cdot}{\underset{\cdot\cdot}{O}}\cdot$

   (2) $R\text{–}\overset{\cdot\cdot}{\underset{\cdot\cdot}{O}}\cdot + Cl\text{–}CCl_3 \longrightarrow R\text{–}\overset{\cdot\cdot}{\underset{\cdot\cdot}{O}}\text{–}Cl + \cdot CCl_3$

       Chain-propagating steps

         CH_3 CH_3

(3) $CH_3CH_2\overset{|}{C}=CH_2 + \cdot CCl_3 \longrightarrow CH_3CH_2\overset{|}{\underset{\cdot}{C}}\text{–}CH_2CCl_3$

         CH_3 CH_3

(4) $CH_3CH_2\overset{|}{C}CH_2CCl_3 + CCl_4 \longrightarrow CH_3CH_2\overset{|}{\underset{\underset{Cl}{|}}{C}}CH_2CCl_3 + \cdot CCl_3$

then (3), (4), (3), (4), etc.

6.13

(a) $CH_3CH_2\overset{|}{\underset{\underset{Br}{|}}{C}}HCH_2Br$
      (b)   (c) Same as (b)

cis

(d)   (e) $CH_3CH_2\text{–}O\text{–}\overset{\overset{O}{\|}}{\underset{\underset{O}{\|}}{S}}\text{–}OH$ (f) CH_3CH_2OH

trans

         CH_3 CH_3 CH_3 H

(g) $(CH_3\overset{|}{C}HCH_2)_3B$ (h) $CH_3\overset{|}{C}HCH_2OH$ (i) $\overset{}{C}=C$ (major product)

                                                  H CH_3

                                                        CH_3

(j)   (k) $CH_2=CHCH_2CH_2CH_3$ (l) $CH_3\overset{|}{\underset{\underset{Br}{|}}{C}}CH(CH_3)_2$

(m) $CH_3\overset{\underset{\displaystyle Cl}{|}}{C}HCH_2CH_2CH_2CH_3$ (n) $BrCH_2CH_2CH_2CH_2CH_2CH_3$

(o) $CH_3CH_2CH\overset{\displaystyle O}{\underset{\displaystyle O—O}{\diagup\diagdown}}CHCH_2CH_3$ (p) $2\ CH_3CH_2CHO$

(q) $2CH_3CH_2\overset{\displaystyle O}{\overset{\|}{C}}-O^-$ (r)   (s) $CH_3\overset{\displaystyle O}{\overset{\|}{C}}CH_2CH_2CH_2CHO$

(t) $CH_3\overset{\displaystyle O}{\overset{\|}{C}}CH_2CH_2CH_2\overset{\displaystyle O}{\overset{\|}{C}}-OH$ (u)   (v)

(w) $CH_3\overset{\underset{\displaystyle I}{|}}{\overset{\displaystyle CH_3}{\overset{|}{C}}}CH_2CH_2CH_3$ (x)

6.14

(a) $CH_3CH{=}CH_2 + H_3O^+ \ \rightleftarrows \ CH_3\overset{+}{C}HCH_3 + H_2O$

$\underset{\displaystyle CH_3}{\overset{\displaystyle CH_3}{\diagup}}\overset{+}{C}H + CH_2{=}CHCH_3 \longrightarrow CH_3\overset{\underset{\displaystyle |}{CH_3}}{\overset{|}{C}}HCH_2\overset{+}{C}HCH_3$

(b) $CH_3\overset{\underset{\displaystyle |}{CH_3}}{\overset{|}{C}}HCH{=}CHCH_3$ (more substituted alkene)

6.15

$CH_3CH{=}CHCH_3 + HCl \rightleftarrows CH_3CH_2\overset{+}{C}HCH_3 + \ :\overset{..}{\underset{..}{Cl}}:^-$

$CH_3CH_2\overset{+}{C}HCH_3 + :\overset{..}{O}{-}CH_2CH_3 \longrightarrow CH_3CH_2\overset{\underset{\displaystyle |}{H}}{\overset{+:\overset{..}{O}-CH_2CH_3}{C}}HCH_3 \overset{-H^+}{\longrightarrow} CH_3CH_2\overset{\underset{}{OCH_2CH_3}}{\overset{|}{C}}HCH_3$

6.16

(a)   +   (* D may also be axial)

(b)

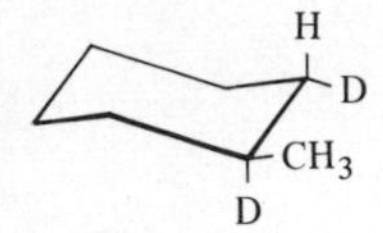

(c)

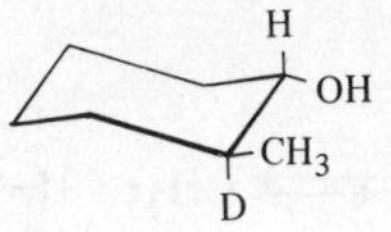

6.17

$$\left(2CH_3\overset{O}{\underset{\|}{C}}CH_3\right); \quad 4\left(O=CHCH_2CH_2\overset{CH_3}{\underset{|}{C}}=O\right), \quad O=CHCH_2CH_2CH=O$$

6.18

$$CH_3\underset{\underset{CH_3}{|}}{C}=CH_2 > CH_3CH=CH_2 > CH_2=CH_2$$

The order is the same as the order of stability of the carbocations formed by protonation of the alkenes.

$$CH_3\underset{\underset{CH_3}{|}}{\overset{+}{C}}-CH_3 > CH_3\overset{+}{CH}-CH_3 > \overset{+}{CH_2}-CH_3$$

6.19

(a) $CH_3CH=CHCH_3 + H^+ \rightleftharpoons CH_3\overset{+}{CH}CH_2CH_3$

(*cis* or *trans*)

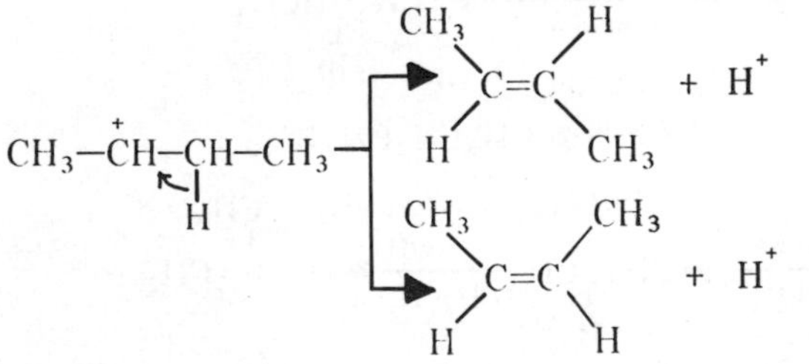

$$CH_3-CH_2-\overset{+}{CH}\overset{\overset{H}{|}}{\curvearrowleft}CH_2 \longrightarrow CH_3CH_2CH=CH_2 + H^+$$

The most stable (most substituted) alkene is formed in greatest amount; i.e., 2-butenes > 1-butene, and *trans*-2-butene > *cis*-2-butene.

(b) 1-Butene, on protonation, gives the same intermediate carbocation: $CH_3CH_2\overset{+}{C}HCH_3$.

(c) The carbocation, $CH_3CH_2\overset{+}{C}HCH_3$, cannot easily rearrange to the branched chain compound because to do so would require the formation of an intermediate primary carbocation, $\overset{+}{C}H_2\underset{\underset{CH_3}{|}}{C}HCH_3$.

6.20

$$CH_3-\underset{\underset{OH}{|}}{CH}-\underset{\underset{CH_3}{|}}{\overset{\overset{CH_3}{|}}{C}}-CH_3 + H-Cl \rightleftharpoons CH_3-\underset{\underset{^+OH_2CH_3}{|}}{CH}-\underset{\underset{}{|}}{\overset{\overset{CH_3}{|}}{C}}-CH_3 + :\ddot{C}l:^-$$

$$\longrightarrow CH_3-\overset{+}{C}H-\underset{\underset{CH_3}{|}}{\overset{\overset{CH_3}{|}}{C}}-CH_3 + H_2O$$

$$CH_3-\overset{+}{\underset{\underset{CH_3}{|}}{C}H}-\underset{\underset{CH_3}{|}}{\overset{\overset{CH_3}{|}}{C}}-CH_3 \longrightarrow CH_3-\underset{\underset{CH_3}{|}}{CH}-\overset{+}{\underset{\underset{CH_3}{|}}{C}}-CH_3 \xrightarrow{:\ddot{C}l:^-} CH_3-\underset{\underset{CH_3}{|}}{CH}-\underset{\underset{CH_3}{|}}{\overset{\overset{CH_3\ Cl}{|\ |}}{C}}-CH_3$$

6.21

(structure: methylcyclohexene ring with CH$_3$ substituent)

6.22

(a) $CH_3CH_2CH_3 + Br_2 \xrightarrow[\text{light}]{CCl_4}$ $\left.\begin{array}{c} CH_3CH_2CH_2Br \\ + \\ CH_3\underset{\underset{Br}{|}}{CH}CH_3 \end{array}\right\}$ $\xrightarrow[CH_3CH_2OH]{KOH} CH_3CH=CH_2$
  (excess)

(b) $CH_3CH=CH_2$ (above) $+ HBr \xrightarrow[\text{inhibitor}]{\text{free-radical}} CH_3\underset{\underset{Br}{|}}{CH}CH_3$

(c) $CH_3CH=CH_2$ (from (a)) $+ HBr \xrightarrow{\text{Peroxides}} CH_3CH_2CH_2Br$

(d) $CH_3\underset{\underset{CH_3}{|}}{CH}CH_3 + Br_2 \xrightarrow[\text{light}]{Br_2,\ \text{heat}} CH_3\underset{\underset{Br}{|}}{\overset{\overset{CH_3}{|}}{C}}CH_3 \xrightarrow[CH_3CH_2OH]{KOH} CH_3\overset{\overset{CH_3}{|}}{C}=CH_2$
  (excess)

(e) $CH_3\overset{\overset{CH_3}{|}}{C}=CH_2$ (from (d)) $+ H_2O \xrightarrow{H_3O^+} CH_3\underset{\underset{OH}{|}}{\overset{\overset{CH_3}{|}}{C}}CH_3$

(f) $CH_3CH_2CH_2CH_2Cl + KOH \xrightarrow{CH_3CH_2OH} CH_3CH_2CH=CH_2 \xrightarrow[\text{dark}]{Cl_2 \atop CCl_4} CH_3CH_2\underset{\underset{Cl}{|}}{CH}CH_2Cl$

(g) $CH_3CH_2Br + KOH \xrightarrow{CH_3CH_2OH} CH_2=CH_2 \xrightarrow{Br_2\ +\ H_2O} \underset{\underset{OH\ \ Br}{|\ \ \ |}}{CH_2CH_2}$

(h) $CH_3\overset{\overset{CH_3}{|}}{C}=CH_2$ (from (d)) $+ HBr \xrightarrow{\text{Peroxides}} CH_3\overset{\overset{CH_3}{|}}{C}HCH_2Br \xrightarrow{Li} CH_3\overset{\overset{CH_3}{|}}{C}HCH_2Li$

$\underset{\underset{CH_3}{|}}{CH_3}CHCH_2CH_2\underset{\underset{CH_3}{|}}{CH}CH_3 \xleftarrow{CH_3\overset{\overset{CH_3}{|}}{C}HCH_2Br.} (CH_3\overset{\overset{CH_3}{|}}{C}HCH_2)_2CuLi \xleftarrow{CuI}$

(or use Wurtz reaction as in (k).)

(i) $+ Br_2 \xrightarrow[\text{light}]{CCl_4}$ $Br \xrightarrow[CH_3CH_2OH]{KOH}$

(excess)

$+ Cl_2 + H_2O \longrightarrow$

(j) $CH_3CH_2CH_2CH_2Br \xrightarrow[CH_3CH_2OH]{KOH} CH_3CH_2CH=CH_2 \xrightarrow[\substack{\text{free radical}\\\text{inhibitor}}]{HBr} CH_3CH_2\underset{\underset{Br}{|}}{C}HCH_3$

(k) $CH_3CH_2CH_2CH_2Cl \xrightarrow[CH_3CH_2OH]{KOH} CH_3CH_2CH=CH_2 \xrightarrow{HCl} CH_3CH_2\underset{\underset{Cl}{|}}{C}HCH_3$

$2CH_3CH_2\underset{\underset{CH_3}{|}}{C}H-Cl + Na \xrightarrow{(Wurtz)} CH_3CH_2\underset{\underset{CH_3}{|}}{C}H-\underset{\underset{CH_3}{|}}{C}HCH_2CH_3$

(or use $(R)_2CuLi$ as in (h).)

(l) $CH_3\underset{\underset{CH_3}{|}}{\overset{\overset{CH_3}{|}}{C}}-OH \xrightarrow[\text{heat}]{H_2SO_4} CH_3\underset{}{\overset{\overset{CH_3}{|}}{C}}=CH_2 \xrightarrow{(BH_3)_2} (CH_3\overset{\overset{CH_3}{|}}{C}HCH_2)_3B$

$(CH_3\overset{\overset{CH_3}{|}}{C}HCH_2)_3B \xrightarrow[OH^-]{H_2O_2} CH_3\overset{\overset{CH_3}{|}}{C}HCH_2OH$

6.23

(a) $CH_3CH_2CH_2CH=CH_2 + Br_2 \longrightarrow CH_3CH_2CH_2\underset{\underset{Br}{|}}{C}HCH_2Br$

MW=70.12   MW=159.8

159.8 g Br_2 will react with 70.12 g pentene. Therefore ~ 16 g Br_2 will react with 7.0 g pentene.

(b) Since bromine and an alkene react in equimolar proportions:

$$\frac{3.20\text{ g}}{160\text{ g/mole}} = 0.02 \text{ mole } Br_2 = 0.02 \text{ mole alkene}$$

2.24 g alkene = (0.02 mole) (Mol. Wt.)

$$\therefore \text{Mol. Wt.} = \frac{2.24\text{ g}}{0.02\text{ mole}} = 112 \text{ g/mole alkene}$$

6.24

Rewriting the starting compound, we can better see the required reaction:

6.25

The intermediate I is competitively attacked by Cl^-, Br^- and H_2O.

6.26

The rate of addition of a proton to an alkene is faster the more stable the carbocation that is produced. The order of carbocation stability is:

$$\overset{+}{C}H_2-CH_3 < CH_3\overset{+}{C}HCH_3 < CH_3\overset{+}{C}HCH_2CH_3 < CH_3\overset{+}{\underset{CH_3}{C}}CH_3$$

Proton addition is the rate-limiting step in the hydration reaction.

6.27

6.28

7

STEREOCHEMISTRY

7.1
Chiral — (a) screw, (e) foot, (f) ear, (g) shoe, (h) spiral staircase
Achiral — (b) plain spoon, (c) fork, (d) cup

7.2
(b) Yes.　(c) No.　(d) No.

7.3
(b) No.　(c) No.

7.4
(a)　1-Chloropropane, (c) 2-methyl-1-chloropropane, (d) 2-methyl-2-chloropropane, (f) 1-chloropentane, and (h) 3-chloropentane are all achiral.

(b)

I　　　　II　　　　　(e)　　I　　　　II

(g)

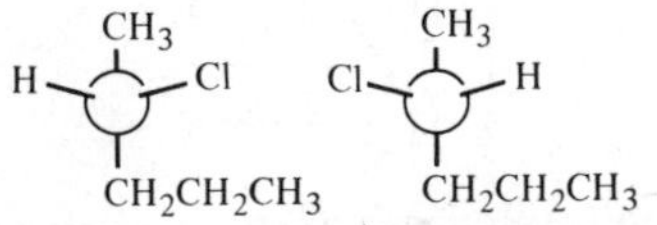

I　　　　II

7.5

(a)　1. 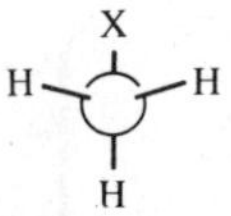　2. 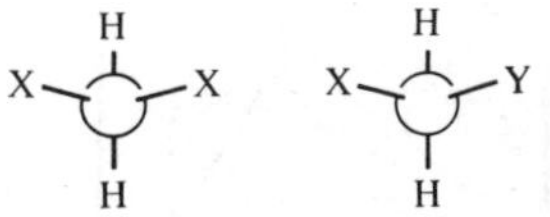　3.

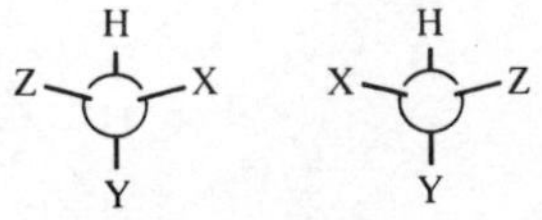

(b)　1. One

2. Two:

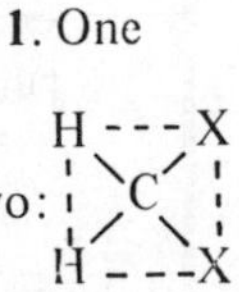

and 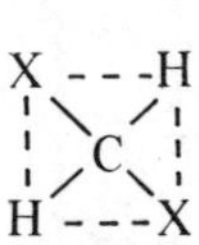　　　　　, or 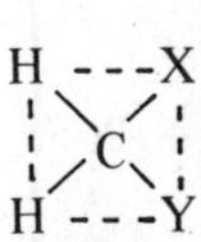　　　　and

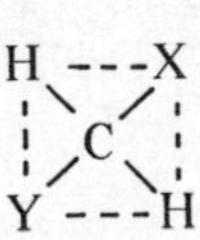

55

3. Three:

$$\text{H---X} \quad \text{C} \quad \text{Z---Y} \quad , \quad \text{H---Y} \quad \text{C} \quad \text{Z---X} \quad , \quad \text{and} \quad \text{H---X} \quad \text{C} \quad \text{Y---Z}$$

(c) 1. One

2. Three:

$$\text{H------X} \; \text{C} \; \text{H------X} \; , \quad \text{H------X} \; \text{C} \; \text{X------H} \; , \quad \text{X------X} \; \text{C} \; \text{H------H} \; ,$$

or

$$\text{H------X} \; \text{C} \; \text{H------Y} \; , \quad \text{H------X} \; \text{C} \; \text{Y------H} \; , \quad \text{X------Y} \; \text{C} \; \text{H------H}$$

3. Six:

$$\text{H------Y} \; \text{C} \; \text{X------Z} \; , \quad \text{H------X} \; \text{C} \; \text{Y------Z} \; , \quad \text{H------Z} \; \text{C} \; \text{Y------X} \; ,$$

$$\text{H------Y} \; \text{C} \; \text{Z------X} \; , \quad \text{H------Z} \; \text{C} \; \text{X------Y} \; , \quad \text{H------X} \; \text{C} \; \text{Z------Y}$$

(d) 1. One

2. Two or three:

$$\text{H------X} \; \text{C} \; \text{H------X} \; , \; \text{H------X} \; \text{C} \; \text{X------H} \quad \text{or} \quad \left\{ \begin{array}{l} \text{H------Y} \; \text{C} \; \text{H------X} \\ \text{H------X} \; \text{C} \; \text{H------Y} \; , \; \text{H------X} \; \text{C} \; \text{Y------H} \end{array} \right.$$

3. Six:

$$\left. \begin{array}{ccc} \underset{1}{\text{H------Z}\;\text{C}\;\text{X------Y}} & \underset{2}{\text{H------Y}\;\text{C}\;\text{Z------X}} & \underset{3}{\text{H------Y}\;\text{C}\;\text{X------Z}} \\ \underset{4}{\text{X------Y}\;\text{C}\;\text{H------Z}} & \underset{5}{\text{Z------X}\;\text{C}\;\text{H------Y}} & \underset{6}{\text{X------Z}\;\text{C}\;\text{H------Y}} \end{array} \right\} \text{enantiomeric pairs}$$

7.6

(b) plain spoon, (c) fork, (d) cup all possess a plane of symmetry.

7.7

(a)

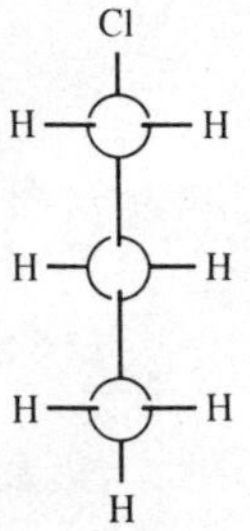

The plane of symmetry is perpendicular to page and passes through Cl and 3C's

(c)

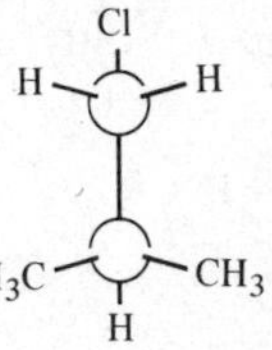

The plane of symmetry is perpendicular to page and passes through Cl and 2C's

(d)

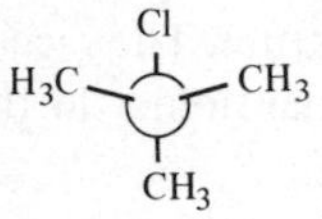

A vertical plane perpendicular to page passes through Cl, tertiary C, and CH₃ at bottom

(f)

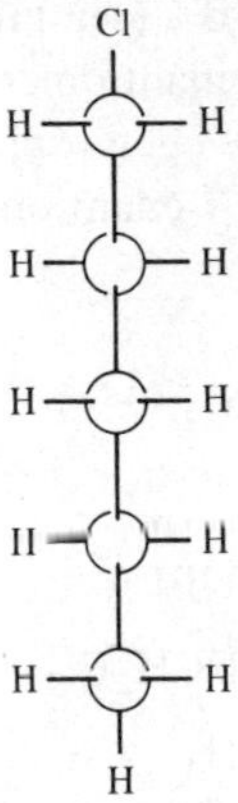

A plane perpendicular to page passes through Cl and 5C's

(h)

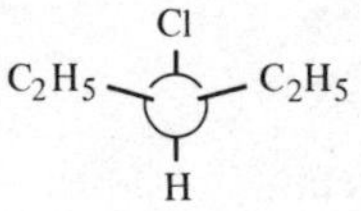

A plane perpendicular to page passes through Cl, C, and H

7.8

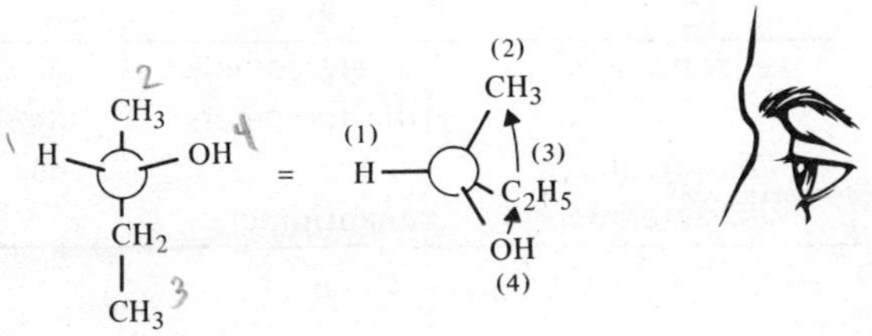

From priority 4 to 3 to 2, the direction is counterclockwise, therefore II is S−2−butanol

7.9

(b)

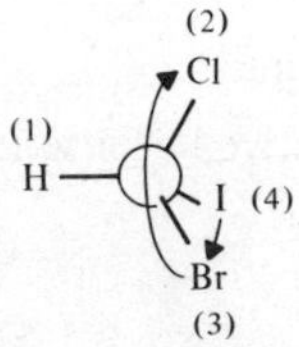

I = R 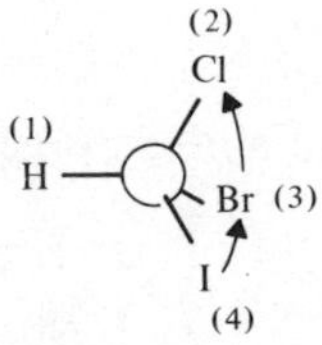   II = S

(e)

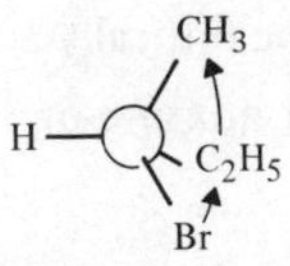

I = S 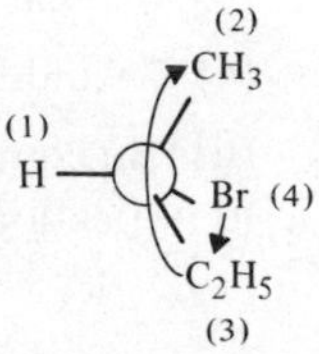   II = R

(g)

I = S   II = R

7.10
(a) *R*   (b) *R*   (c) *R*

7.11
The optical purity is 50% (see previous paragraph in text). That means that the sample contains 50% of the *S*-enantiomer and 50% of the racemic mixture. The racemic mixture is 50% *S*- and 50% *R*. Therefore the total percentage of *S*-enantiomer in the sample is 75%, the percentage of *R*-enantiomer is 25%.

7.12
(a) (±) $CH_3CH_2\underset{|}{C}HCH_3$ (b) Same as (a)
            OH

   (Racemic modification)

(c) (±) $CH_3CH_2CH_2\underset{|}{C}HCH_2CH_3$ (d) Same as (a)
                CH_3

   (Racemic modification)   (e) Same as (a)

7.13
(a) Diastereomers   (b) Diastereomers   (c) Diastereomers

(d)

	1	2	3	4
1		enantiomers	diastereomers	diastereomers
2	enantiomers		diastereomers	diastereomers
3	diastereomers	diastereomers		enantiomers
4	diastereomers	diastereomers	enantiomers	

(e) Yes.   (f) No.

7.14
(a) **5** alone would be optically active.

(b) **6** alone would be optically active.

(c) **7** would not be optically active because it is a meso compound.

(d) An equimolar mixture of **5** and **6** would not be optically active because it is a racemic modification.

7.15

(a)

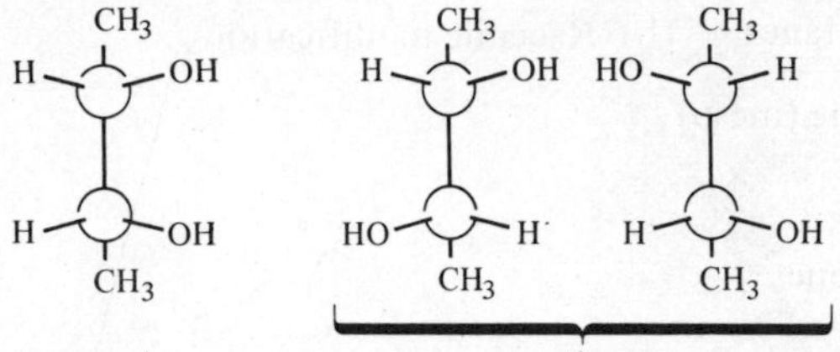

CH₃ — H—OH / H—OH — CH₃ *(meso)*

CH₃ — H—OH / HO—H — CH₃   CH₃ — HO—H / H—OH — CH₃

enantiomers

(b)

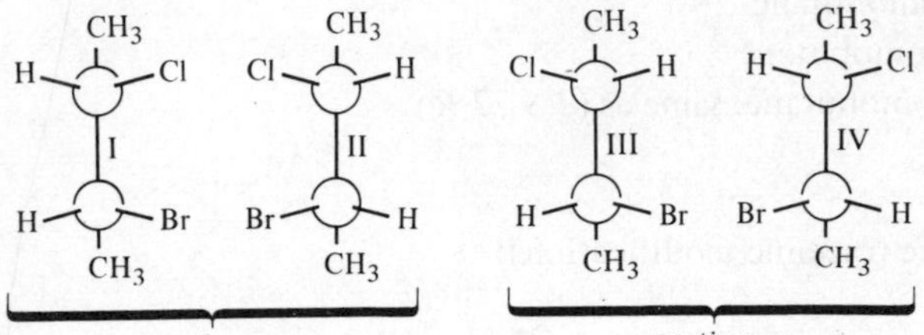

CH₃ / H—Cl / H—Br / CH₃ (I)   CH₃ / Cl—H / Br—H / CH₃ (II)   CH₃ / Cl—H / H—Br / CH₃ (III)   CH₃ / H—Cl / Br—H / CH₃ (IV)

enantiomers   enantiomers

Diastereomers are I and III, I and IV, II and III, and II and IV.

(c)

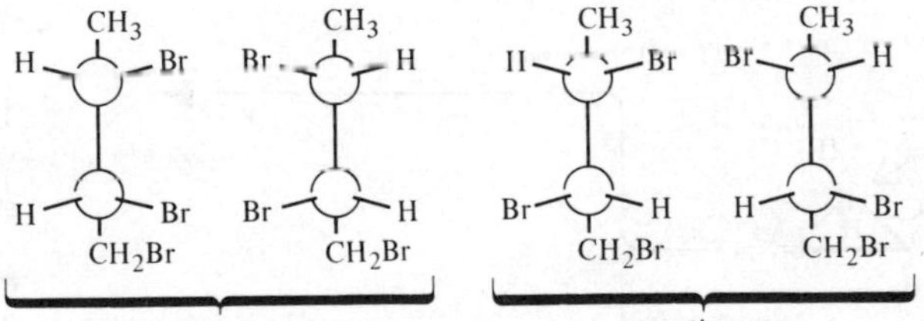

CH₃ / H—Br / H—Br / CH₂Br   CH₃ / Br—H / Br—H / CH₂Br   CH₃ / H—Br / Br—H / CH₂Br   CH₃ / Br—H / H—Br / CH₂Br

enantiomers   enantiomers

(d)

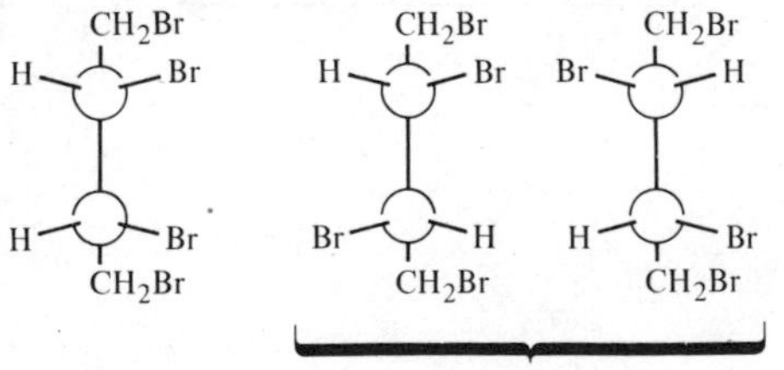

CH₂Br / H—Br / H—Br / CH₂Br *meso*   CH₂Br / H—Br / Br—H / CH₂Br   CH₂Br / Br—H / H—Br / CH₂Br

enantiomers

(e)

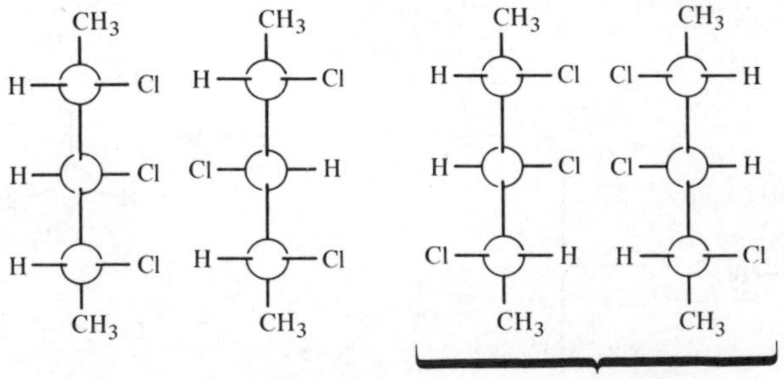

CH₃ / H—Cl / H—Cl / H—Cl / CH₃ *meso*   CH₃ / H—Cl / Cl—H / H—Cl / CH₃ *meso*   CH₃ / H—Cl / H—Cl / Cl—H / CH₃   CH₃ / Cl—H / Cl—H / H—Cl / CH₃

enantiomers

7.16

(a) No   (b) Yes   (c) No   (d) No   (e) Diastereomers   (f) Diastereomers

7.17

(a) *Trans*-1,2-dibromocyclopentanes   (b) Racemic modification

(c) *Cis*-1,2-dibromocyclopentane (meso)

7.18

2     $2R, 3S$ -2, 3-dibromopentane
3     $2S, 3S$ -2, 3-dibromopentane
4     $2R, 3R$ -2, 3-dibromopentane
5     $2S, 3S$ -2, 3-dibromobutane
6     $2R, 3R$ -2, 3-dibromobutane
7     $2R, 3S$ -2, 3-dibromobutane; same as $(2S, 3R)$

7.19

(a) 2, 3-Dibromobutane (racemic modification):

(b) *Meso* -2, 3-dibromobutane:

The enantiomer gives the same result in
this step as well.

7.20

(a)

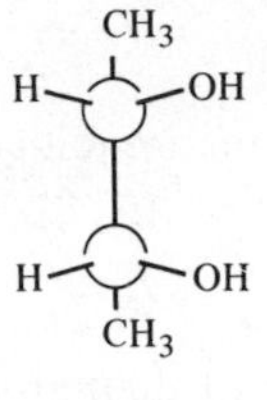

(meso)

(b)

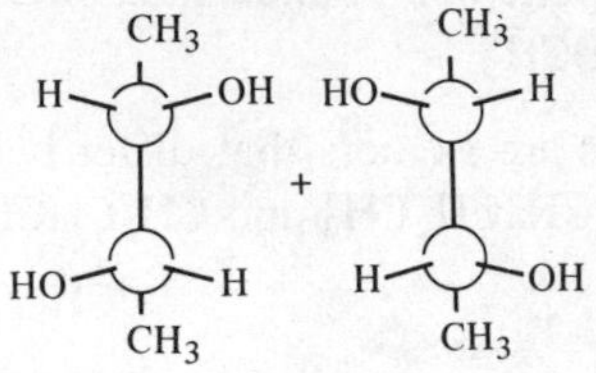

+

(racemic modification)

7.21

(a) (b)

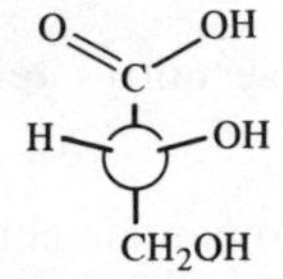

(R)–(−)–glyceric acid

(a)(c)

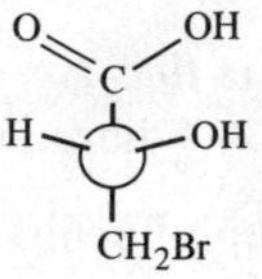

(S)–(−)–3−bromo−
2−hydroxypropanoic
acid

(d)

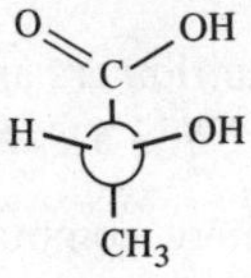

(R)–(−)–lactic acid

7.22

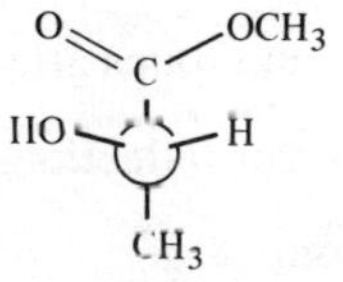

(S)–(−)–methyl lactate

7.23

(a)

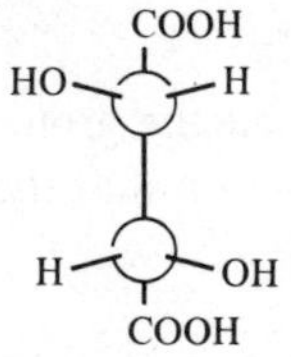

(−)−tartaric acid

(b)

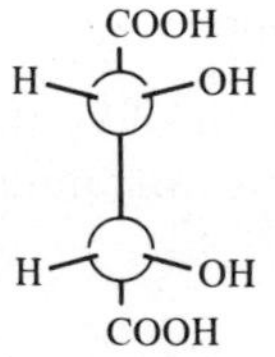

(meso)−tartaric acid

(c) No, *meso*-tartaric acid would *not* be optically active.

7.24

They are diastereomers.

7.25

(a) *(Z)* - 1-Bromo-1-chloro-1-butene

(b) *(Z)* - 2-Bromo-1-chloro-1-iodopropene

(c) *(E)* - 3-Ethyl-4-methyl-2-pentene

(d) *(E)* - 1-Chloro-1-fluoro-2-methyl-1-butene

7.26

(a) Isomers are different compounds that have the same molecular formula. C_2H_6O: CH_3CH_2OH and CH_3OCH_3

(b) Structural isomers are isomers that differ because their atoms are joined in a different order. C_4H_{10}: $CH_3CH_2CH_2CH_3$ and $CH_3\underset{\underset{\displaystyle CH_3}{|}}{CH}CH_3$

(c) Stereoisomers are isomers that differ only in the arrangement of their atoms in space: *cis*- and *trans* -2-butene.

(d) Diastereomers are stereoisomers that are not mirror reflections of each other: *cis*- and *trans* -2-butene, or $(2\,S, 3\,S)$- and $(2\,S, 3\,R)$- 2, 3-dibromobutane.

(e) Enantiomers are stereoisomers that are **non-superposable** mirror reflections of each other: $(2\,S, 3\,S)$- and $(2\,R, 3\,R)$- 2, 3-dibromobutane.

(f) A meso compound is made up of achiral molecules that contain chiral centers: $(2\,S, 3\,R)$- 2, 3-dibromobutane.

(g) A racemic modification is an equimolar mixture of a pair of enantiomers.

(h) A plane of symmetry is an imaginary plane that bisects a molecule in such a way that the two halves of the molecule are mirror reflections of each other. (See Fig. 7.7.)

(i) A chiral center is any tetrahedral atom that has four different groups attached to it.

(j) A chiral molecule is one that is not superposable on its mirror reflection.

(k) An achiral molecule is superposable on its mirror reflection.

(l) Optical activity is the rotation of the plane of polarization of plane polarized light by a substance placed in the light path.

(m) A dextrorotatory substance is one that rotates the plane of polarization of plane polarized light in a clockwise direction.

(n) A reaction occurs with retention of configuration when all the groups around the chiral atom retain the same relative configuration after the reaction that they had before the reaction.

7.27

(a) Enantiomers   (b) Same   (c) Enantiomers   (d) Diastereomers   (e) Same

(f) Structural isomers   (g) Same   (h) Diastereomers   (i) Same

(j) Enantiomers   (k) Same   (l) Enantiomers   (m) Same   (n) Structural isomers   (o) Same   (p) Diastereomers   (q) Enantiomers

7.28

(a)

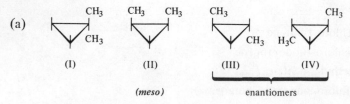

(b) III and IV   (c) II   (d) Three: I, II, and a mixture of III and IV.   (e) None, since the only chiral molecules are III and IV, and they would be obtained in the same amounts as a racemic modification.

7.29

(a)
```
     CH3
 H ──┤── OH
          and
 H ──┤── OH    enantiomer,
     CH2CH3
```

(b)
```
     CH3
 H ──┤── OH
          and
HO ──┤── H     enantiomer,
     CH2CH3
```

(c)
```
     CH3
 H ──┤── OH
          and
HO ──┤── H     enantiomer,
     CH2CH3
```

(d)
```
     CH3
 H ──┤── OH
          and
 H ──┤── OH    enantiomer,
     CH2CH3
```

(e)
```
     CH3
 H ──┤── Br
          and
 H ──┤── Br    enantiomer,
     CH2CH3
```

(f)
```
     CH3
 H ──┤── Br
          and
Br ──┤── H     enantiomer,
     CH2CH3
```

7.30

(a) 2S, 3R - (the enantiomer is 2R,3S)   (b) 2 S, 3 S - (the enantiomer is 2 R, 3 R-)

(c) Same as (b)   (d) Same as (a)   (e) 2 S, 3 R- (the enantiomer is 2 R, 3 S-)

(f) 2 S, 3 S- (the enantiomer is 2 R, 3 R-)

7.31

(b) and (c) *must* occur with retention of configuration because no bonds to the chiral carbon are broken in either reaction.

7.32

(a)
```
              O
              ‖
     CH3     O─C              CH3     O─C
 H ──┤── CH2  H ──┤── OH   H ──┤── CH2  HO ──┤── H
     CH2CH3    CH3            CH2CH3      CH3
      (S)       (R)    and     (S)        (S)
```

(b) They are diasteromers.

(c) The boiling points of these esters *will* be different. If there is a large enough difference in boiling points, then separation by fractional distillation will be possible.

(d) Yes.

(e) After separation of the diastereomeric esters by fractional distillation, they could each be hydrolyzed to yield the separate enantiomeric acids.

7.33

(a) 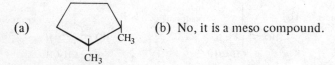   (b) Yes.

7.34

(a)   (b) No, it is a meso compound.

7.35

(a)

(b)

$+ (BH_3)_2 \longrightarrow$ and

The BH_3 group may attack
the double bond from either
side of the ring

$+$

$\downarrow H_2O_2,\ OH^-$

8 SPECIAL TOPICS I

8.1

(a) $X \longrightarrow Y$, $K_{eq} = \dfrac{[Y]}{[X]} = 10$

Initial $[X] = 1.0$
Equilibrium $[Y] = a$
Equilibrium $[X] = 1.0 - a$

then $K_{eq} = 10 = \dfrac{a}{1.0 - a}$

$10 - 10a = a$

$-11a = -10$

$a = \dfrac{10}{11} = 0.91$ mole/liter

At equilibrium, $[Y] = 0.91$ mole/liter,
$\qquad\qquad\quad [X] = 0.09$ mole/liter,
and 91% of X is converted to product, Y.

(b) If $K_{eq} = 1$, $1 = \dfrac{a}{1.0 - a}$

$1 - a = a$

$-2a = -1$

$a = 0.5$

$\therefore$ At equilibrium, $[Y] = 0.5$ mole/liter,
$\qquad\qquad\qquad [X] = 0.5$ mole/liter,
and 50% of X is converted to product, Y.

(c) If $K_{eq} = 10^{-3}$, $10^{-3} = \dfrac{a}{1 - a}$

$10^{-3} - 10^{-3} a = a$

$-1.001a = -10^{-3}$

$a = \dfrac{10^{-3}}{1.001} \cong 10^{-3}$

At equilibrium, $[Y] = 10^{-3}$ mole/liter,
$\qquad\qquad\quad [X] = 0.999$ mole/liter,
and 0.1% of X is converted to product, Y.

8.2

(a) $\Delta G = \Delta H - T\Delta S$

$\Delta G = -41700 \text{ cal/mole} - 300 \text{deg}(-26.6 \text{ cal/deg mole})$

$\Delta G = -41,700 + 7980 = -33,720 \text{ cal/mole}$

or $\Delta G = -33.72 \text{ kcal/mole}$

(b) Yes, because a negative value of ΔG tells us that the products are favored at equilibrium.

(c) No, a negative entropy tells us that the products are more ordered, and therefore less favored than the reactants.

(d) There are fewer degrees of freedom in the product molecule, ethene, than in the separate and independent molecules, ethyne and hydrogen.

8.3

(a) Yes, as shown by the negative value of ΔG.

(b) Yes, as shown by the positive value of ΔS.

(c) The extra degrees of freedom associated with rotation about the C—C single bond of ethane are not possible in ethene.

8.4

The reaction will proceed through the most stable free radical that can be produced. Head-to-head polymerization, as shown, will lead to a primary radical,

$$R-CH_2-\underset{\underset{CH_3}{|}}{CH} \cdot \; + \; \underset{\underset{CH_3}{|}}{CH}=CH_2 \longrightarrow R-CH_2-\underset{\underset{CH_3}{|}}{CH}-\underset{\underset{CH_3}{|}}{CH}-CH_2 \cdot,$$

which is less stable than the secondary radical that is produced by head-to-tail polymerization:

$$R-CH_2-\underset{\underset{CH_3}{|}}{CH} \cdot \; + \; CH_2=\underset{\underset{CH_3}{|}}{CH} \longrightarrow R-CH_2-\underset{\underset{CH_3}{|}}{CH}-CH_2-\underset{\underset{CH_3}{|}}{CH} \cdot$$

8.5

(a) $n\,CH_2=\underset{\underset{F}{|}}{CH} \xrightarrow[\text{peroxide}]{\text{organic}} \{CH_2-\underset{\underset{F}{|}}{CH}\}_n$

(b) $n\,CF_2=\underset{\underset{Cl}{|}}{CF} \xrightarrow[\text{peroxide}]{\text{organic}} \{CF_2-\underset{\underset{Cl}{|}}{CF}\}_n$

(c) $n\,CF_2=\underset{\underset{CF_3}{|}}{CF} \; + \; m\,CH_2=CF_2 \xrightarrow[\text{peroxide}]{\text{organic}} \{CF_2-\underset{\underset{CF_3}{|}}{CF}\}_n \{CH_2-CF_2\}_m$

Note that the units are randomly ordered, and not necessarily joined to their own kind as shown.

8.6

Polymerization will occur to produce the most stable carbocation possible. The scheme shown in this problem involves formation of the primary carbocations,

$$\underset{\underset{CH_3}{|}}{\overset{\overset{CH_3}{|}}{CH-CH_2}}{}^{+}\ ,\ \underset{\underset{CH_3}{|}}{\overset{\overset{CH_3}{|}}{CH-CH_2-C-CH_2}}{}^{+}\ ,\ etc.$$

instead of the tertiary carbocations,

$$CH_3-\underset{\underset{CH_3}{|}}{\overset{\overset{CH_3}{|}}{C}}-CH_2-\underset{\underset{CH_3}{|}}{\overset{\overset{CH_3}{|}}{C}}{}_+$$

8.7

(a) By proton transfer from water to the strongly basic carbanion,

$$R-CH_2-\underset{\underset{CN}{|}}{CH}{\,}^{\!\!-} + H-\underset{\underset{H}{|}}{\ddot{O}}: \longrightarrow R-CH_2-\underset{\underset{CN}{|}}{CH_2} + :\ddot{O}H^{-}$$

(b) $(CH_2CH)_n-CH_2CH{\,}^{\!\!-} + (m+1)CH_2-CH_2 \longrightarrow$
with R and R substituents, and epoxide $\overset{}{O}$

$$(CH_2CH)_{n+1}(CH_2-CH_2-O)_m CH_2-CH_2-\ddot{O}:{}^{-}$$
with R substituent

$$\xrightarrow{H_2O} (CH_2CH)_{\overline{n+1}}(CH_2-CH_2-O)_m CH_2-CH_2-OH$$
with R substituent

In this polymer, each chain consists of a long uninterrupted segment of the first repeating unit, $(CH_2CH)_{n+1}$ with R substituent, followed by a long uninterrupted segment of the second repeating unit, $(CH_2-CH_2-O)_m$.

8.8

(a)

(b)

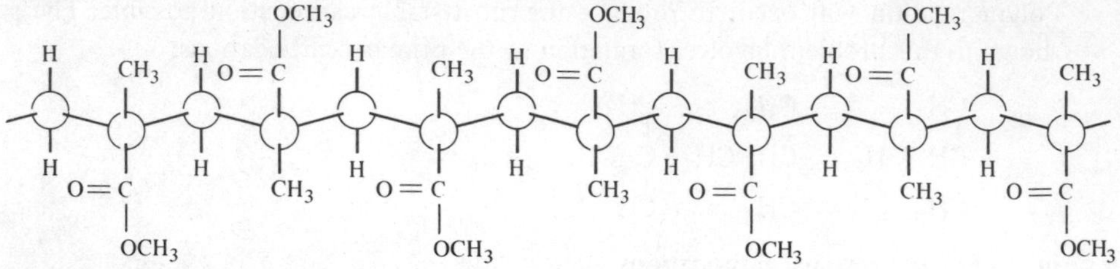

(c)

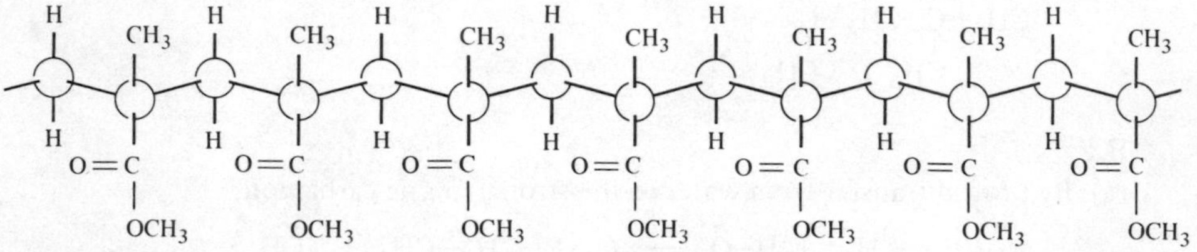

8.9

The singlet methylene reacts with the double bond at the same rate from either side as shown:

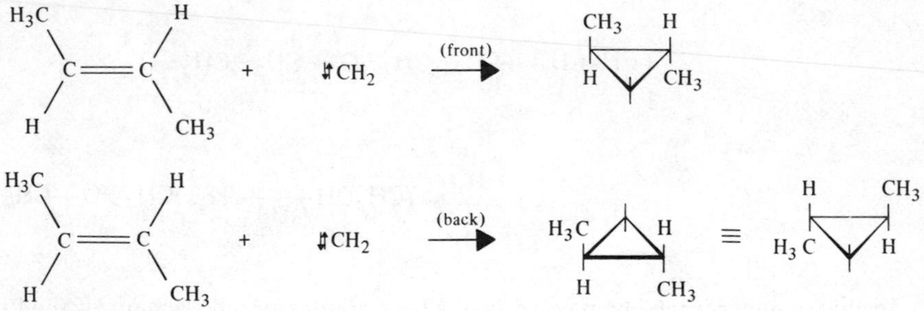

8.10

(a) + (racemic modification)

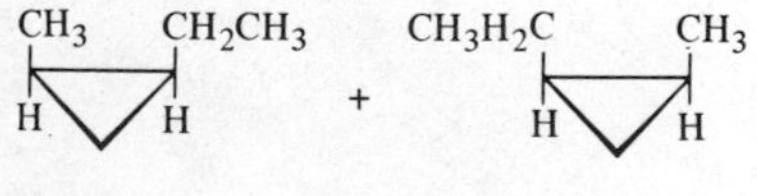

(b) + (racemic modification)

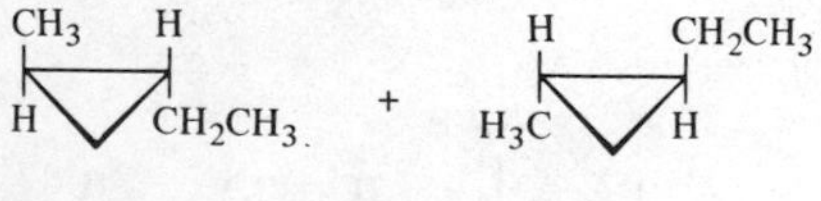

(c) + and +

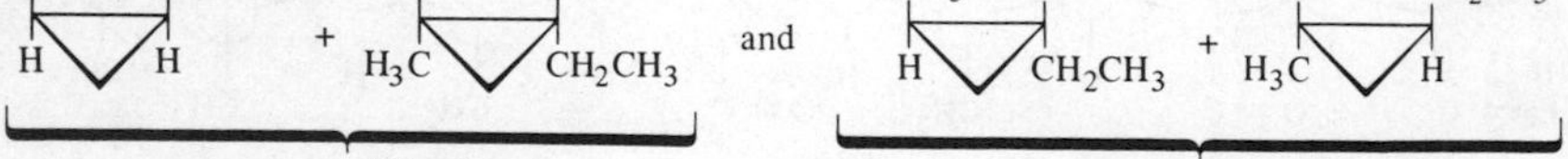

racemic modification   racemic modification

(d) Same mixture as in (c).

8.11
Recombination of radicals should yield ethane, butane, hexane, isobutane, 2-methylpentane, 2,3-dimethylbutane.

8.12

(a)

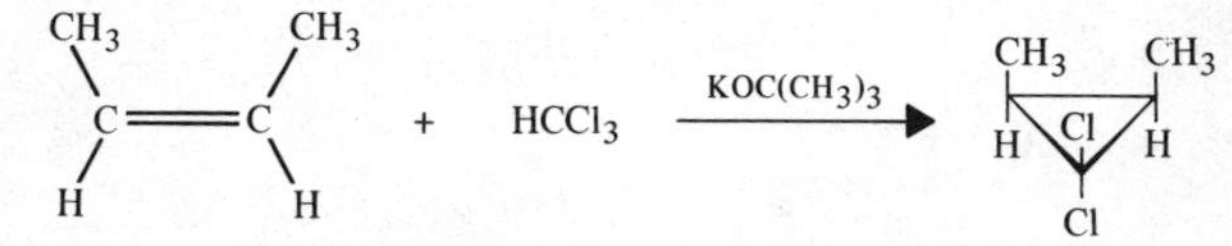

(b)

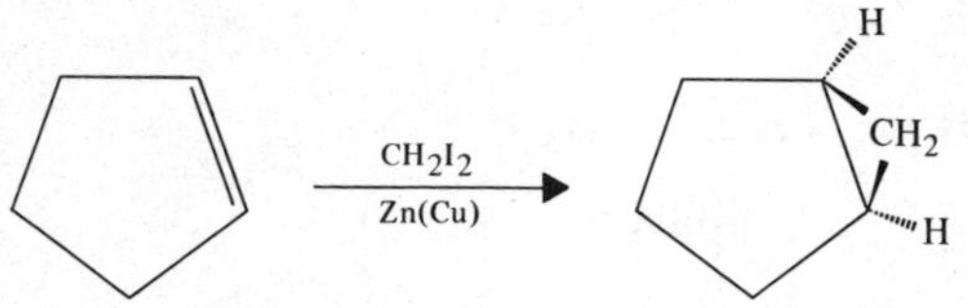

(c)

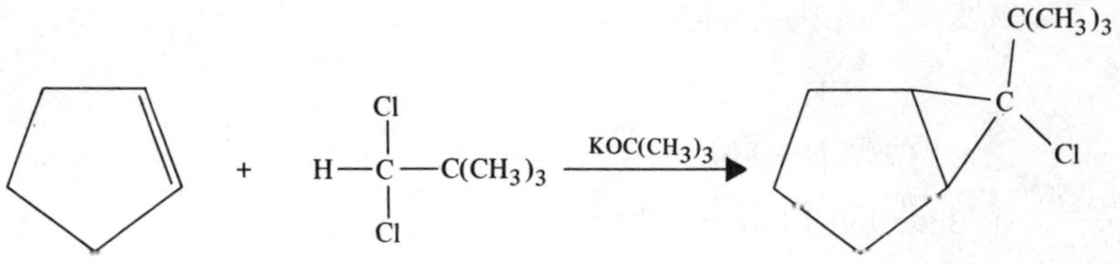

9 ALKYNES

9.1

(a) C_4H_6 : $CH_3CH_2C{\equiv}CH$, $CH_3C{\equiv}CCH_3$

                     1-Butyne   2-Butyne

(b) C_5H_8 : $CH_3CH_2CH_2C{\equiv}CH$ $CH_3CH_2C{\equiv}CHCH_3$

                     1-Pentyne   2-Pentyne

$$\underset{\text{3-Methyl-1-butyne}}{CH_3\overset{\overset{\displaystyle CH_3}{|}}{C}HC{\equiv}CH}$$

(c) $CH_3CH_2CH_2CH_2C{\equiv}CH$ $CH_3CH_2CH_2C{\equiv}CCH_3$

               1-Hexyne   2-Hexyne

$CH_3CH_2C{\equiv}CCH_2CH_3$ $CH_3\overset{\overset{\displaystyle }{}}{C}HCH_2C{\equiv}CH$

         3-Hexyne $|$
                                         CH_3

                               4-Methyl-1-pentyne

$CH_3C{\equiv}C\overset{\overset{\displaystyle }{}}{C}HCH_3$ $HC{\equiv}C\overset{\overset{\displaystyle }{}}{C}HCH_2CH_3$

            $|$ $|$
           CH_3 CH_3

   4-Methyl-2-pentyne   3-Methyl-1-pentyne

            CH_3
              $|$
$HC{\equiv}CCCH_3$
              $|$
            CH_3

   3,3-Dimethyl-
   1-Butyne

9.2

(a) $HC{\equiv}CH + :\overset{..}{N}H_2^{-} \rightleftharpoons HC{\equiv}C:^{-} + :NH_3$

   stronger   stronger   weaker   weaker
   acid   base   base   acid

$\left(\begin{array}{l} \text{No appreciable amount of re-} \\ \text{actants are present at equi-} \\ \text{librium.} \end{array} \right)$

(b) $CH_2=CH_2$ + $:\overset{..}{N}H_2^-$ $\longleftrightarrow$ $CH_2=\overset{..}{C}H^-$ + $:NH_3$ $\left.\begin{array}{l}\text{No appreciable amount of pro-}\\ \text{ducts are present at equilibrium.}\end{array}\right|$

     weaker   weaker   stronger   stronger
     acid   base   base   acid

(c) CH_3CH_3 + $:\overset{..}{N}H_2^-$ $\longleftrightarrow$ $CH_3\overset{..}{C}H_2^-$ + $:NH_3$ $\left.\begin{array}{l}\text{No appreciable amount of pro-}\\ \text{ducts are present at equilibrium.}\end{array}\right|$

     weaker   weaker   stronger   stronger
     acid   base   base   acid

(d) $HC\equiv C:^-$ + $CH_3CH_2\overset{..}{O}H$ $\longleftrightarrow$ $HC\equiv CH$ + $CH_3CH_2\overset{..}{O}:^-$ $\left.\begin{array}{l}\text{No appreciable amount of}\\ \text{reactants are present at equi-}\\ \text{librium.}\end{array}\right|$

    stronger   stronger   weaker   weaker
    base   acid   acid   base

(e) $HC\equiv C:^-$ + $H-\overset{\overset{\displaystyle..}{|}}{\underset{\displaystyle H}{O}}:$ $\longleftrightarrow$ $HC\equiv CH$ + $:\overset{..}{O}H^-$ $\left.\begin{array}{l}\text{No appreciable amount of}\\ \text{reactants are present at equi-}\\ \text{librium.}\end{array}\right|$

    stronger   stronger   weaker   weaker
    base   acid   acid   base

9.3

$$\overset{\displaystyle CH_3 \quad\ CH_3}{\underset{\displaystyle\qquad}{CH_3CH-CH-OH}} \text{ from the Sia}_2BH.$$

9.4

(a) $CH_3CH_2C\equiv CH \xrightarrow[\substack{HgSO_4 \\ H_2SO_4}]{H_2O} CH_3CH_2\overset{\overset{\displaystyle O}{||}}{C}CH_3$

(b) $-C\equiv CH$ + $Sia_2BH \xrightarrow{O^\circ}$ $-CH=CH-BSia_2 \xrightarrow[OH^-]{H_2O_2}$ $-CH_2\overset{\overset{\displaystyle O}{||}}{C}H$

(c) $3CH_3C\equiv CCH_3$ + $\frac{1}{2}(BD_3)_2 \xrightarrow{O^\circ}$ $\left(\underset{\displaystyle D}{\overset{\displaystyle CH_3}{}}C=C\underset{\displaystyle B}{\overset{\displaystyle CH_3}{}}\right)_3 \xrightarrow[O^\circ]{CH_3COOD}$ $\underset{\displaystyle D}{\overset{\displaystyle CH_3}{}}C=C\underset{\displaystyle D}{\overset{\displaystyle CH_3}{}}$

or $CH_3C\equiv C-CH_3$ + $D_2 \xrightarrow{Ni_2B(P\text{-}2)}$ $\underset{\displaystyle D}{\overset{\displaystyle CH_3}{}}C=C\underset{\displaystyle D}{\overset{\displaystyle CH_3}{}}$

(d) $CH_3CH_2C\equiv CH$ + $Sia_2BH \longrightarrow$ $\underset{\displaystyle H}{\overset{\displaystyle CH_3CH_2}{}}C=C\underset{\displaystyle BSia_2}{\overset{\displaystyle H}{}} \xrightarrow{CH_3COOD}$ $\underset{\displaystyle H}{\overset{\displaystyle CH_3CH_2}{}}C=C\underset{\displaystyle D}{\overset{\displaystyle H}{}}$

9.5

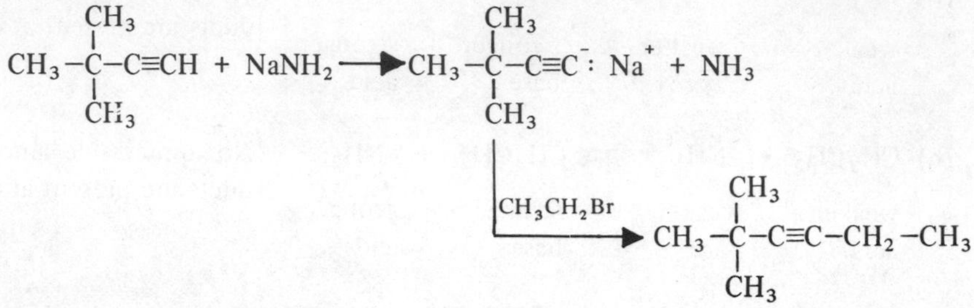

$$CH_3-\underset{\underset{CH_3}{|}}{\overset{\overset{CH_3}{|}}{C}}-C\equiv CH + NaNH_2 \longrightarrow CH_3-\underset{\underset{CH_3}{|}}{\overset{\overset{CH_3}{|}}{C}}-C\equiv C^-\!\!: Na^+ + NH_3$$

$$\xrightarrow[]{CH_3CH_2Br} CH_3-\underset{\underset{CH_3}{|}}{\overset{\overset{CH_3}{|}}{C}}-C\equiv C-CH_2-CH_3$$

A reaction between $CH_3CH_2C\equiv C^-\!\!:Na^+$ and $CH_3-\underset{\underset{CH_3}{|}}{\overset{\overset{CH_3}{|}}{C}}-Br$ would result in elimination to pro-

duce $CH_2\!\!=\!\!\underset{\underset{CH_3}{|}}{C}-CH_3 + CH_3CH_2C\equiv CH.$

9.6

(a)

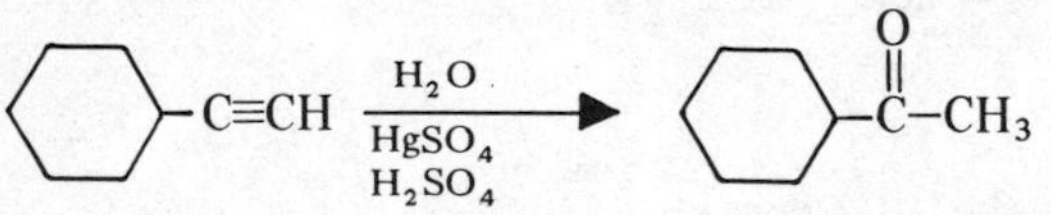

(b)

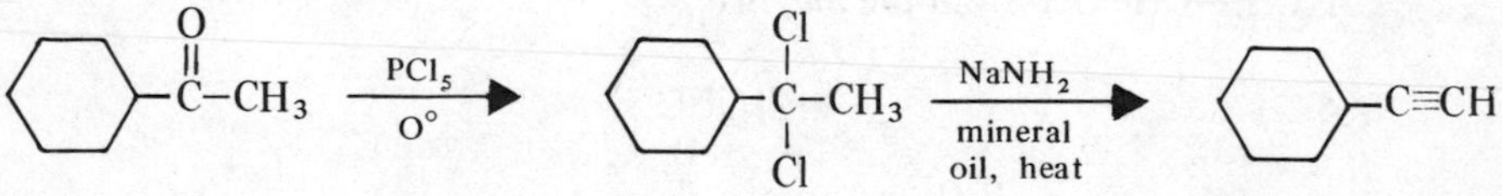

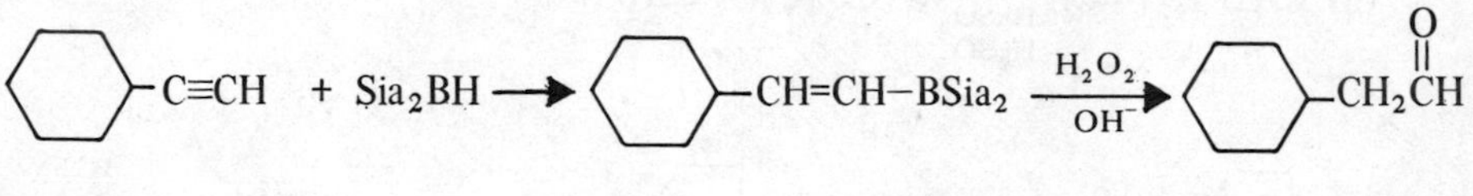

9.7

(a) $CH_3\underset{\underset{CH_3}{|}}{C}HC\equiv CCH_2CH_3$

2-Methyl-3-hexyne

(b)

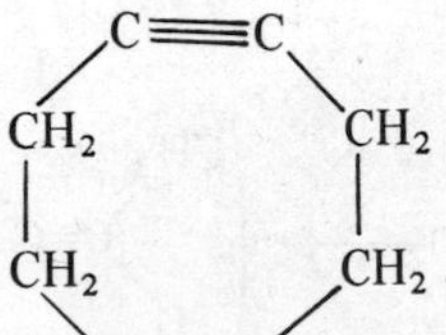

Cyclooctyne

(c) $HC\equiv CCH_2CH_2CH_2CH_2CH_3$

1-Heptyne

9.8

(a) 3-Methyl-1-butyne

(b) 2,2-Dimethyl-3-hexyne

(c) 3-Nonyne

(d) 3-Hexyne

(e) 2,2,5,5-Tetramethyl-3-hexyne

(f) 2,5-Dimethyl-3-hexyne

(g) 2-Hexyne

(h) 4-Methyl-2-hexyne

(i) 2,7-Dimethyl-4-octyne

(j) 1-Octyne

9.9
a, j

9.10
(a) d, g   (b) None

9.11
(a) $3\,C + CaO \xrightarrow{2500°} CaC_2 + CO$
(coke) (lime)

$CaC_2 + 2H_2O \xrightarrow[\text{temperature}]{\text{room}} HC{\equiv}CH + Ca(OH)_2$

(b) $HC{\equiv}CH + H_2 \xrightarrow{Ni_2B} CH_2{=}CH_2$

(c) $HC{\equiv}CH + NaNH_2 \xrightarrow[NH_3]{\text{liquid}} HC{\equiv}C\!:^-Na^+ + NH_3$

$CH_4 + Br_2 \xrightarrow[\text{heat}]{\text{light}} CH_3Br + HBr$
(excess)

$HC{\equiv}C\!:^-Na^+ + CH_3Br \longrightarrow HC{\equiv}C{-}CH_3 + Na^+Br^-$

(d) $CH_3C{\equiv}CH + H_2 \xrightarrow{Ni_2B} CH_3CH{=}CH_2$

(e) $CH_3C{\equiv}CH + H_2O \xrightarrow[H_2SO_4]{HgSO_4} CH_3\overset{\overset{\displaystyle O}{\|}}{C}CH_3$

(f) $CH_3C{\equiv}CH + NaNH_2 \xrightarrow[NH_3]{\text{liquid}} CH_3C{\equiv}C\!:^-Na^+ + NH_3$
$\xrightarrow{CH_3Br} CH_3C{\equiv}CCH_3 + Na^+Br^-$

(g) $CH_2{=}CH_2 + HBr \longrightarrow CH_3CH_2Br \xrightarrow{HC{\equiv}\bar{C}:Na^+} HC{\equiv}CCH_2CH_3 + Na^+Br^-$

(h) $CH_3C{\equiv}CCH_3 + H_2O \xrightarrow[H_2SO_4]{HgSO_4} CH_3\overset{\overset{\displaystyle O}{\|}}{C}CH_2CH_3$

(i) See (g) above.

(j) $CH_3C{\equiv}C\!:^-Na^+ + CH_3CH_2Br \longrightarrow CH_3C{\equiv}CCH_2CH_3 + Na^+Br^-$

(k) $CH_2{=}CH_2 + H_2O \xrightarrow{H_2SO_4} CH_3CH_2OH$

(l) $CH_2{=}CH_2 + Br_2 \xrightarrow[\text{dark}]{CCl_4} CH_2BrCH_2Br$

(m) $CH_3C{\equiv}CH + 2HCl \longrightarrow CH_3{-}\underset{\underset{\displaystyle Cl}{|}}{\overset{\overset{\displaystyle Cl}{|}}{C}}{-}CH_3$

(n) $CH_3C{\equiv}CCH_3$ + H_2 $\xrightarrow{Ni_2B}$

$$\underset{H}{\overset{CH_3}{C}}{=}\underset{H}{\overset{CH_3}{C}}$$

(o) $CH_3C{\equiv}CCH_3$ $\xrightarrow[-78°]{Li\ +\ C_2H_5NH_2}$

$$\underset{H}{\overset{CH_3}{C}}{=}\underset{CH_3}{\overset{H}{C}}$$

(p) $CH_3CH_2C{\equiv}CH$ + H_2 $\xrightarrow{Ni_2B}$ $CH_3CH_2CH{=}CH_2$

(q) $CH_3CH_2CH{=}CH_2$ + HBr $\xrightarrow[inhibitor]{peroxide}$ $CH_3CH_2\overset{Br}{\underset{|}{C}}HCH_3$

(r) $CH_3CH_2CH{=}CH_2$ + HBr $\xrightarrow{peroxide}$ $CH_3CH_2CH_2CH_2Br$

(s) $CH_3CH_2CH{=}CH_2$ + $(BH_3)_2$ $\longrightarrow$ $(CH_3CH_2CH_2CH_2)_3B$ $\xrightarrow[OH^-]{H_2O_2}$

$CH_3CH_2CH_2CH_2OH$

(t) $CH_3CH_2CH{=}CH_2$ + Hg(OAc)2 $\longrightarrow$ $CH_3CH_2\underset{HgOAc}{\overset{|}{C}}HCH_3$ $\xrightarrow[OH^-]{NaBH_4}$ $CH_3CH_2\underset{OH}{\overset{|}{C}}HCH_3$

or $CH_3CH_2CH{=}CH_2$ + H_2O $\xrightarrow{H_2SO_4}$ $CH_3CH_2\underset{OH}{\overset{|}{C}}H{-}CH_3$

(u)

$$\underset{H}{\overset{H_3C}{C}}{=}\underset{CH_3}{\overset{H}{C}}$$ + Br_2 $\xrightarrow[dark]{CCl_4}$

$$H{-}\overset{CH_3}{\underset{}{\bigcirc}}{-}Br$$
$$H{-}\underset{CH_3}{\overset{}{\bigcirc}}{-}Br$$

(v)

$$\underset{H}{\overset{H_3C}{C}}{=}\underset{H}{\overset{CH_3}{C}}$$ + Br_2 $\xrightarrow[dark]{CCl_4}$

$$H{-}\overset{CH_3}{\underset{}{\bigcirc}}{-}Br$$
$$Br{-}\underset{CH_3}{\overset{}{\bigcirc}}{-}H$$ + enantiomer

(w) $CH_3C{\equiv}CCH_3$ + HCl $\xrightarrow[CH_3COOH]{Cl^-,\ 25°}$

$$\underset{Cl}{\overset{CH_3}{C}}{=}\underset{CH_3}{\overset{H}{C}}$$

(x) $CH_3CH{=}CH_2$ + HBr $\xrightarrow{peroxide}$ $CH_3CH_2CH_2Br$

(y) $CH_3CH_2C{\equiv}CH$ + $NaNH_2$ $\xrightarrow[NH_3]{liquid}$ $CH_3CH_2C{\equiv}C\!:^-\overset{+}{Na}$ + NH_3

$\xrightarrow{CH_3CH_2Br}$ $CH_3CH_2C{\equiv}CCH_2CH_3$

(z) $CH_3CH_2C{\equiv}CH$ + HBr $\xrightarrow{\text{peroxide}}$ $CH_3CH_2CH{=}CHBr$

9.12

(a)
$$\begin{array}{c} CH_3CH_2CH_2 \qquad Br \\ {\diagdown}C{=}C{\diagup} \\ Br \qquad\qquad H \end{array}$$

(b)
$$\begin{array}{c} CH_3CH_2CH_2 \\ {\diagdown}C{=}CH_2 \\ Cl \end{array}$$

(c)
$$CH_3CH_2CH_2\overset{\displaystyle Cl}{\underset{\displaystyle Cl}{\overset{|}{\underset{|}{C}}}}CH_3$$

(d) $CH_3CH_2CH_2CH{=}CHBr$

(e)
$$CH_3CH_2CH_2\overset{\displaystyle O}{\overset{\|}{C}}CH_3$$

(f) $CH_3CH_2CH_2CH{=}CH_2$

(g) $CH_3CH_2CH_2CH{=}CH_2$

(h)
$$CH_3CH_2CH_2CH_2\overset{\displaystyle O}{\overset{\|}{C}}H$$

(i) $CH_3CH_2CH_2C{\equiv}C\overset{..}{}{:}Na^+$

(j) $CH_3CH_2CH_2C{\equiv}CCH_3$

(k) $CH_3CH_2CH_2C{\equiv}C{-}C{\equiv}CCH_2CH_2CH_3$

(l) $CH_3CH_2CH_2C{\equiv}CAg$

(m) $CH_3CH_2CH_2C{\equiv}CCu$

(n) $CH_3CH_2CH_2COOH$
$\qquad\qquad\qquad + CO_2$

9.13

(a)
$$\begin{array}{c} CH_3CH_2 \qquad H \\ {\diagdown}C{=}C{\diagup} \\ Cl \qquad\qquad CH_2CH_3 \end{array}$$

(b)
$$CH_3CH_2\overset{\displaystyle Cl}{\underset{\displaystyle Cl}{\overset{|}{\underset{|}{C}}}}CH_2CH_2CH_3$$

(c)
$$\begin{array}{c} CH_3CH_2 \qquad Br \\ {\diagdown}C{=}C{\diagup} \\ Br \qquad\qquad CH_2CH_3 \end{array}$$

(d)
$$CH_3CH_2\overset{\displaystyle Br}{\underset{\displaystyle Br}{\overset{|}{\underset{|}{C}}}}{-}\overset{\displaystyle Br}{\underset{\displaystyle Br}{\overset{|}{\underset{|}{C}}}}CH_2CH_3$$

(e)
$$\begin{array}{c} CH_3CH_2 \qquad CH_2CH_3 \\ {\diagdown}C{=}C{\diagup} \\ H \qquad\qquad H \end{array}$$

(f) Same as (e)

(g)
$$\begin{array}{c} CH_3CH_2 \qquad H \\ {\diagdown}C{=}C{\diagup} \\ H \qquad\qquad CH_2CH_3 \end{array}$$

(h)
$$CH_3CH_2\overset{\displaystyle O}{\overset{\|}{C}}CH_2CH_2CH_3$$

(i) No reaction

(j) No reaction

(k) Same as (e)

(l) Same as (h)

(m) No reaction

(n) $CH_3CH_2CH_2CH_2CH_2CH_3$

(o) $2CH_3CH_2COOH$

(p) $2CH_3CH_2COOH$

(q) No reaction

9.14

(a) $CH_3CH_2CH_2CH{=}CH_2$ + $Br_2\longrightarrow$ $CH_3CH_2CH_2\underset{\displaystyle Br}{\overset{|}{C}HCH_2Br}$

$\qquad\qquad\qquad\qquad\qquad\qquad\overset{\displaystyle 2NaNH_2}{\underset{\displaystyle liq\ NH_3}{\xrightarrow{\hspace{2cm}}}}$ $CH_3CH_2CH_2C{\equiv}CH$

(b) $CH_3CH_2CH_2CH_2CH_2Cl \xrightarrow[CH_3CH_2OH]{KOH} CH_3CH_2CH_2CH=CH_2$

then proceed as in (a) above.

(c) $CH_3CH_2CH_2CH=CHCl \xrightarrow[NH_3]{NaNH_2} CH_3CH_2CH_2C\equiv CH$

(d) $CH_3CH_2CH_2CH_2CHCl_2 \xrightarrow[NH_3]{2NaNH_2} CH_3CH_2CH_2C\equiv CH$

(e) $HC\equiv CH \xrightarrow[liq.\ NH_3]{NaNH_2} HC\equiv C:^- Na^+ \xrightarrow{CH_3CH_2CH_2Br} HC\equiv CCH_2CH_2CH_3$

9.15

(a) Propyne is soluble in cold, concentrated H_2SO_4; propane is not. Other tests are Br_2/CCl_4 and $KMnO_4/H_2O$.

(b) $Ag(NH_3)_2^+ OH^-$ gives a precipitate with propyne, not with propene.

(c) Dilute $KMnO_4$ oxidizes 1-bromopropene and not 2-bromopropane.

(d) $Ag(NH_3)_2^+ OH^-$ gives a precipitate with 1-butyne, not with 2-bromo-2-butene.

(e) Sodium fusion followed by acidification with dilute HNO_3 and addition of $AgNO_3$ gives a AgBr precipitate with 2-bromo-2-butene, not with 2-butyne.

(f) Br_2/CCl_4 is decolorized by 2-butyne, not by *n*-butyl alcohol.

(g) $AgNO_3/C_2H_5OH$ gives a AgBr precipitate wth 2-bromobutane, not with 2-butyne.

(h) Br_2/CCl_4 is decolorized by $CH_3C\equiv CCH_2OH$, not by $CH_3CH_2CH_2CH_2OH$.

(i) Br_2/CCl_4 is decolorized by $CH_3CH=CHCH_2OH$, not by $CH_3CH_2CH_2CH_2OH$.

(In many cases above other tests are possible.)

9.16

(a) A = $CH_3CH_2CH_2C\equiv CH$, B = $CH_3CH_2C\equiv CCH_3$, C =

(b) Yes, B may also be $CH_3CH=CH-CH=CH_2$ or $CH_2=CH-CH_2-CH=CH_2$.

  C may also be , or .

(c) B = $CH_3CH_2C\equiv CCH_3$

(d)

9.17

(a)

(b)

(c) Product of (a) $\xrightarrow{H_2/Ni}$ $\underset{\underset{Cl}{|}}{\overset{\overset{CH_3}{|}}{CH_3CHCHCH_3}}$

(d) Product of (a) $\xrightarrow[dark]{Cl_2/CCl_4}$ $CH_3\overset{\overset{CH_3}{|}}{CH}-\overset{\overset{Cl}{|}}{\underset{\underset{Cl}{|}}{C}}-CH_2Cl$

(e) Product of (a) $\xrightarrow[\substack{peroxide \\ inhibitor}]{HBr}$ $CH_3\overset{\overset{CH_3}{|}}{CH}-\overset{\overset{Cl}{|}}{\underset{\underset{Br}{|}}{C}}-CH_3$

(f) $\underset{\underset{CH_3}{\overset{\overset{CH_3}{|}}{}}}{CH_3CHC{\equiv}CH}\xrightarrow[then\quad H^+]{KMnO_4/OH^-,}$ $CH_3\overset{\overset{CH_3}{|}}{CH}COOH + CO_2$

9.18

(a) $CH_3CH_2CH_2C{\equiv}CH + D_2 \xrightarrow{Ni_2B}$ $\underset{D}{\overset{CH_3CH_2CH_2}{\diagdown}}C{=}C\underset{D}{\overset{H}{\diagup}}$

(b) $CH_3CH_2CH_2C{\equiv}CH + Sia_2BH \longrightarrow$ $\underset{H}{\overset{CH_3CH_2CH_2}{\diagdown}}C{=}C\underset{BSia_2}{\overset{H}{\diagup}}$

$\xrightarrow{CH_3COOD}$ $\underset{H}{\overset{CH_3CH_2CH_2}{\diagdown}}C{=}C\underset{D}{\overset{H}{\diagup}}$

(c) $CH_3CH_2CH_2C{\equiv}CH + DCl \longrightarrow$ $\underset{Cl}{\overset{CH_3CH_2CH_2}{\diagdown}}C{=}C\underset{H}{\overset{D}{\diagup}}$

(d) $CH_3CH_2CH_2C{\equiv}CH + Sia_2BD \longrightarrow$ $\underset{D}{\overset{CH_3CH_2CH_2}{\diagdown}}C{=}C\underset{BSia_2}{\overset{H}{\diagup}}$

$\xrightarrow[OH^-]{H_2O_2}$ $CH_3CH_2CH_2\overset{\overset{}{|}}{\underset{\underset{D}{|}}{C}}H\overset{\overset{O}{\|}}{C}H$

9.19

(a) These results indicate a preference for abstraction of an internal hydrogen.

(b) Elimination of HBr must be anti. (Compound 3 has no hydrogen anti to Br.)

(c) 2-Decyne is more stable than 1-decyne, but it is produced at a slower rate. After only 9 hours, the more rapidly formed isomer is produced; after 21 hours, enough time for equilibrium to be established, the more stable 2-decyne predominates.

10

CONJUGATED UNSATURATED SYSTEMS. VISIBLE-ULTRAVIOLET SPECTROSCOPY

10.1

(a) $^{14}CH_2=CHCH_2X$ and $CH_2=CH^{14}CH_2X$

(b) The reaction proceeds through the resonance stabilized free radical,

$$^{14}\overset{\bullet}{C}H_2=CH-CH_2 \longleftrightarrow {}^{14}CH_2-CH=\overset{\bullet}{C}H_2 \text{ or } {}^{14}\overset{\delta\bullet}{C}H_2 \text{=\!=\!=} CH \text{=\!=\!=} \overset{\delta\bullet}{C}H_2$$

Thus attack on X_2 can occur by the carbon at either end of the chain since they are equivalent.

(c) 50:50

10.2

(a)
$$\overset{4}{CH_3}-\overset{3}{\underset{+}{CH}}\overset{2}{\diagup}\overset{CH}{\diagdown}\overset{1}{CH_2}\longleftrightarrow \overset{4}{CH_3}-\overset{3}{CH}\overset{2}{\diagup}\overset{CH}{\diagdown}\overset{1}{\underset{+}{CH_2}} \text{ or } \overset{4}{CH_3}-\overset{3}{\underset{\delta+}{CH}}\overset{2}{\diagup}\overset{CH}{\diagdown}\overset{1}{\underset{\delta+}{CH_2}}$$

D	E	F

(b) We know that the allyl cation is almost as stable as a tertiary carbocation. Here we find not only the resonance stabilization of an allyl cation but the additional stabilization that arises from contributor **D** in which the plus charge is on a secondary carbon.

(c)
$$CH_3-\overset{\overset{\displaystyle Cl}{|}}{CH}-CH=CH_2 \text{ and } CH_3-CH=CH-CH_2-Cl$$

10.3

(a) *Cis*-1,3-pentadiene, *trans,trans*-2,4-hexadiene, *cis,trans*-2,4-hexadiene, and 1,3-cyclo-hexadiene are conjugated dienes.

(b) 1,4-Cyclohexadiene is an isolated diene.

(c) 1-Penten-4-yne is an isolated enyne.

10.4

(a) $CH_3CH_2\underset{\underset{\displaystyle Cl}{|}}{CH}CH=CHCH_3$ and $CH_3CH_2CH=CH\underset{\underset{\displaystyle Cl}{|}}{CH}CH_3$

(b) The most stable cation is a hybrid of equivalent forms: $CH_3\underset{+}{C}HCH=CHCH_3 \longleftrightarrow$

79

$CH_3CH=CHCHCH_3$. Thus 1,4 and 1,2- addition yield the same product,

$$CH_3\overset{+}{C}HCH=CHCH_3$$
$$\underset{Cl}{|}$$

10.5

(a) Addition of the proton gives the resonance hybrid of

$$CH_3-\overset{+}{C}H-CH=CH_2 \longleftrightarrow CH_3-CH=CH-\overset{+}{C}H_2$$
$$\qquad\qquad I \qquad\qquad\qquad\qquad\qquad II$$

The inductive effect of the methyl group in I stabilizes the positive charge on the adjacent carbon. Such stabilization of the positive charge does not occur in II. Because I contributes more heavily to the resonance hybrid than does II, C-2 bears a greater positive charge and reacts faster with the chloride ion.

(b) In the 1,4-addition product, the double bond is more highly substituted than in the 1,2-addition product, hence it is the more stable alkene.

10.6

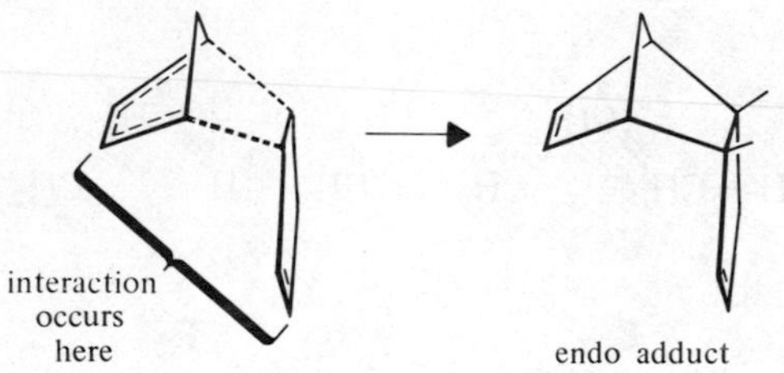

interaction
occurs
here

endo adduct

10.7

(a)

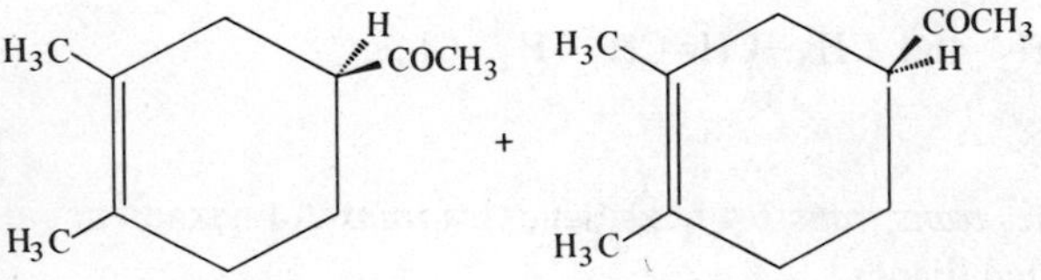

(b)

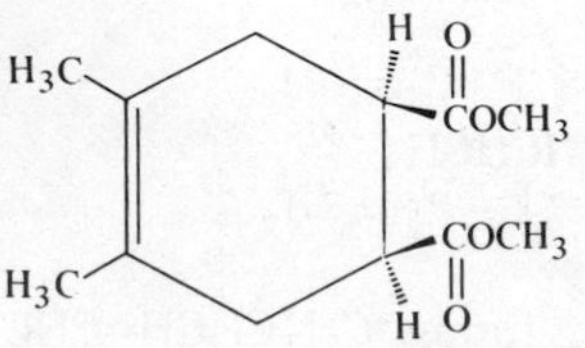

(c) 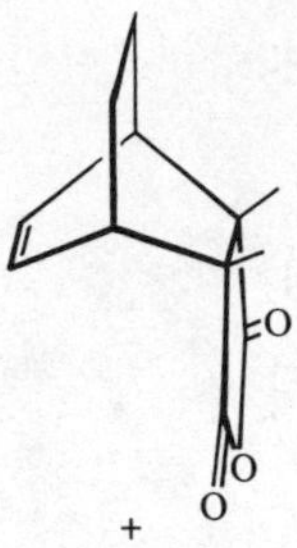 (major product)

+

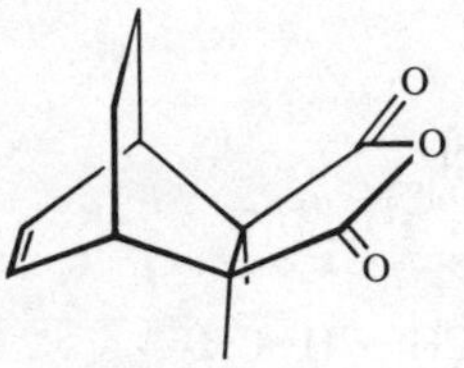

(minor product)

10.8

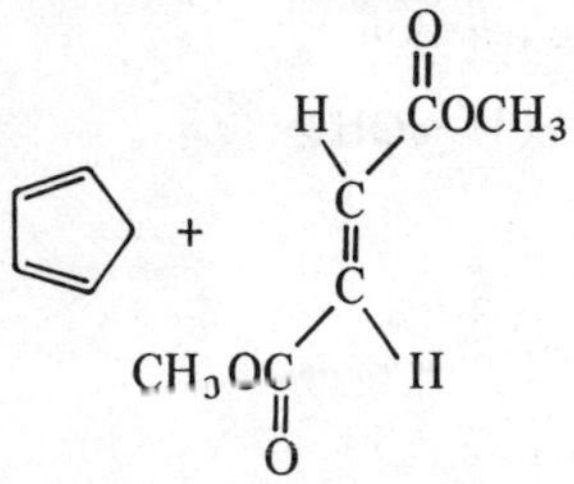

10.9

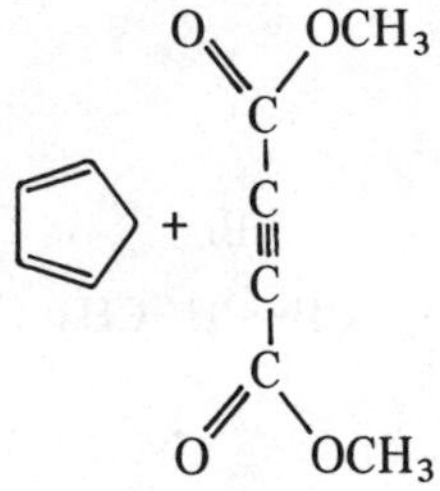

10.10

(a) $BrCH_2CH_2CH_2CH_2Br \xrightarrow[\text{(CH}_3\text{)}_3\text{COH}]{\text{(CH}_3\text{)}_3\text{COK}} CH_2=CH-CH=CH_2$

(b) $HOCH_2CH_2CH_2CH_2OH \xrightarrow[\text{heat}]{\text{conc.\,H}_2\text{SO}_4} CH_2=CH-CH=CH_2$

(c) $CH_2=CH-CH_2CH_2-OH \xrightarrow[\text{heat}]{\text{conc.\,H}_2\text{SO}_4} CH_2=CH-CH=CH_2$

(d) $CH_2=CH-CH_2CH_2-Cl \xrightarrow[\text{(CH}_3\text{)}_3\text{COH}]{\text{(CH}_3\text{)}_3\text{COK}} CH_2=CH-CH=CH_2$

(e) $CH_2=CH-\underset{\underset{Cl}{|}}{CH}-CH_3 \xrightarrow[\text{(CH}_3\text{)}_3\text{COH}]{\text{(CH}_3\text{)}_3\text{COK}} CH_2=CH-CH=CH_2$

(f) $CH_2=CH-CH-CH_3$ $\xrightarrow[\text{heat}]{\text{conc. } H_2SO_4}$ $CH_2=CH-CH=CH_2$
            | OH

(g) $HC\equiv C-CH=CH_2$ + H_2 $\xrightarrow[\text{quinoline}]{\text{Pd. } BaSO_4}$ $CH_2=CH-CH=CH_2$

10.11

$CH_2=C\underset{\underset{CH_3}{|}}{}\;\;\underset{\underset{CH_3}{|}}{C}=CH_2$

10.12

(a) $Cl-CH_2CHCH=CH_2$ + $Cl-CH_2-CH=CH-CH_2-Cl$
             | Cl

(b) $CH_2-CH-CH-CH_2$ (c) $CH_2-CH-CH-CH_2$
   | Cl | Cl | Cl | Cl | Br | Br | Br | Br

(d) $CH_3-CH_2-CH_2-CH_3$ (e) No reaction

(f) $Cl-CH_2-CH-CH=CH_2$ + $Cl-CH_2-CH=CH-CH_2-OH$
           | OH

(g) $4CO_2$ (Note: $KMnO_4$ oxidizes
          $HOOC-COOH$ to $2CO_2$)

(h) $CH_3-CH-CH=CH_2$ + $CH_3-CH=CH-CH_2OH$
        | OH

10.13

(a) $CH_2=CH-CH_2-CH_3$ + (NBS) $\xrightarrow{CCl_4}$ $CH_2=CH-\underset{\underset{}{\overset{\overset{Br}{|}}{}}}{C}H-CH_3$

$\xrightarrow[(CH_3)_3COH]{(CH_3)_3COK}$ $CH_2=CH-CH=CH_2$

(b) $CH_2=CH-CH_2CH_2CH_3$ + NBS $\xrightarrow{CCl_4}$ $CH_2=CH-\overset{\overset{Br}{|}}{C}HCH_2CH_3$

$\xrightarrow[(CH_3)_3COH]{(CH_3)_3COK}$ $CH_2=CH-CH=CH-CH_3$

(c) $CH_3CH_2CH_2CH_2OH \xrightarrow[\text{heat}]{\text{conc. } H_2SO_4} CH_3CH_2CH=CH_2 \xrightarrow{\text{[as in (a)]}} CH_2=CH-CH=CH_2$

$CH_2-CH=CH-CH_2 \xleftarrow[\text{heat}]{Br_2}$
| Br | Br

(d) $CH_3-CH=CH-CH_3$ + NBS $\xrightarrow{\;CCl_4\;}$ $CH_3-CH=CH-CH_2-Br$

(e) [cyclopentane] + Br_2 $\xrightarrow[\text{heat}]{\text{light}}$ [bromocyclopentane, Br] $\xrightarrow[(CH_3)_3COH]{(CH_3)_3COK}$ [cyclopentene] $\xrightarrow[CCl_4]{NBS}$ [bromocyclopentene, Br]

(excess)

(f) [bromocyclopentene, Br] $\xrightarrow[(CH_3)_3COH]{(CH_3)_3COK}$ [cyclopentadiene] $\left(\text{same as } \text{[cyclopentadiene]}\right)$

10.14

$$R-\ddot{\underset{..}{O}}-\ddot{\underset{..}{O}}-R \xrightarrow[\text{or heat}]{\text{light}} 2R-\ddot{\underset{..}{O}}\cdot$$

$$R-\ddot{\underset{..}{O}}\cdot + H-\ddot{\underset{..}{Br}}: \longrightarrow R-\ddot{\underset{..}{O}}-H + \cdot\ddot{\underset{..}{Br}}:$$

$$CH_2=CH-CH=CH_2 + \cdot\ddot{\underset{..}{Br}}: \longrightarrow \left[CH_2=CH-\overset{\cdot}{C}H-\underset{Br}{C}H_2 \longleftrightarrow \overset{\cdot}{C}H_2-CH=CH-\underset{Br}{C}H_2\right]$$

$$\xrightarrow{\;HBr\;} CH_2=CH-\underset{\underset{Br}{|}}{\underset{\overset{|}{H}}{C}}H-\underset{\underset{Br}{|}}{\underset{\overset{|}{H}}{C}}H_2 + CH_2-CH=CH-\underset{\underset{Br}{|}}{\overset{\overset{|}{H}}{C}}H_2 + \cdot\ddot{\underset{..}{Br}}:$$

(*cis* and *trans*)

10.15

(a) $Ag(NH_3)_2OH$ gives a precipitate with 1-butyne only.

(b) 1,3-Butadiene decolorizes bromine solution; *n*-butane does not.

(c) CrO_3/H_2SO_4 oxidizes the alcohol. The solution changes from orange to green. No color change with 1,3-butadiene.

(d) $AgNO_3$ in C_2H_5OH gives a AgBr precipitate with $CH_2=CHCH_2CH_2Br$. No reaction with 1,3-butadiene.

(e) $AgNO_3$ in C_2H_5OH gives a AgBr precipitate with $BrCH_2CH=CHCH_2Br$ (it is an allylic bromide), but not with $CH_3\underset{\underset{Br}{|}}{C}H=\underset{\underset{Br}{|}}{C}HCH_3$ (a vinyl bromide).

10.16

(a) Because a highly resonance-stabilized free radical is formed:

$$CH_2=CH-\overset{\cdot}{C}H-CH=CH_2 \longleftrightarrow \overset{\cdot}{C}H_2=CH-CH=CH-\overset{\cdot}{C}H_2 \longleftrightarrow \overset{\cdot}{C}H_2-CH=CH-CH=CH_2$$

(b) Because the carbanion is more stable:

$$CH_2=CH-\overset{..}{C}H-CH=CH_2 \longleftrightarrow CH_2=CH-CH=CH-\overset{..}{C}H_2 \longleftrightarrow \overset{..}{C}H_2-CH=CH-CH=CH_2$$

i.e., we can write more resonance structures of nearly equal energies.

10.17

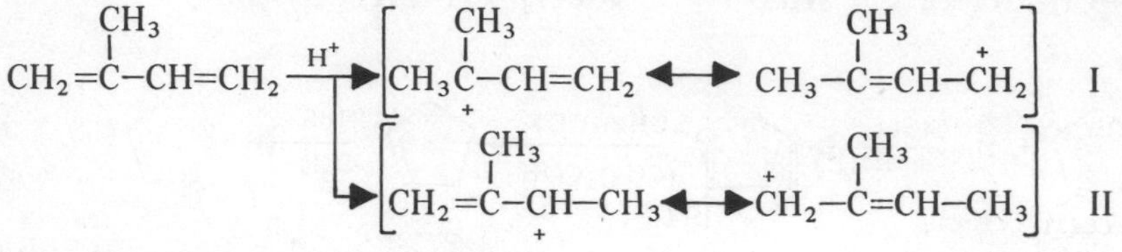

$$CH_2=\overset{\overset{\textstyle CH_3}{|}}{C}-CH=CH_2 \xrightarrow{H^+} \left[CH_3\overset{\overset{\textstyle CH_3}{|}}{\underset{+}{C}}-CH=CH_2 \longleftrightarrow CH_3-\overset{\overset{\textstyle CH_3}{|}}{C}=CH-\overset{+}{C}H_2 \right] \quad I$$

$$\left[CH_2=\overset{\overset{\textstyle CH_3}{|}}{\underset{+}{C}}-CH-CH_3 \longleftrightarrow \overset{+}{C}H_2-\overset{\overset{\textstyle CH_3}{|}}{C}=CH-CH_3 \right] \quad II$$

The resonance hybrid, I, has the positive charge, in part, on the tertiary carbon; in II, the positive charge is on primary and secondary carbons only. Therefore hybrid I is more stable, and will be the intermediate carbocation. 1,4-addition to I gives

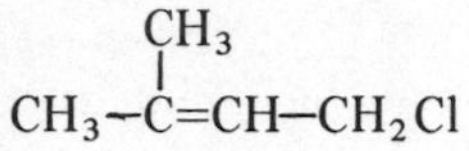

$$CH_3-\overset{\overset{\textstyle CH_3}{|}}{C}=CH-CH_2Cl$$

10.18

(a)

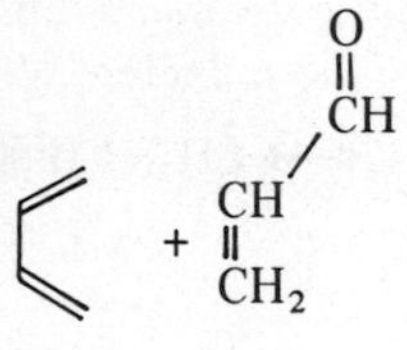

(d)

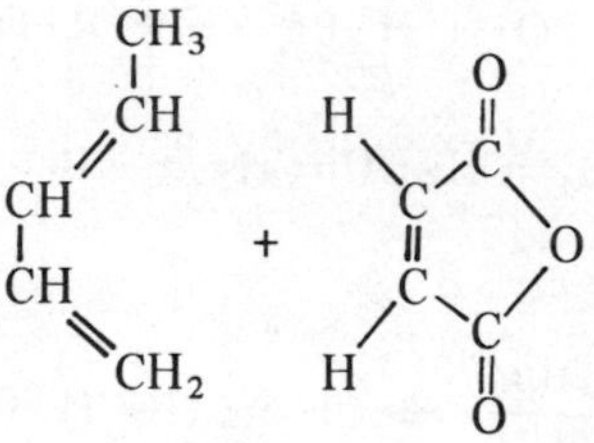

(b)

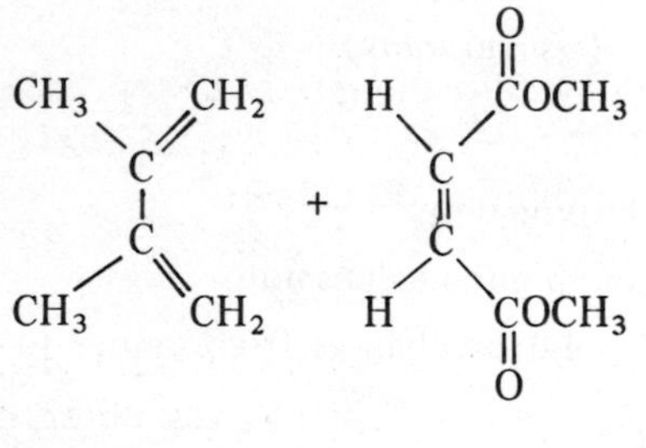

(e)

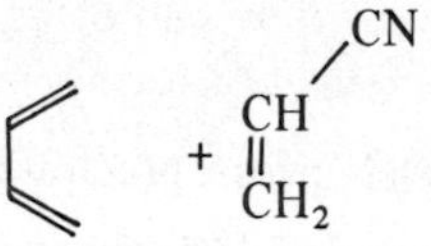

(c)

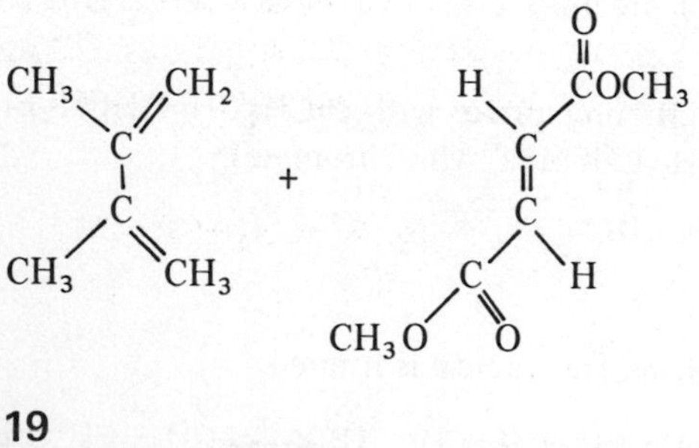

10.19

Neither compound can assume the s-*cis*-conformation. 1,3-Butadiyne is linear, and

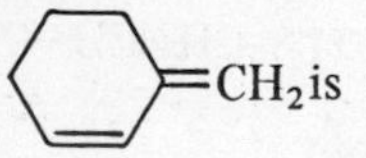

$=CH_2$ is forced into the s-*trans* conformation by the requirements of the ring.

10.20

(a)

(b)

10.21

The formula, C_6H_8, tells us that **A** and **B** have six hydrogens less than an alkane. This unsaturation may be due to three double bonds, one triple bond and one double bond, or combinations of two double bonds and a ring, or one triple bond and a ring. Since both **A** and **B** react with two moles of H_2 to yield cyclohexane, they are either cyclohexyne or cyclohexadienes. The absorption maximum of 256 nm for **A** tells us that it is conjugated. **B**, with no absorption maximum beyond 200, possesses isolated double bonds. We can rule out cyclohexyne because of ring strain caused by the requirement of linearity of the $-C\equiv C-$ system. Therefore **A** is 1,3-cyclohexadiene; **B** is 1,4-cyclohexadiene.

10.22

In the dimer, the remaining double bonds are isolated:

The rate of dimerization can be followed by observing the rate of disappearance of the absorption maximum at 239 nm that is due to the conjugated double bonds in cyclopentadiene (see Table 10.3). The product does not absorb at 239 nm.

10.23

All three compounds have an unbranched five-carbon chain. The formula, C_5H_6, suggests that they have one double bond and one triple bond. **D**, **E**, and **F** must differ, therefore, in the way the multiple bonds are distributed in the chain. **E** and **F** have a terminal $-C\equiv CH$ (reaction with $Ag(NH_3)_2{}^+OH^-$). The absorption maximum near 230 nm for **D** and **E** suggests that in these compounds, the multiple bonds are conjugated. The structures are:

$$CH_3-C\equiv C-CH=CH_2 \qquad HC\equiv C-CH=CH-CH_3 \qquad HC\equiv C-CH_2-CH=CH_2$$

$$\textbf{D} \qquad\qquad\qquad \textbf{E} \qquad\qquad\qquad \textbf{F}$$

10.24

The C_2-C_3 bond has partial double bond character.

10.25

The *endo* adduct is less stable than the *exo*, but is produced at a faster rate at 25°. At 90°. equilibrium is established, and the more stable *exo* adduct predominates.

10.26

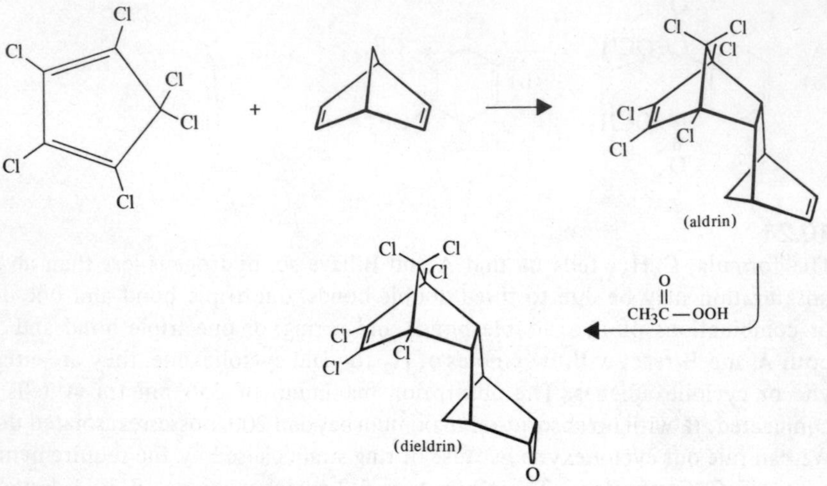
(aldrin)
$CH_3C—OOH$
(dieldrin)

10.27

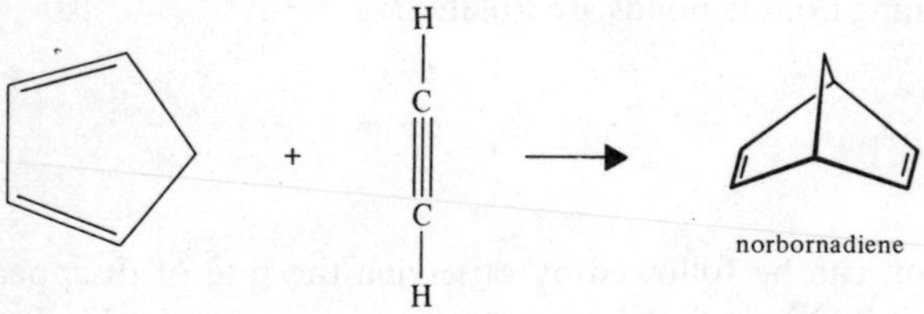
norbornadiene

10.28

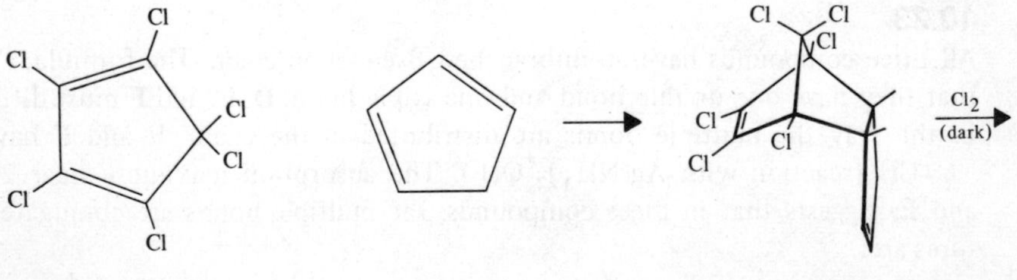
Cl_2
(dark)

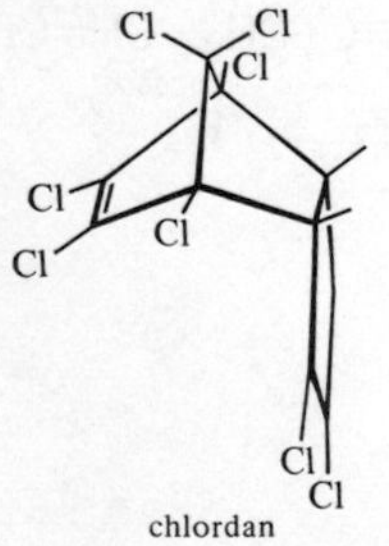
chlordan

Note: The other double bond is less reactive because of the presence of the two chlorine substituents.

heptachlor

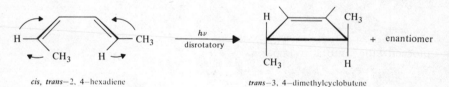

11
SPECIAL
TOPICS II

11.1
Conrotatory motion of the type shown would lead to increasingly unfavorable interaction of the methyl groups as the transition state is approached. Thus this path is not followed to any appreciable extent.

11.2
According to the Woodward-Hoffmann rule for electrocyclic reactions of $4n$ π electron systems (p. 390), the photochemical cyclization of *cis,trans*-2,4-hexadiene should proceed with *disrotatory motion*. Thus it should yield *trans*-3,4-dimethylcyclobutene:

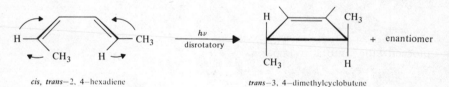

cis, trans–2, 4–hexadiene   *trans*–3, 4–dimethylcyclobutene

11.3

(a)

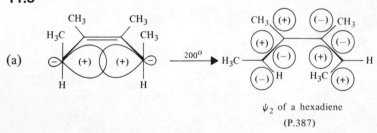

ψ_2 of a hexadiene
(P.387)

(b) This is a thermal electrocyclic reaction of a $4n$ π electron system; it should, *and does*, proceed with conrotatory motion.

11.4

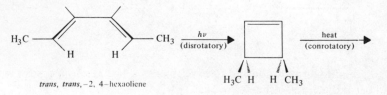

trans, trans,–2, 4–hexaoliene H_3C H   H CH_3
cis–3, 4–dimethylcyclobutene

cis, trans–2, 4–hexadiene

Here we find that two consecutive electrocyclic reactions (the first photochemical, the second thermal), provide a stereospecific synthesis of *cis,trans*-2,4-hexadiene from *trans,trans*-2,4-hexadiene.

11.5

(a) This is a photochemical electrocyclic reaction of an eight π electron system—a $4n\ \pi$ system where $n = 2$. It should, therefore, proceed with disrotatory motion.

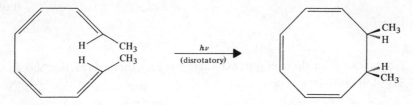

cis–7, 8–dimethyl–1, 3, 5–cyclooctatriene

(b) This is a thermal electrocyclic reaction of the eight π electron system. It should proceed with conrotatory motion.

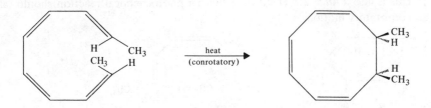

cis–7, 8–dimethyl–1, 3, 5–cyclooctatriene

11.6

(a) This is conrotatory motion and since this is a $4n\ \pi$ electron system (where $n = 1$) it should occur under the influence of heat.

(b) This is conrotatory motion and since this is also a $4n\ \pi$ electron system (where $n = 2$) it should occur under the influence of heat.

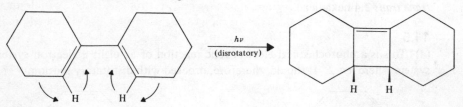

(c) This is disrotatory motion. This, too is a $4n$ π electron system (where $n = 1$), thus it should occur under the influence of light.

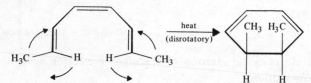

11.7

(a) This is a $4n + 2$ π electron system (where $n = 1$); a thermal reaction should take place with disrotatory motion:

(b) This is also a $4n + 2$ π electron system; a photochemical reaction should take place with conrotatory motion.

11.8

Here we need a conrotatory ring-opening of *trans*-5,6-dimethyl-1,3-cyclohexadiene (to produce *trans,cis,trans*-2,4,6-octatriene), then we need a disrotatory cyclization to produce *cis*-5,6-dimethyl-1,3-cyclohexadiene.

trans−5, 6−dimethyl−1, 3−
cyclohexadiene

trans, cis, trans−2, 4, 6−
octatriene

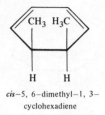

cis−5, 6−dimethyl−1, 3−
cyclohexadiene

Since both reactions involve $4n + 2$ π electron systems we apply light to accomplish the first step and heat to accomplish the second. It would also be possible to use heat to produce *trans,cis,cis*-2,4,6-octatriene then use light to produce the desired product.

11.9
The first electrocyclic reaction is a thermal, conrotatory ring opening of a $4n$ π electron system. The second electrocyclic reaction is a thermal, disrotatory ring closure of a $4n + 2$ π electron system.

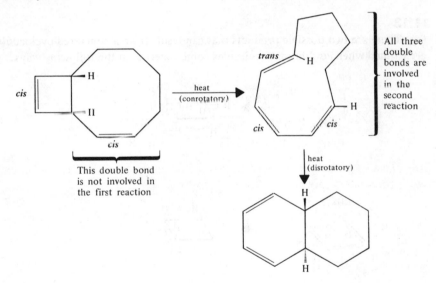

11.10
(a) This reaction involves two π electrons, thus it is a $4n + 2$ π system where $n = 0$.

(b) An allylic cation is formed.

(c) The cyclopropyl anion is a $4n$ π system (where $n = 1$), thus a thermal reaction should take place with conrotatory motion.

11.11
(a) This is a $4n + 2$ π electron system undergoing disrotatory motion. Heat is required.

(b) This is a $4n + 2$ π electron system undergoing conrotatory motion. Light is required.

(c) This is a $4n + 2$ π electron system undergoing conrotatory motion. Light is required.

(d) This is a $4n + 2$ π electron system undergoing disrotatory motion. Heat is required.

11.12

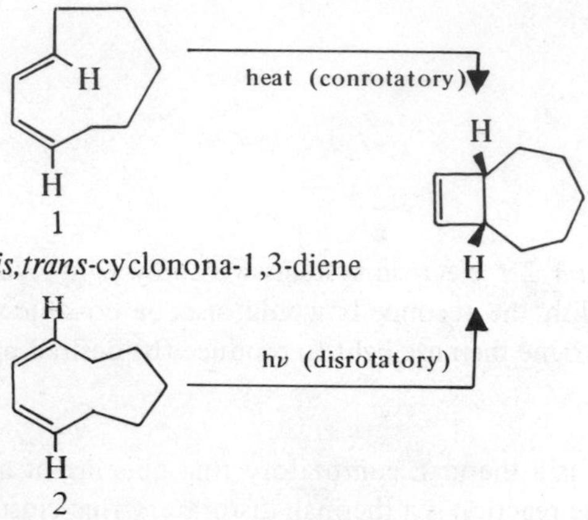

heat (conrotatory)

Cis,trans-cyclonona-1,3-diene

hν (disrotatory)

Cis,cis-cyclonona-1,3-diene

11.13

(a) There are two possible products that can result from a concerted cycloaddition. They are formed when *cis*-2-butene molecules come together in the following ways:

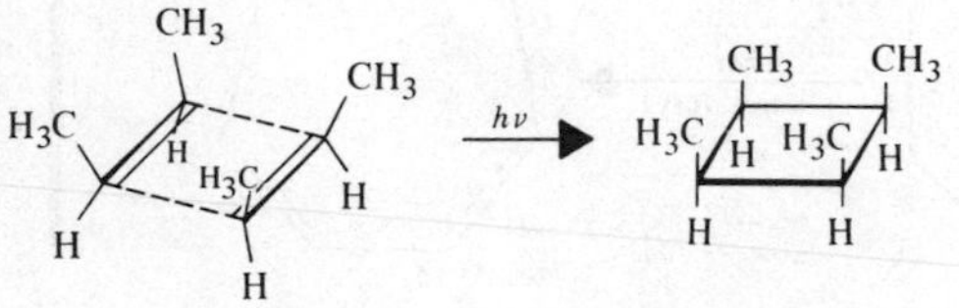

and

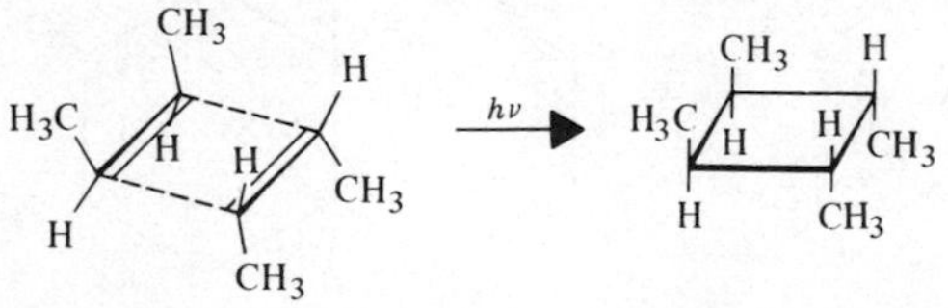

(b) There are two possible products that can be obtained from *trans*-2-butene as well.

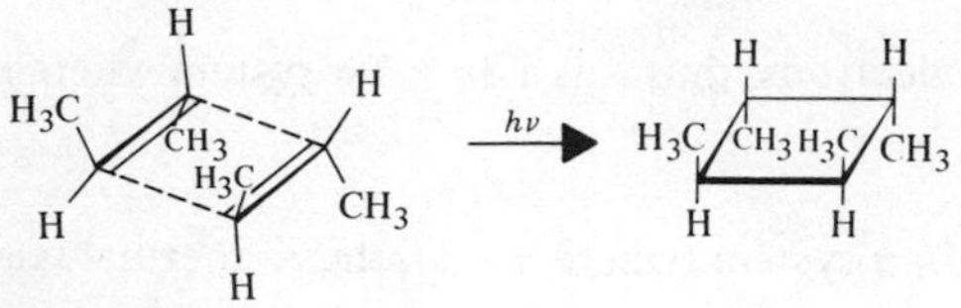

and

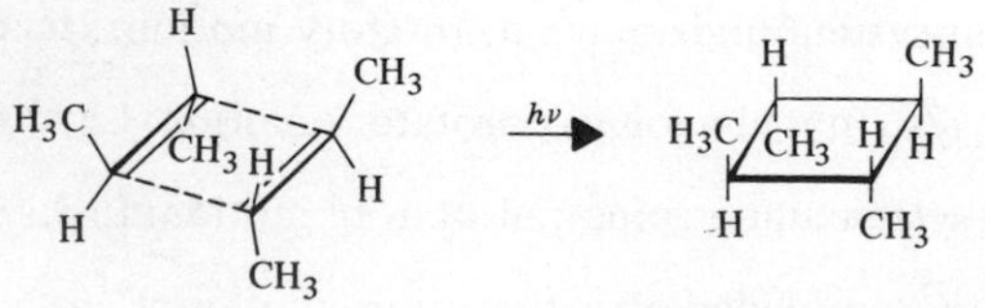

11.14

This is an intramolecular [2 + 2] cycloaddition.

11.15

(a) No. A thermal [2 + 2] concerted cycloaddition is symmetry forbidden and therefore, is likely to have a very high activation energy.

(b) The formation of a relatively stable intermediate in which both unpaired electrons are involved in allylic systems explains the preference for head-to-head cyclization.

$$CH_2 \!-\! C\!=\!CH_2 \longrightarrow \text{diallylic diradical} \longrightarrow CH_2\!-\!C, \; CH_2\!-\!C$$

diallylic diradical

Cyclization in a head-to-tail manner would require the formation of a less stable diradical intermediate — only one electron would be involved in an allylic system:

$$CH_2\!=\!C\!-\!CH_2 \longrightarrow \text{vinylic-allylic diradical}$$

vinylic-allylic diradical

11.16

(a)

$$\begin{array}{c} CN \quad CN \\ CN \quad NC \\ H \quad H \\ CH_3 \quad CH_3 \end{array}$$

(b)

$$\begin{array}{c} CN \quad CN \\ CN \quad NC \\ CH_3 \quad H \\ H \quad CH_3 \end{array} \quad + \quad \begin{array}{c} CN \quad CN \\ CN \quad NC \\ H \quad H_3C \\ CH_3 \quad H \end{array}$$

enantiomers

11.17

$$\underset{CH_2}{\overset{CH_2}{\big|}} \quad + \quad \text{(maleic anhydride)} \quad \xrightarrow{\text{room temp.}} \quad \text{3}$$

3

$$\Big\downarrow 150°$$

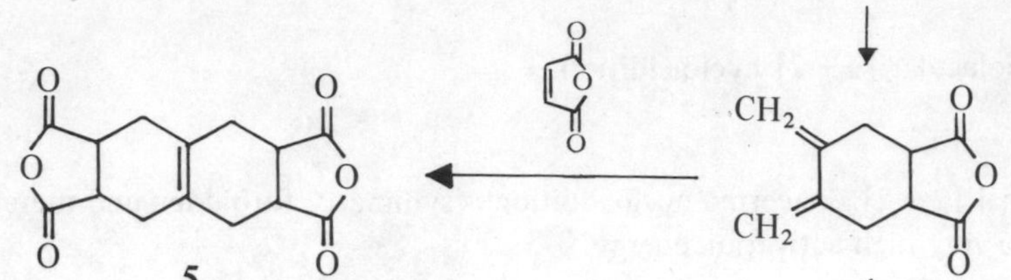

11.18

Compound **7** (below) results from a conrotatory ring opening; it then reacts as the diene component of a Diels-Alder reaction.

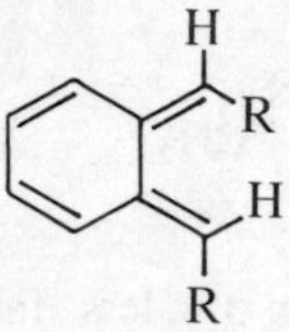

7

11.19

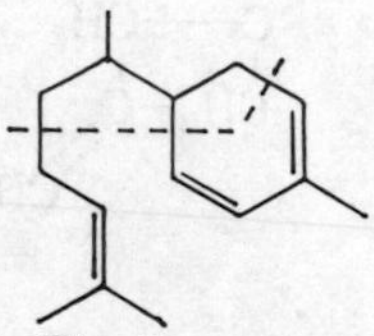

Zingiberene
(a sesquiterpene)

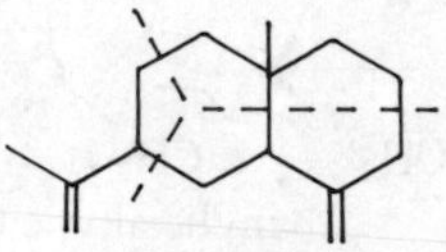

β-Selinene
(a sesquiterpene)

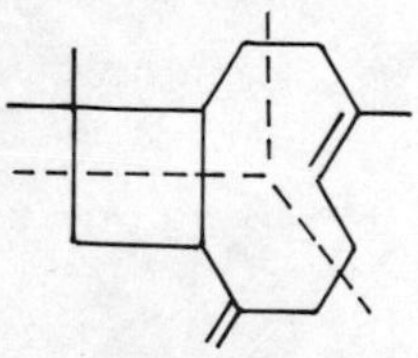

caryophyllene
(a sesquiterpene)

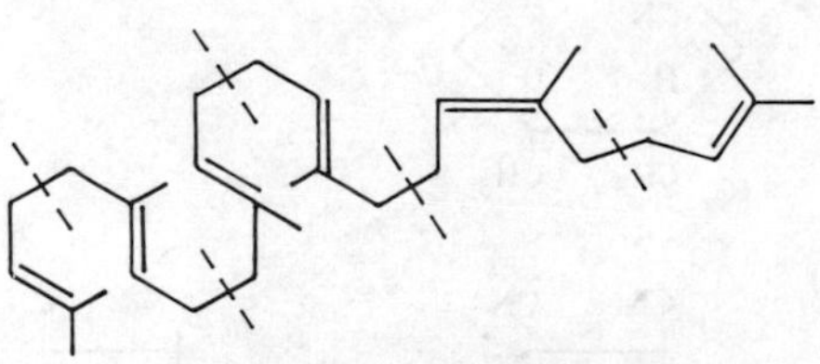

squalene
(a triterpene)

11.20

(a)

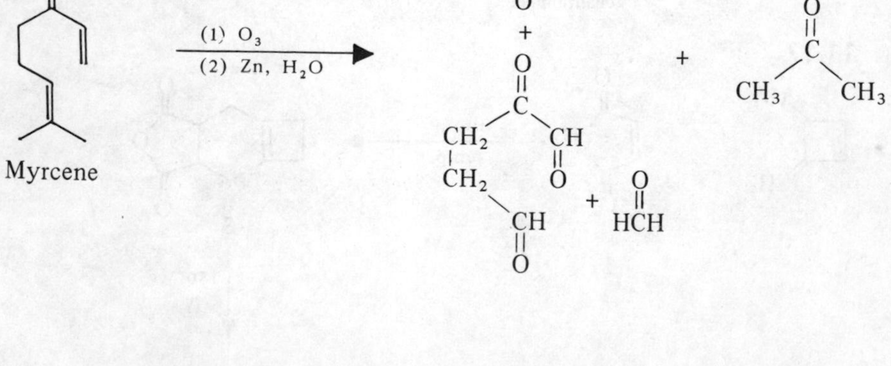

Myrcene

$\xrightarrow[\text{(2) Zn, H}_2\text{O}]{\text{(1) O}_3}$

(b)

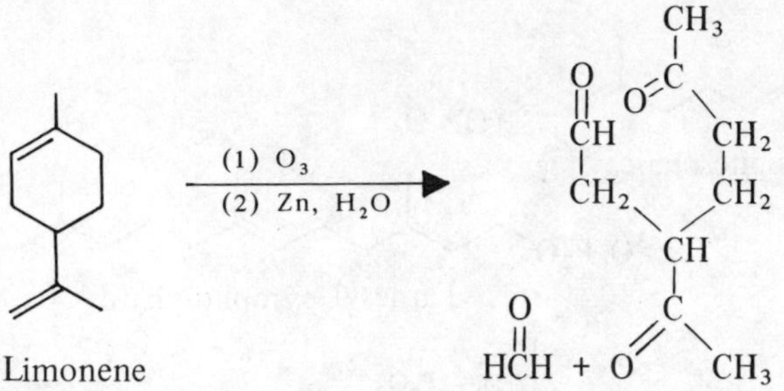

Limonene

$\xrightarrow[\text{(2) Zn, H}_2\text{O}]{\text{(1) O}_3}$

(c) α-Farnesene $\xrightarrow[\text{(2) Zn, H}_2\text{O}]{\text{(1) O}_3}$ (See p. 404)

$$CH_3\overset{O}{\overset{\|}{C}}CH_3 + H\overset{O}{\overset{\|}{C}}CH_2CH_2\overset{O}{\overset{\|}{C}}CH_3$$

$$+ H\overset{O}{\overset{\|}{C}}CH_2\overset{O}{\overset{\|}{C}}H + H\overset{O}{\overset{\|}{C}}-\overset{O}{\overset{\|}{C}}CH_3 + H\overset{O}{\overset{\|}{C}}H$$

(d) Geraniol $\xrightarrow[\text{(2) Zn, H}_2\text{O}]{\text{(1) O}_3}$ (See p. 405)

$$CH_3\overset{O}{\overset{\|}{C}}CH_3 + H\overset{O}{\overset{\|}{C}}CH_2CH_2\overset{O}{\overset{\|}{C}}CH_3$$

$$+ H\overset{O}{\overset{\|}{C}}CH_2OH$$

(e) Squalene $\xrightarrow[\text{(2) Zn, H}_2\text{O}]{\text{(1) O}_3}$ (See p. 405)

$$2CH_3\overset{O}{\overset{\|}{C}}CH_3 + H\overset{O}{\overset{\|}{C}}CH_2CH_2\overset{O}{\overset{\|}{C}}H$$

$$+ 4CH_3\overset{O}{\overset{\|}{C}}CH_2CH_2\overset{O}{\overset{\|}{C}}H$$

11.21

(a)

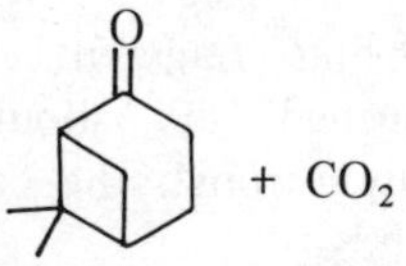

+ CO$_2$

(c)

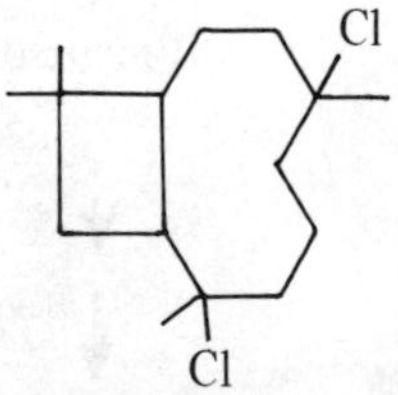

(b)

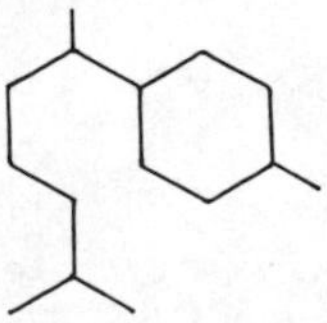

(d)

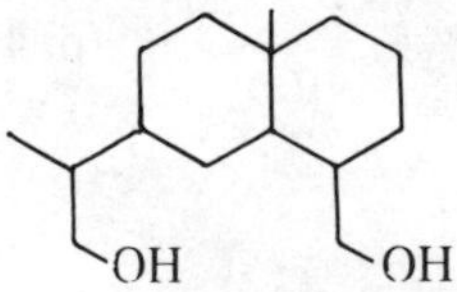

11.22

Br$_2$ in CCl$_4$ or KMnO$_4$ in H$_2$O. Either reagent would give a positive result with geraniol and a negative result with menthol.

11.23

(a)

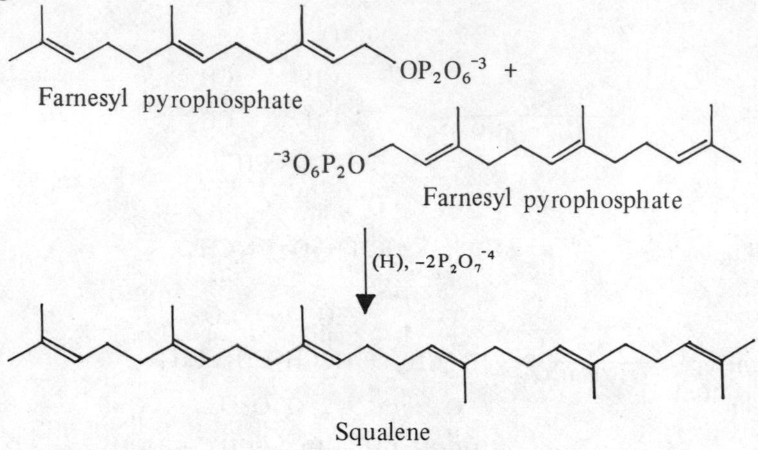

Farnesyl pyrophosphate

$OP_2O_6^{-3}$ +

$^{-3}O_6P_2O$

Farnesyl pyrophosphate

$(H), -2P_2O_7^{-4}$

Squalene

(b)

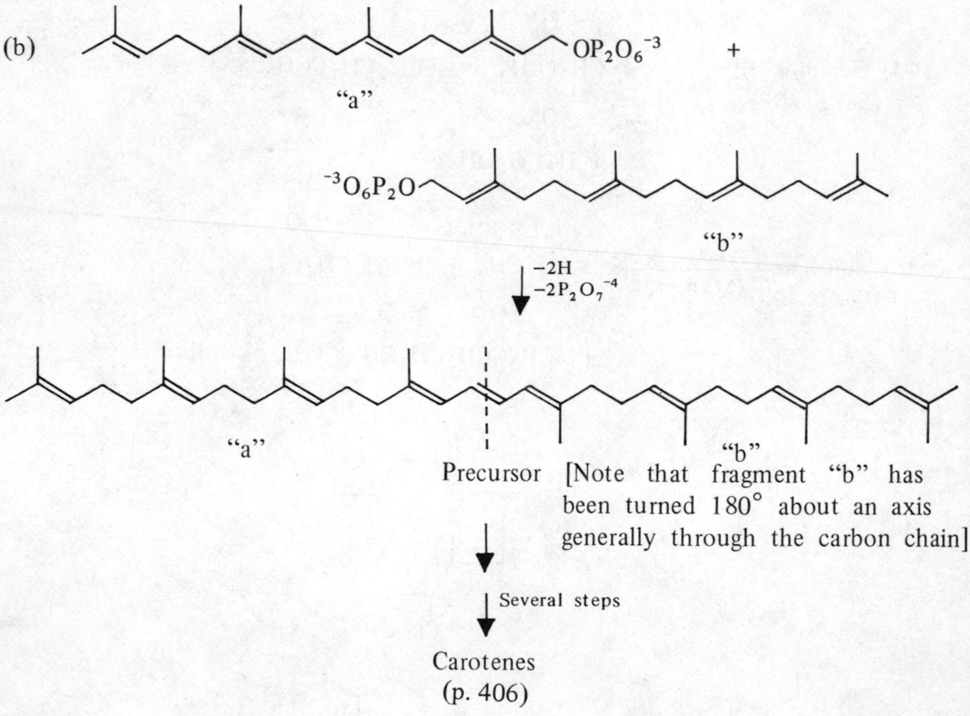

"a"

$OP_2O_6^{-3}$ +

$^{-3}O_6P_2O$

"b"

$-2H$
$-2P_2O_7^{-4}$

"a"

"b"

Precursor [Note that fragment "b" has
been turned 180° about an axis
generally through the carbon chain]

Several steps

Carotenes
(p. 406)

11.24

Farnesol

12 AROMATIC COMPOUNDS I: THE PHENOMENON OF AROMATICITY

12.1

(a) d (b) None

12.2

C=C = 1.33 Å C−C = 1.47 Å (See page 354 of text for sp^2−sp^2 carbon-carbon single bond length.)

12.3

(a)

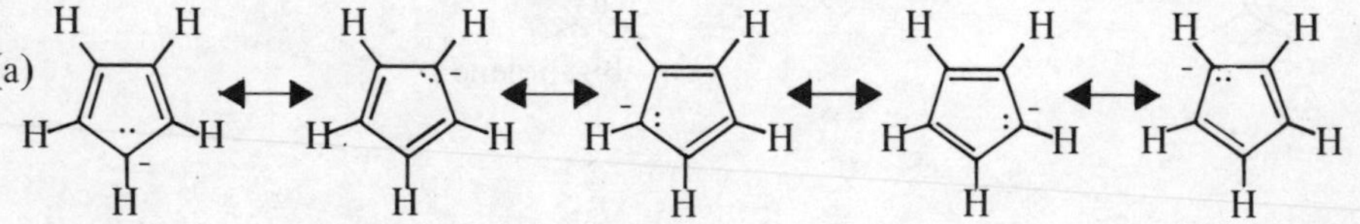

(b) Yes, all of the five resonance structures are equivalent, and all five hydrogen atoms are equivalent.

12.4

(a)

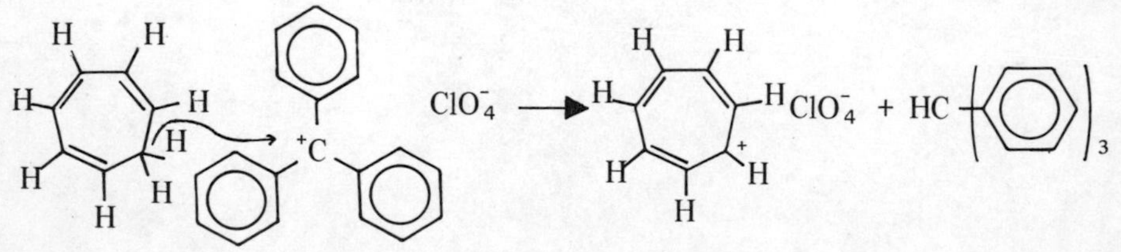

(b) Triphenylmethane (See (a) above).

(c) ClO_4^-

12.5

Tropylium bromide is ionic and has the structure, 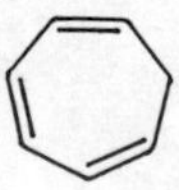 + Br^-. The ring is aromatic.

12.6

(a) $+CH\begin{smallmatrix}CH=CH_2\\\\CH=CH-CH=CH_2\end{smallmatrix}$ (b) Cycloheptatrienyl cation is aromatic.

12.7

(a) $+CH\begin{smallmatrix}CH=CH_2\\\\CH=CH_2\end{smallmatrix}$ $\xrightarrow[\text{energy increases}]{\pi \text{ electron}}$ $+\langle\ \rangle$ $+\ H_2$

The π electron energy of cyclopentadienyl cation is higher than that of the open chain counterpart.

(b) $+CH_2\begin{smallmatrix}\\CH=CH_2\end{smallmatrix}$ $\xrightarrow[\text{energy decreases}]{\pi \text{ electron}}$ $\overset{+}{\triangle}$ $+\ H_2$

The π electron energy of cyclopropenyl cation is lower than that of the open chain counterpart.

(c) $4n+2 = 2$ when $n = 0$. Hückel's rule predicts that cyclopropenyl cation is aromatic.

(d) The π electron energy of cyclopropenyl anion is higher than that of the open chain counterpart.

12.8

(a)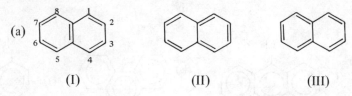

(I)   (II)   (III)

(b) Two of the structures (I and III) have a double bond between the C_1-C_2 carbons, whereas only structure II has a double bond between the C_2-C_3 carbons. Assuming that the three structures contribute nearly equally, the C_1-C_2 bond should be more like a double bond and therefore should be shorter than the C_2-C_3 bond.

12.9

$$C_6H_5\begin{smallmatrix}\ \\C=C\\\\C+\\|\\:O:-\end{smallmatrix}C_6H_5$$

III

III is a more important contributor to the resonance hybrid of I than a corresponding ionic structure of II is to the hybrid of II. III is an important contributor to the hybrid of diphenylcyclopropenone because it resembles the aromatic cyclopropenyl cation (Cf problem 12.7); i.e., the ring in structure III has 2 π electrons and is a $4n + 2$ system where $n = 0$.

12.10

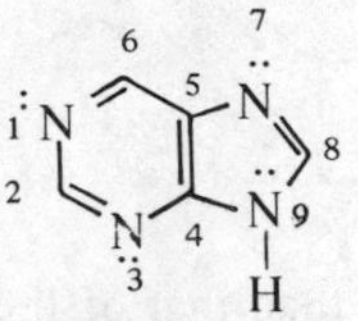

1,3,7 are pyridine type nitrogens; 9 is a pyrrole type nitrogen

12.11

(a) 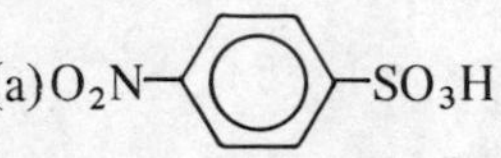 $O_2N-\bigcirc-SO_3H$ (b) 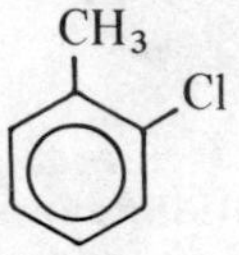 (c) 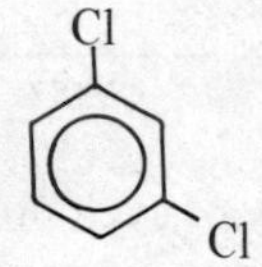 (d)

(e)  (f) 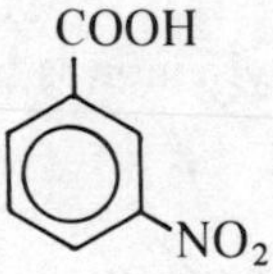 (g)  (h)

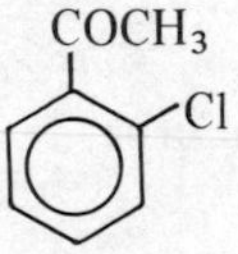

(i) 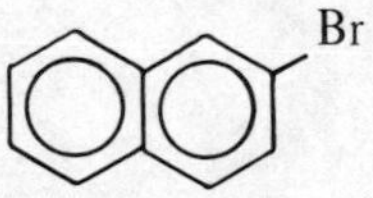 (j) 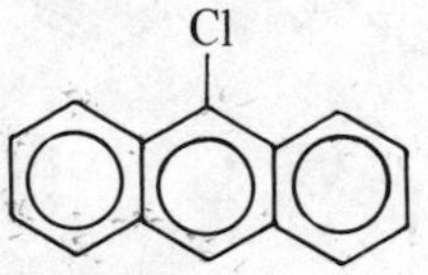 (k) 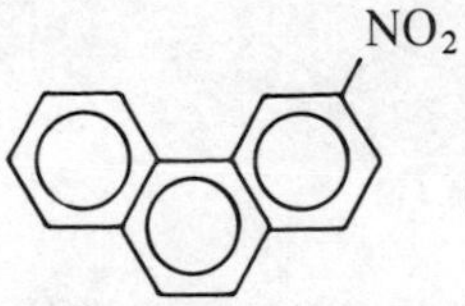 (l)

(m) 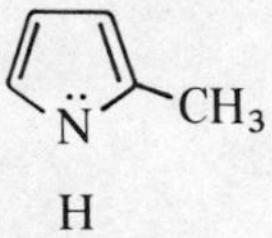 (n) 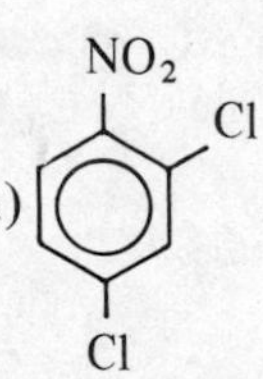 (o) 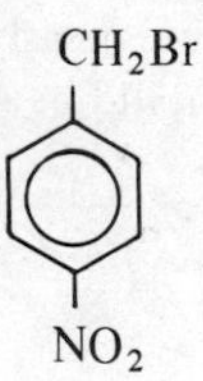 (p)

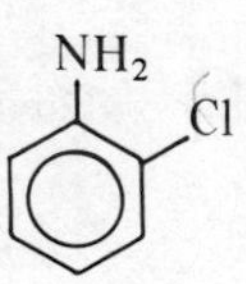

(q) 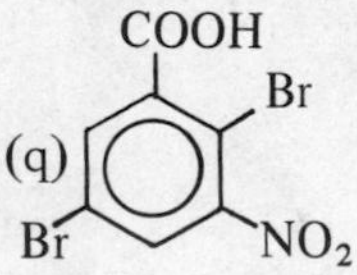 (r) 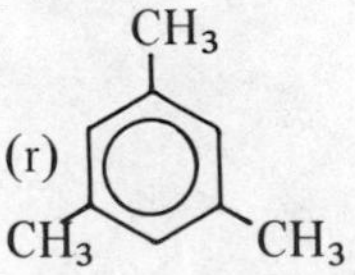 (s)  (t)

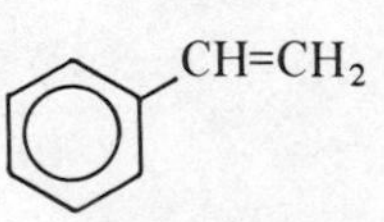

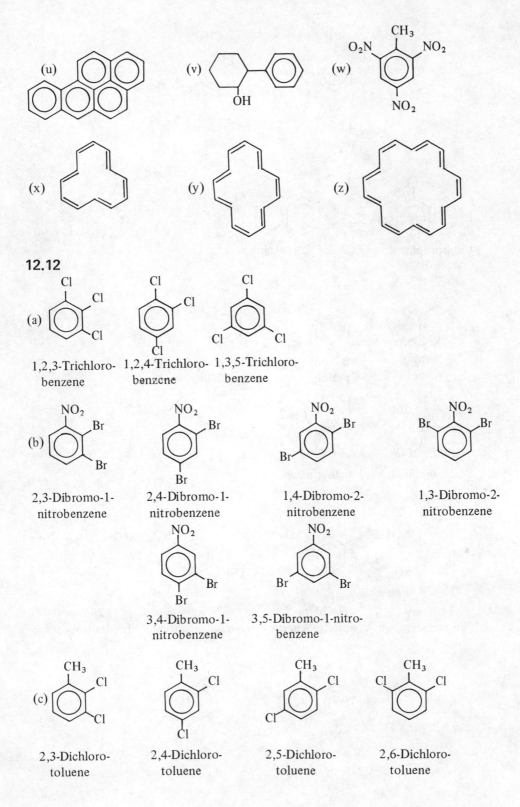

(u)

(v)

(w)

(x)

(y)

(z)

12.12

(a)

1,2,3-Trichloro-
benzene

1,2,4-Trichloro-
benzene

1,3,5-Trichloro-
benzene

(b)

2,3-Dibromo-1-
nitrobenzene

2,4-Dibromo-1-
nitrobenzene

1,4-Dibromo-2-
nitrobenzene

1,3-Dibromo-2-
nitrobenzene

3,4-Dibromo-1-
nitrobenzene

3,5-Dibromo-1-nitro-
benzene

(c)

2,3-Dichloro-
toluene

2,4-Dichloro-
toluene

2,5-Dichloro-
toluene

2,6-Dichloro-
toluene

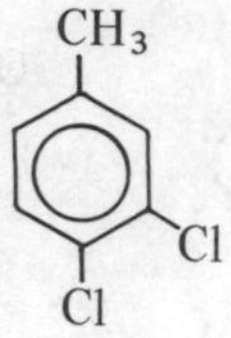

3,4-Dichloro-
toluene

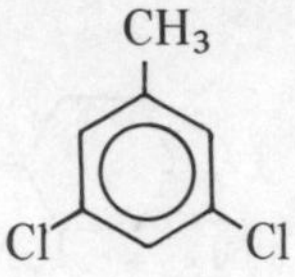

3,5-Dichloro-
toluene

(d)

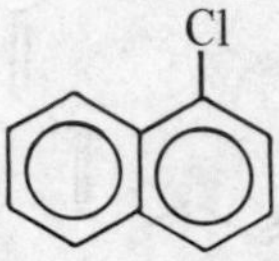

1-Chloronaphthalene

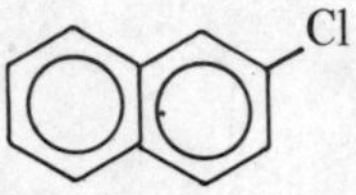

2-Chloronaphthalene

(e)

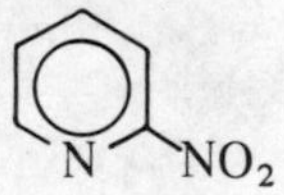

2-Nitro-
pyridine

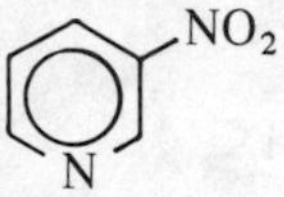

3-Nitro-
pyridine

4-Nitro-
pyridine

(f)

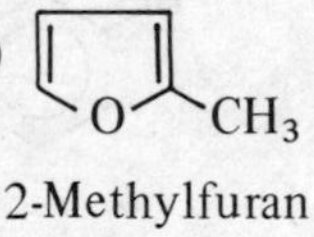

2-Methylfuran

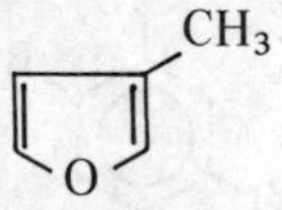

3-Methylfuran

(g)

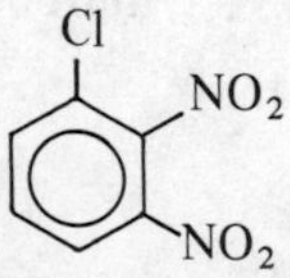

1-Chloro-
2,3-dinitrobenzene

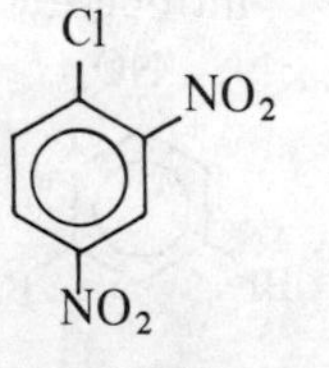

1-Chloro-
2,4-dinitrobenzene

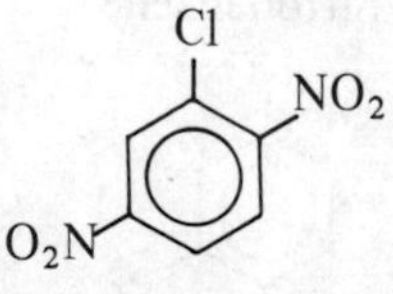

2-Chloro-
1,4-dinitrobenzene

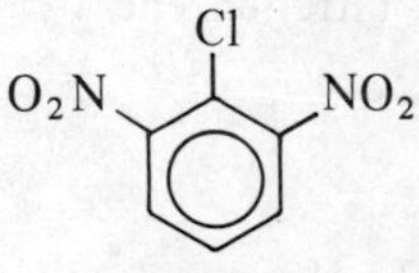

2-Chloro-
1,3-dinitrobenzene

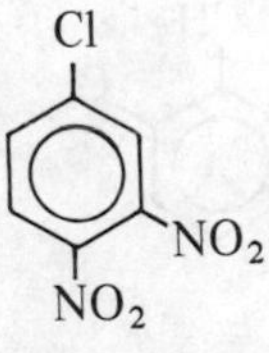

4-Chloro-
1,2-dinitrobenzene

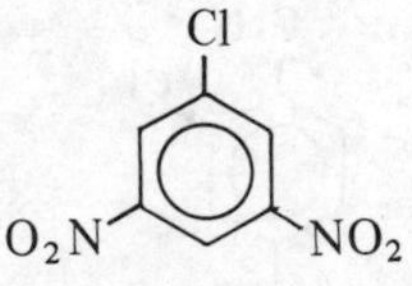

1-Chloro-
3,5-dinitrobenzene

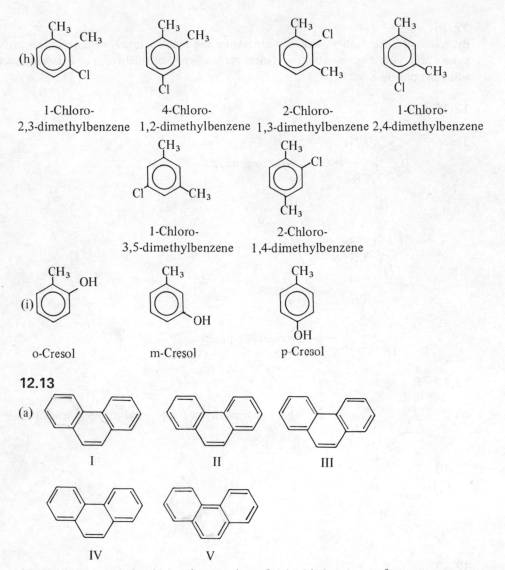

(h)

1-Chloro-
2,3-dimethylbenzene

4-Chloro-
1,2-dimethylbenzene

2-Chloro-
1,3-dimethylbenzene

1-Chloro-
2,4-dimethylbenzene

1-Chloro-
3,5-dimethylbenzene

2-Chloro-
1,4-dimethylbenzene

(i)

o-Cresol

m-Cresol

p-Cresol

12.13

(a)

I

II

III

IV

V

(b) The 9.10 bond should be close to that of a double bond, 1.33Å, since in four of the five contributors it is a double bond.

(c) Almost that of an actual double bond.

(d) Bromine adds to the 9,10 double bond because of its large double bond character and because addition disrupts only one of three aromatic rings.

12.14

(a) (mp + 87°) (b) (mp + 6°) (c) (mp − 7°)

12.15
By mononitrating each xylene and separating the mononitration products. The *ortho*-xylene will yield two mononitroxylenes, *meta*-xylene will yield three, and *para*-xylene will yield only one.

12.16

Three.

CH_3
CH_3
CH_3 → 2 mononitro products

CH_3
CH_3
CH_3 → 3 mononitro products

CH_3
CH_3 — CH_3 → 1 mononitro product

12.17

(a)
CH_3
$+$
CH_3 CH_3
BF_4^-

(b) The trimethylcyclopropenyl cation is aromatic.

12.18

C_6H_5 C_6H_5
$+$
OH
Br^-

or

C_6H_5 C_6H_5
$\oplus$
OH
Br^-

12.19

(a)  + many other equivalent resonance structures

(It may be drawn as ⬡ .)

(b) It has $4n + 2 = 10$ π electrons ($n = 2$), and is therefore aromatic; i.e., it obeys Huckel's rule.

12.20

(a) Cyclononatetraenyl anion: Li^+

(b) 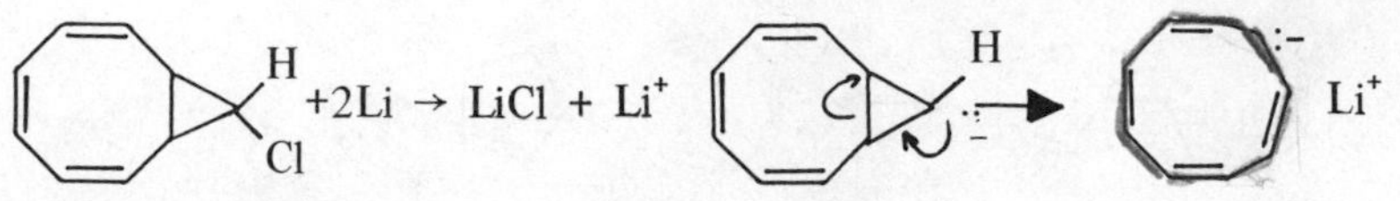   +2Li → LiCl + Li⁺

The product anion has 10 π-electrons and is therefore aromatic.

12.21

(a) 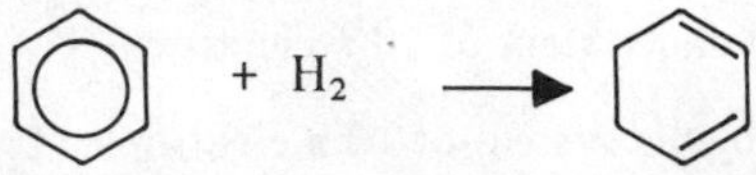 + H₂ ⟶ $\Delta H = (-49.8) - (-55.4) = +5.6$ kcal/mole

(b) 2 ⟶ + $\Delta H = -5.6 + (-55.4 - (-28.6))$
$= -5.6 - 26.8 = -32.4$ kcal/mole

(c) Reaction (a) is endothermic, and the competing reaction (reaction (b)) is exothermic. Therefore we must conclude that reaction (a) is unlikely under conditions (i.e., the presence of a catalyst) that will allow equilibrium to be established.

12.22

(a)

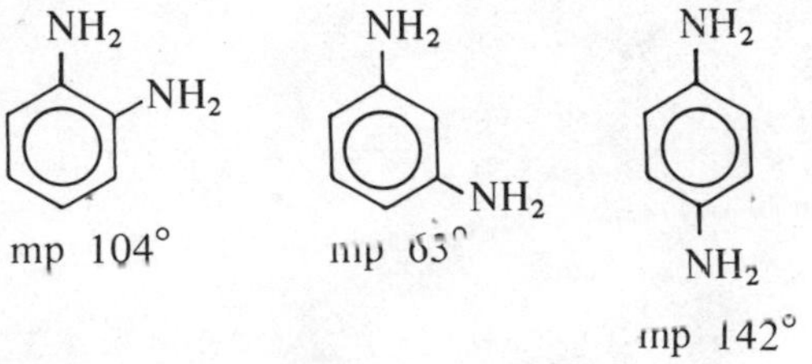

mp 104°   mp 63°   mp 142°

(b) mp 104°:

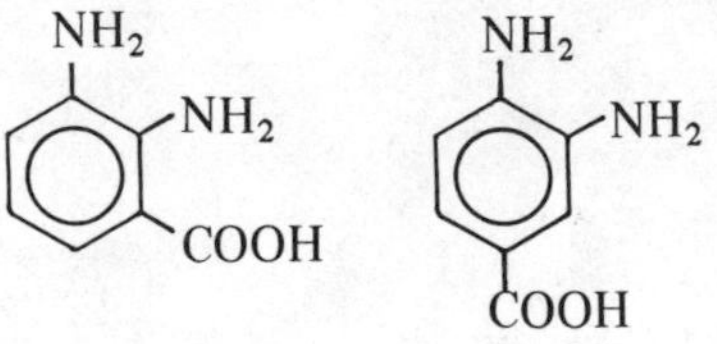

mp 142°:

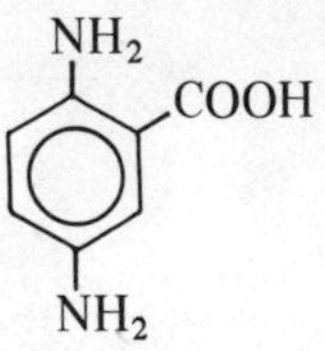

mp 63°:

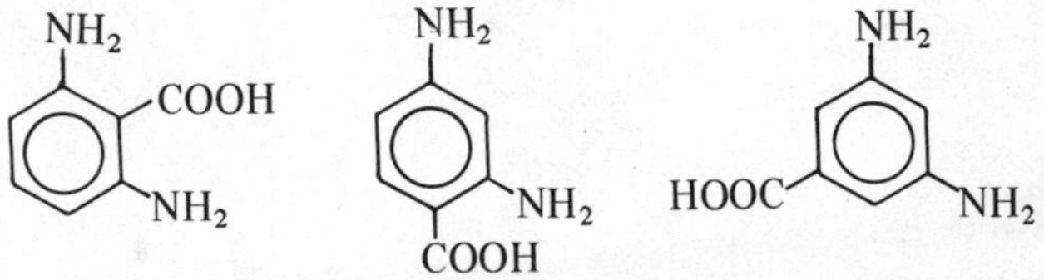

12.23

(a) Would not be aromatic. It is a monocyclic system of 12 π electrons and thus does not obey Huckel's rule.

(b) Would not be aromatic; it is not a conjugated system.

(c) Would not be aromatic; it is an 8 π electron monocyclic system and thus does not obey Huckel's rule.

(d) Would not be aromatic; it is a 16 π electron monocyclic system and thus does not obey Huckels rule.

(e) Would be aromatic because of resonance structures (below) that consist of a cyclo-heptatrienyl cation and cyclopentadienyl anion.

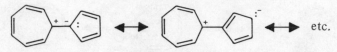

(f) Would be aromatic; it is a planar monocyclic system of 14 π electrons. (We count only two electrons of the triple bond because only two are in p-orbitals that overlap with those of the double bonds on either side.)

(g) Would be aromatic; it is a planar monocyclic system of 10 π electrons.

(h) Would be aromatic; it is a planar monocyclic system of 10 π electrons. (The bridging $-CH_2-$ groups allows the ring system to be planar.)

13 AROMATIC COMPOUNDS II: REACTIONS OF AROMATIC COMPOUNDS WITH ELECTROPHILES

13.1

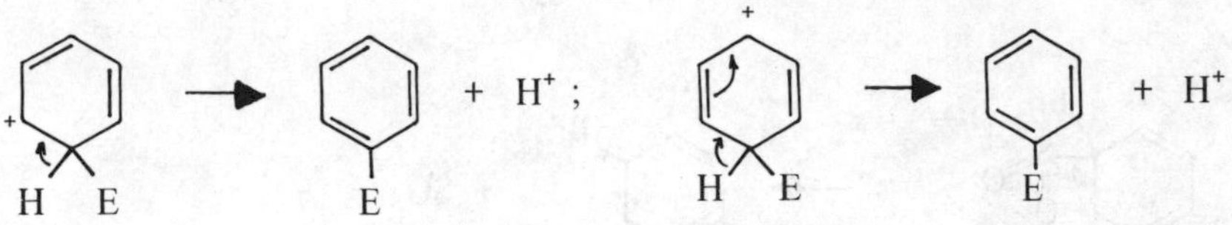

13.2

(a) 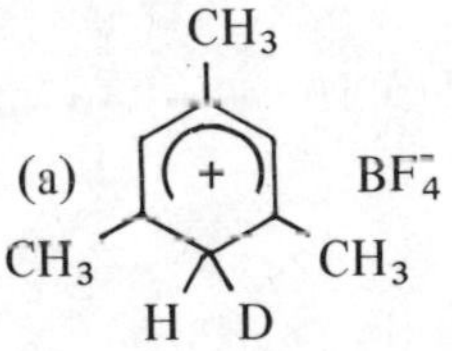

(b) 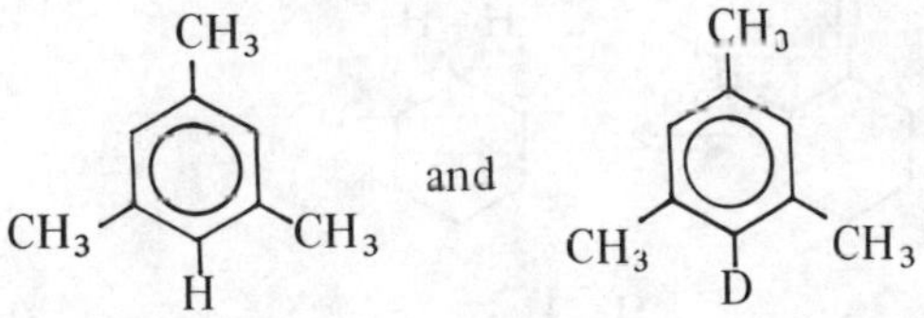

13.3

Oxidation of the Fe by X_2 generates the ferric salt:

$$2Fe + 3X_2 \longrightarrow 2FeX_3$$

13.4

$$H\text{–}\ddot{O}\text{–}NO_2 + H\text{–}\ddot{O}\text{–}NO_2 \longrightarrow H\text{–}\overset{+}{\ddot{O}}(H)\text{–}NO_2 + NO_3^-$$

$$H\text{–}\overset{+}{\ddot{O}}(H)\text{–}NO_2 + HONO_2 \longrightarrow NO_2^+ + H_3O^+ + NO_3^-$$

13.5

(a) 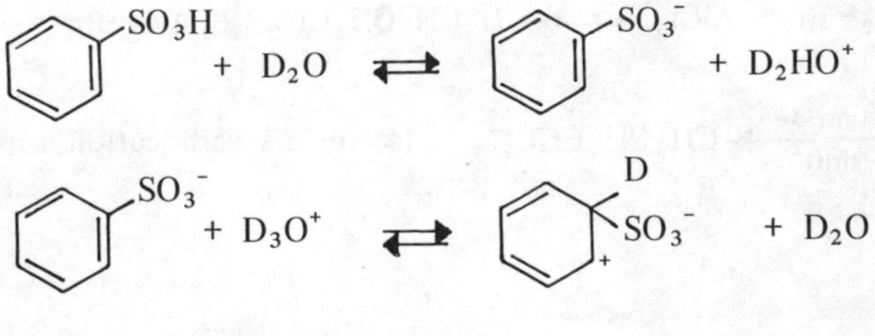

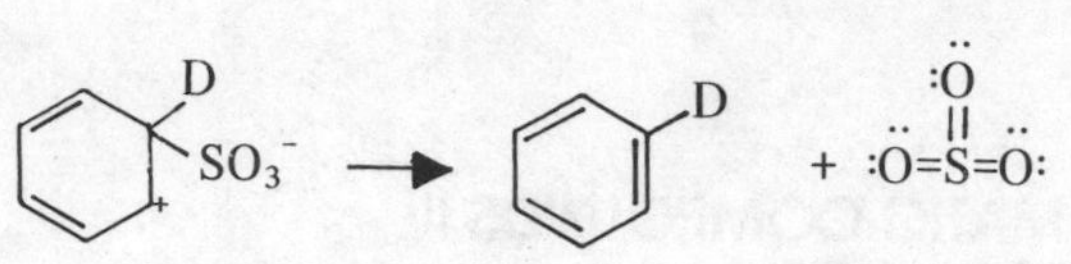

(b) $:\overset{..}{Br}-\overset{..}{Br}-\overset{-}{Fe}Br_3$ is the eletrophile.
$\quad\quad \overset{\delta+}{} \overset{\delta+}{}$

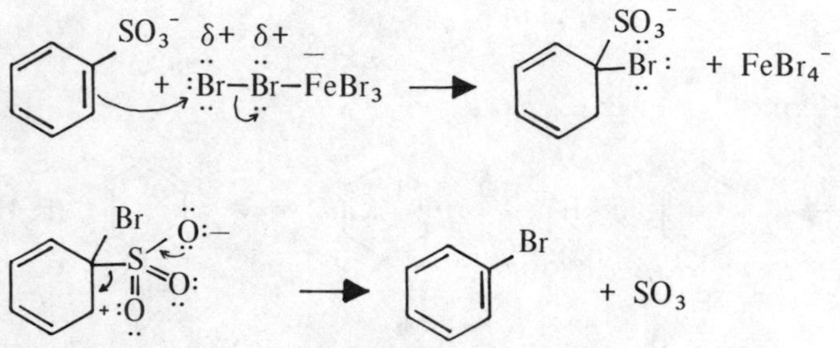

13.6

(a)

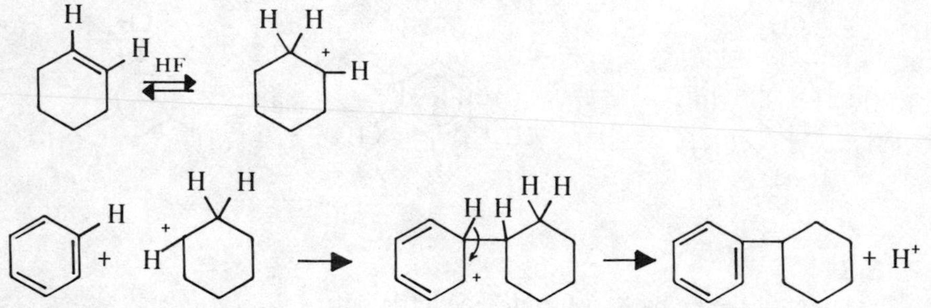

(b)

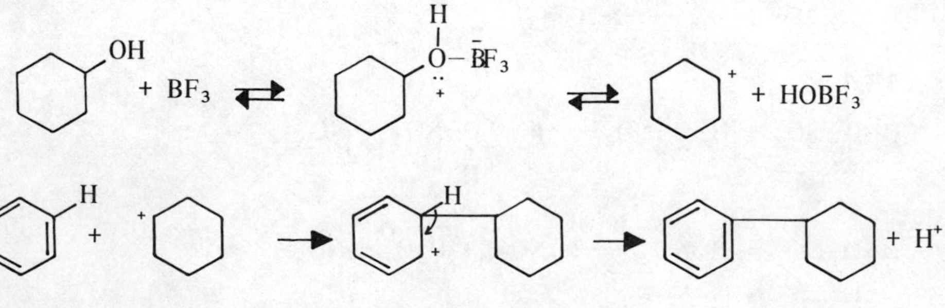

13.7

(a) $CH_3CH_2CH_2CH_2-Br + AlCl_3 \rightleftharpoons CH_3CH_2CH_2CH_2^+ + \overset{-}{A}lCl_3Br$

$CH_3CH_2CH_2CH_2^+ \xrightarrow[\text{shift}]{\text{Hydride}} CH_3CH_2\overset{+}{C}HCH_3$ (secondary carbocation is more stable than primary)

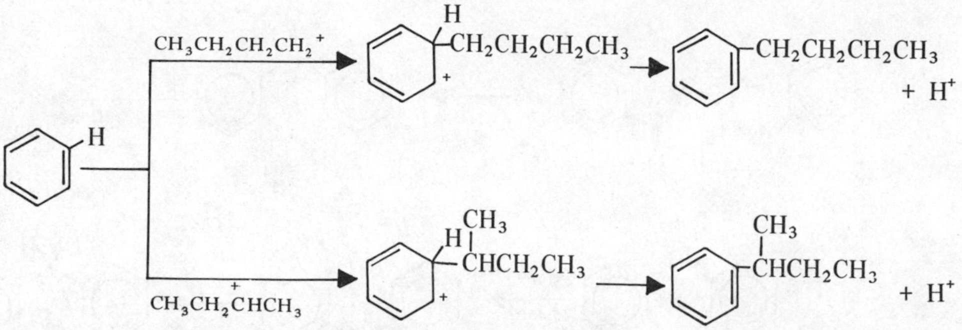

$CH_3CH_2CH_2CH_2^+$ → (ring with H, $-CH_2CH_2CH_2CH_3$, +) → (ring with $-CH_2CH_2CH_2CH_3$) $+ H^+$

(ring with $-H$)

$CH_3CH_2\overset{+}{C}HCH_3$ → (ring with H, $\overset{CH_3}{-CHCH_2CH_3}$, +) → (ring with $\overset{CH_3}{-CHCH_2CH_3}$) $+ H^+$

(b) $CH_3CH_2CH_2-OH + BF_3 \rightleftharpoons CH_3CH_2CH_2^+ + HOBF_3^-$.

The *n*-propyl cation can rearrange to an isopropyl cation:

$$CH_3CH_2\overset{+}{C}H_2 \xrightarrow[\text{Shift}]{\text{Hydride}} CH_3\overset{+}{C}HCH_3$$

Both cations can then attack the benzene ring.

13.8

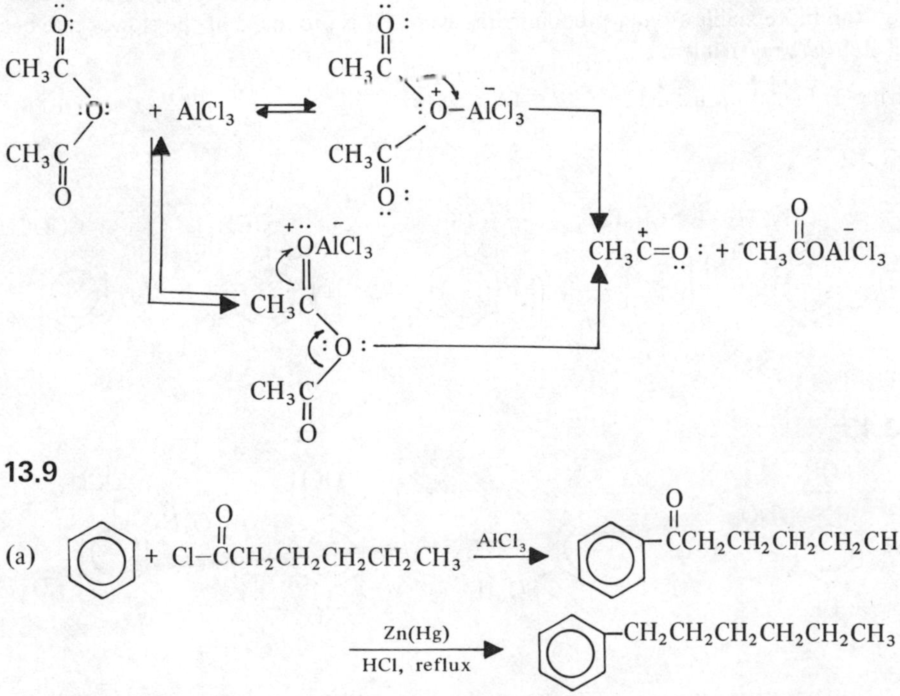

$CH_3C(=O)-O-CH_3C(=O)-O + AlCl_3 \rightleftharpoons CH_3C(=O)\cdots \overset{+}{O}-AlCl_3^- / CH_3C(=O)-O \longrightarrow$

$\overset{+}{O}AlCl_3^- / CH_3C / CH_3C(=O) / :O:$

$CH_3\overset{+}{C}=O: + CH_3\overset{O}{C}OAlCl_3^-$

13.9

(a) 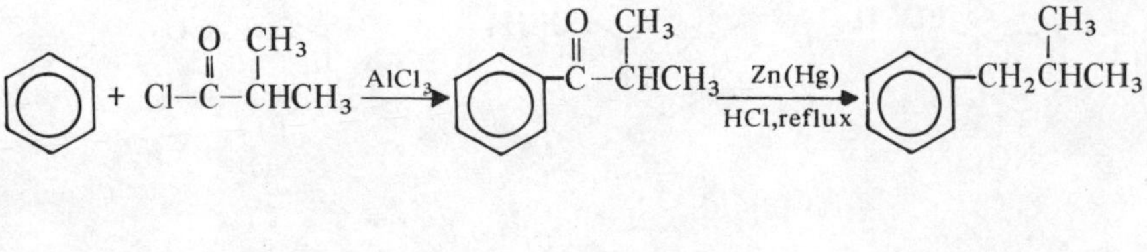 (benzene ring) $+ Cl-\overset{O}{C}CH_2CH_2CH_2CH_2CH_3 \xrightarrow{AlCl_3}$ (ring)$-\overset{O}{C}CH_2CH_2CH_2CH_2CH_3$

$\xrightarrow[\text{HCl, reflux}]{Zn(Hg)}$ (ring)$-CH_2CH_2CH_2CH_2CH_2CH_3$

(b) (benzene ring) $+ Cl-\overset{O}{C}-\overset{CH_3}{C}HCH_3 \xrightarrow{AlCl_3}$ (ring)$-\overset{O}{C}-\overset{CH_3}{C}HCH_3 \xrightarrow[\text{HCl,reflux}]{Zn(Hg)}$ (ring)$-CH_2\overset{CH_3}{C}HCH_3$

(c) [benzene] + [phthalic anhydride] $\xrightarrow{AlCl_3}$ [2-benzoylbenzoic acid] $\xrightarrow[\text{HCl, reflux}]{Zn(Hg)}$

[diphenylmethane-carboxylic acid] $\xrightarrow[\text{heat}]{SOCl_2}$ [acid chloride] $\xrightarrow[\text{CS}_2]{AlCl_3}$ [anthrone]

13.10
40% ortho, 40% meta, 20% para.

13.11
(a) At the lower temperature the proportions are determined by the relative reaction rates. At the higher temperature the proportions are determined by the relative stabilities; i.e., the more stable isomer predominates even if it is produced at the slower rate because all steps are reversible.

(b) p-Toluenesulfonic acid.

13.12

(a) [o-chloroethylbenzene with CH_2CH_3, Cl] + [p-chloroethylbenzene with CH_2CH_3, Cl]

(b) [benzene with CF_3 and Cl meta]

(c) [benzene with $\overset{+}{N}(CH_3)_3$ Cl$^-$ and Cl meta]

(d) [benzene with $COOCH_3$ and Cl meta]

13.13

(a) [phenol with OH, NO_2, CF_3]

(b) [benzene with CN, O_2N, SO_3H]

(c) [benzene with OCH_3, NO_2, NO_2] + [benzene with OCH_3, O_2N, NO_2]

(d) [benzene with $NHCOCH_3$, $COCH_3$, NO_2] + [benzene with $NHCOCH_3$, O_2N, $COCH_3$]

(e) [benzene with CH_3, NO_2, NO_2]

(f)

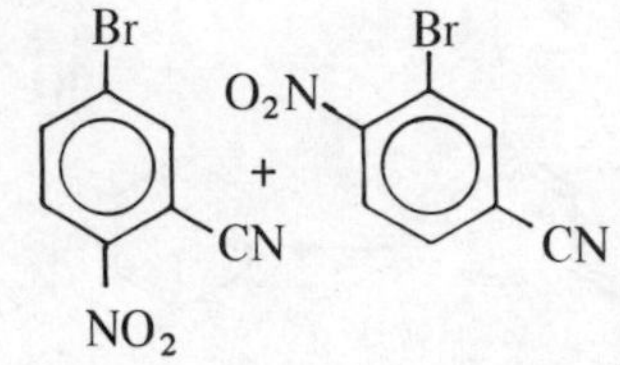

(g)

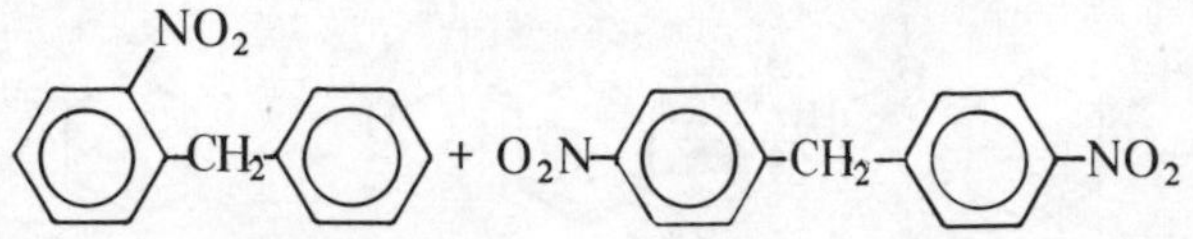

13.14

(a) $Cl-CH_2-CH_2-\overset{+}{N}(CH_3)_3\overset{-}{Cl}$. This is anti-Markovnikov addition because the intermediate carbocation formed by Markovnikov addition would have positive charges on adjacent carbons, i.e.,

$$CH_3-\overset{+}{C}H-\overset{+}{N}(CH_3)_3\overset{-}{Cl}$$

The anti-Markovnikov carbocation is $\overset{+}{C}H_2-CH_2-\overset{+}{N}(CH_3)\overset{-}{Cl}$. Although this is a primary carbocation, it is more stable because of the greater separation of charges.

(b) Slower because it requires formation of an intermediate primary cation with two positive charges.

(c) The positive charge of the trimethylammonium ion makes it a powerful electron-withdrawing group.

(d)

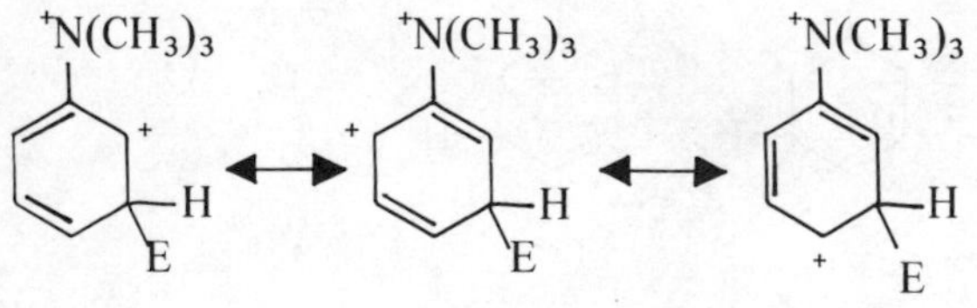

None of these structures is especially unstable, whereas in *ortho* or *para* attack at least one structure would be.

13.15

(a) The electronic influence of the trimethylammonium ion on the ring is an inductive effect that causes *meta* orientation and it is diminished by every methylene group that separates it from the ring.

(b) The increasing number of chlorines makes the methyl group increasingly electron-withdrawing, and hence increasingly *meta*-directing.

13.16

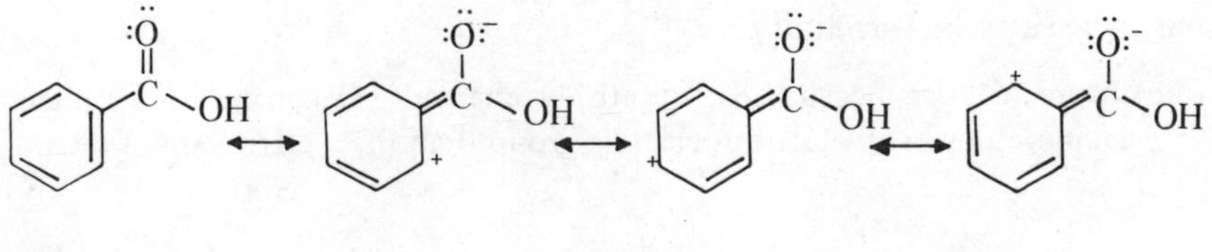

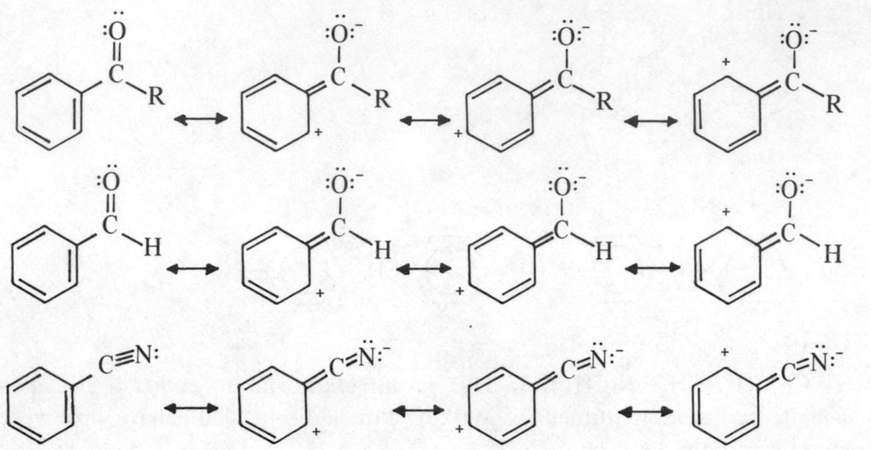

13.17

(a) ortho:

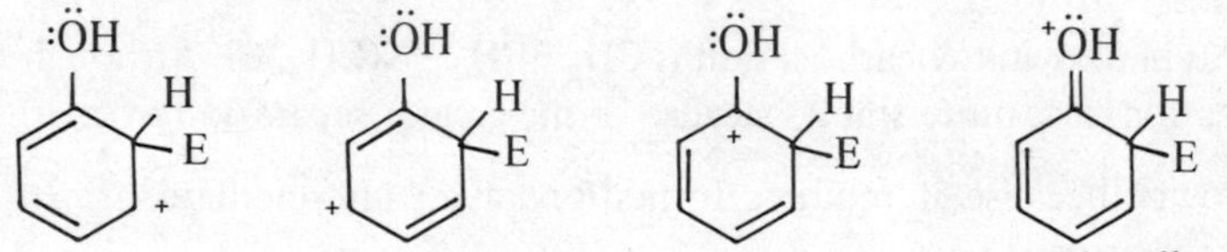

exceptionally
stable

meta:

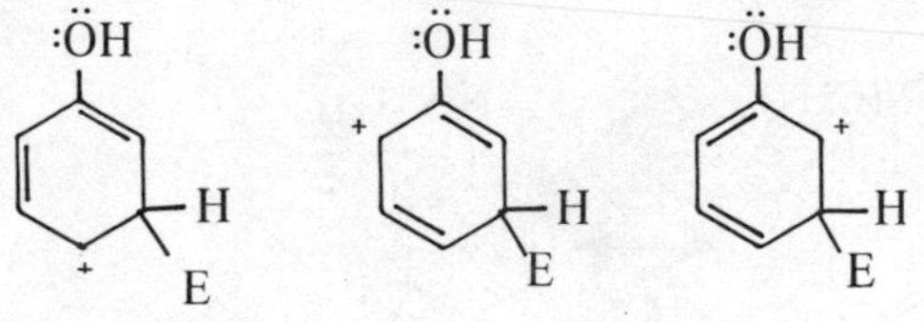

para:

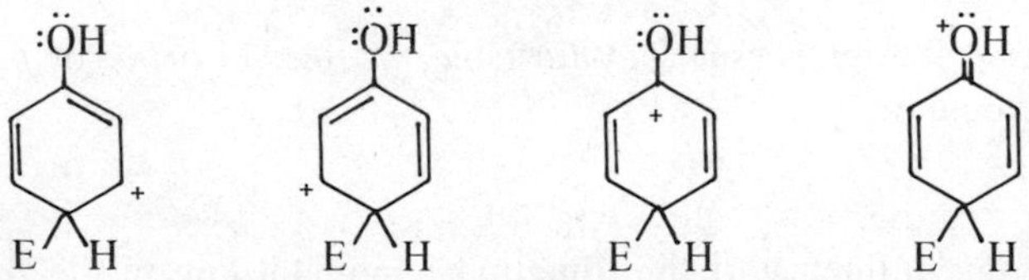

exceptionally
stable

(b) The electron-releasing ability of the —OH group through resonance increases the electron density of the ring, and it stabilizes the positive charge of the intermediate carbocation.

(c) An exceptionally stable structure (above) contributes to the intermediate carbocation only when attack is *ortho* or *para*.

(d,e) More reactive because the negatively charged —O⁻ group of the phenoxide ion is an even more powerful electron-releasing group than the —OH group of phenol.

13.18

(a)

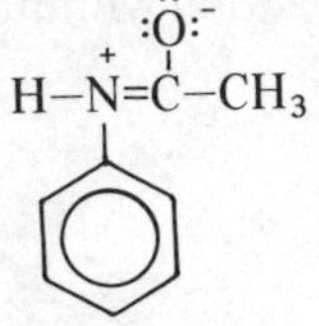

$$H-\overset{+}{N}=\overset{:\ddot{O}:^-}{\underset{|}{C}}-CH_3$$

(b) The $-\overset{O}{\overset{\|}{C}}-CH_3$ group competes with the ring for the electron pair on N, therefore stabilization of the intermediate carbocation is less effective than in aniline.

(c)

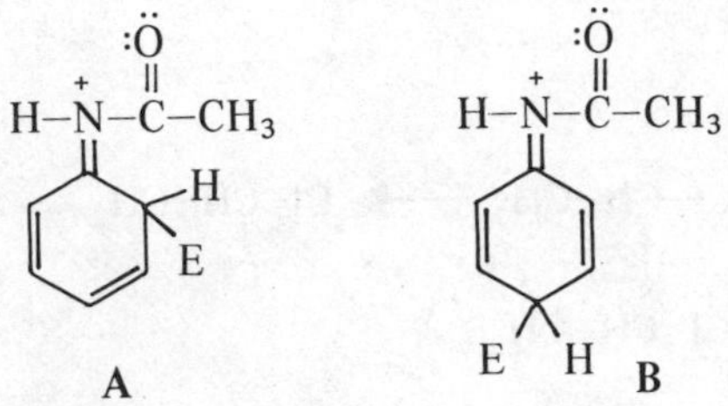

A   **B**

Yes, resonance accounts for an electron release from nitrogen to the ring, and exceptionally stable structures (**A** and **B** above) contribute to the sigma complexes formed when attack takes place at an *ortho* or *para* carbon.

(d) Phenyl acetate should be less reactive than phenol because the $-COCH_3$ group competes with the ring for electrons on oxygen as shown in the following structure,

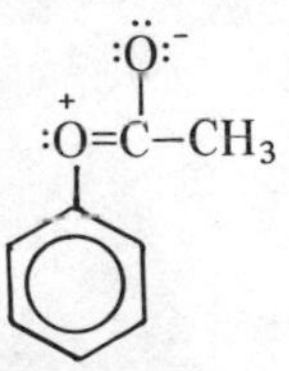

$$:\overset{+}{O}=\overset{:\ddot{O}:^-}{\underset{|}{C}}-CH_3$$

Notice that this structure also places a positive charge on the oxygen attached to the ring.

(e) Ortho-para

(f) More reactive

13.19

In each case the orientation results from the formation of the more stable intermediate carbocation.

With 3,3,3-tri fluoropropene the possibilities are:

$$CF_3-\overset{+}{C}H-CH_3$$
I
(less stable)

$$CF_3-CH=CH_2 + H^+$$

$$CF_3-CH_2-\overset{+}{C}H_2 \xrightarrow{Cl^-} CF_3-CH_2-CH_2Cl$$
II
(more stable)

Although II is a primary carbocation and I is a secondary carbocation, II is more stable because in it the positive charge is separated from the highly electron-withdrawing CF_3 group by an intervening carbon.

With chloroethene the possibilities are:

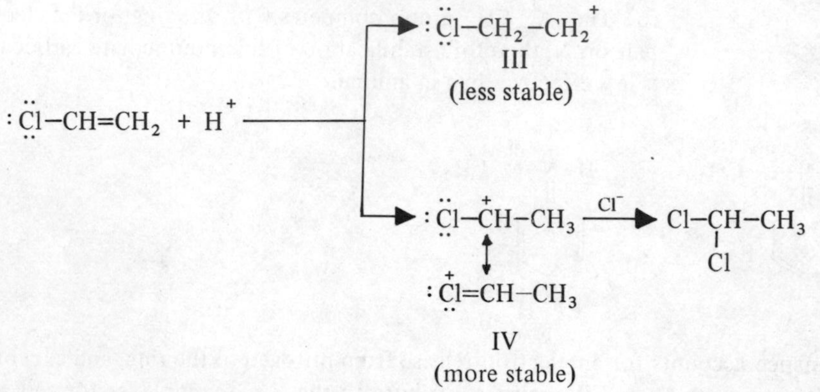

Here carbocation IV is more stable than III because of resonance involving an electron pair of the chlorine atom.

13.20

(a) ortho:

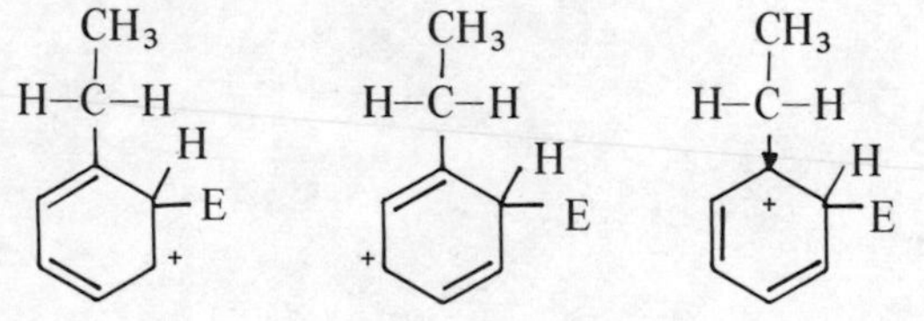

(exceptionally stable)

meta:

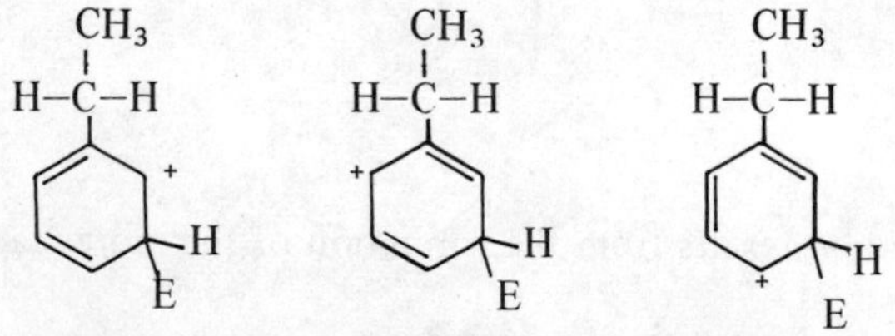

para:

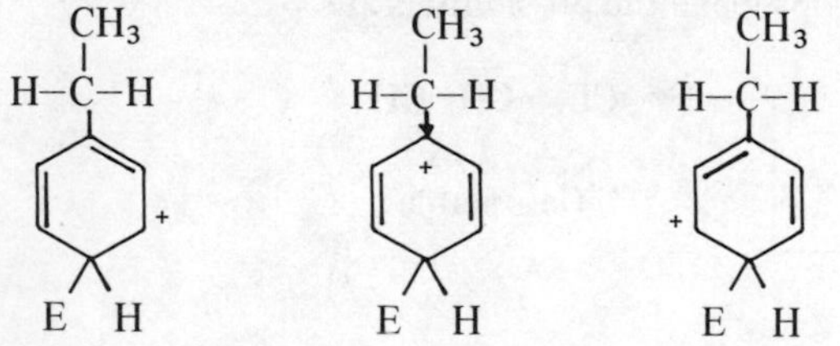

(exceptionally stable)

(b) Yes, substitution at an ortho or para position yields a carbocation that is stabilized by the contribution of an exceptionally stable structure. The exceptionally stable structures (above) have a positive charge on the ring carbon that bears the ethyl group.

(c) Because the ethyl group is electron-releasing it stabilizes the intermediate sigma complexes.

13.21

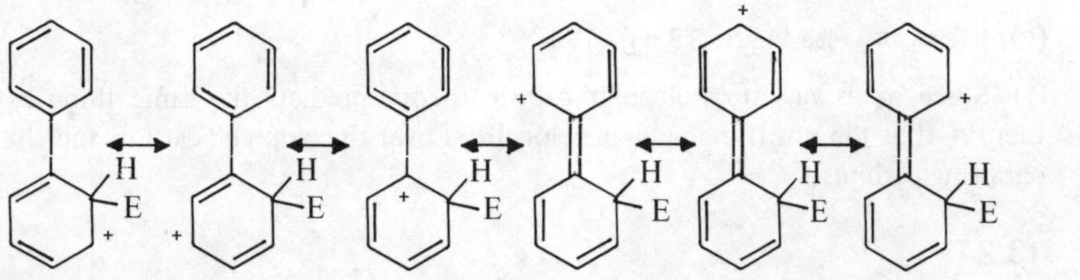

The carbocation above has the positive charge delocalized over both rings and thus it is exceptionally stable. Similar structures can be drawn for the sigma complex formed when substitution takes place at a para position. However, when electrophilic attack takes place at the meta position it produces a carbocation whose positive charge cannot be delocalized over both rings:

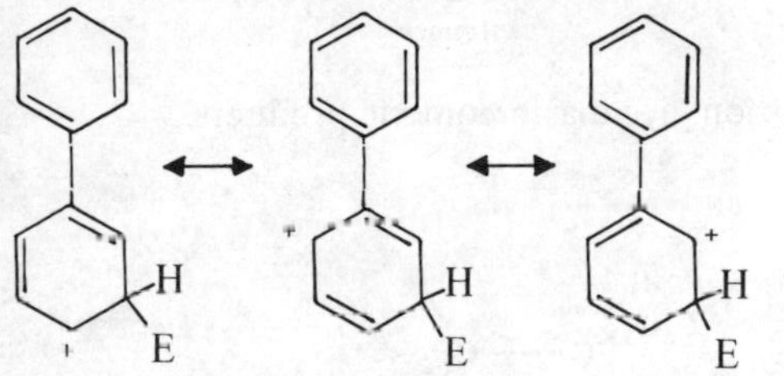

13.22

(a) 1-Chloro-1-phenylethane results from the more stable free radical. The more stable free radical is $\langle\bigcirc\rangle$—$\overset{\cdot}{C}HCH_3$ because the unshared electron pair is conjugated with the ring:

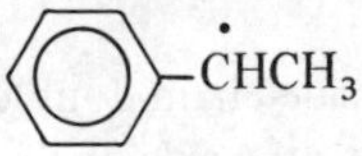

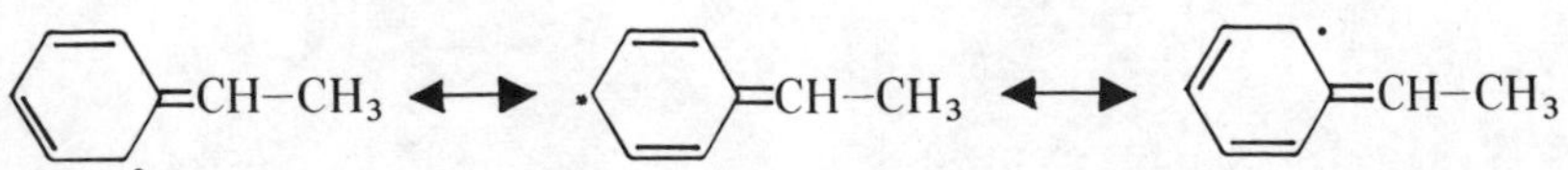

(b) A primary free radical is involved, and the unpaired electron is not conjugated with the ring: $\langle\bigcirc\rangle$—CH_2—$CH_2\cdot$

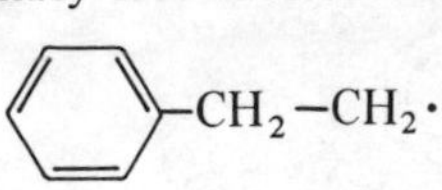

(c) 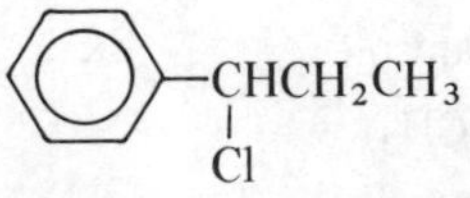 $\langle\bigcirc\rangle$—$\underset{\underset{Cl}{|}}{C}HCH_2CH_3$ (1-chloro-1-phenylpropane)

13.23

(a)

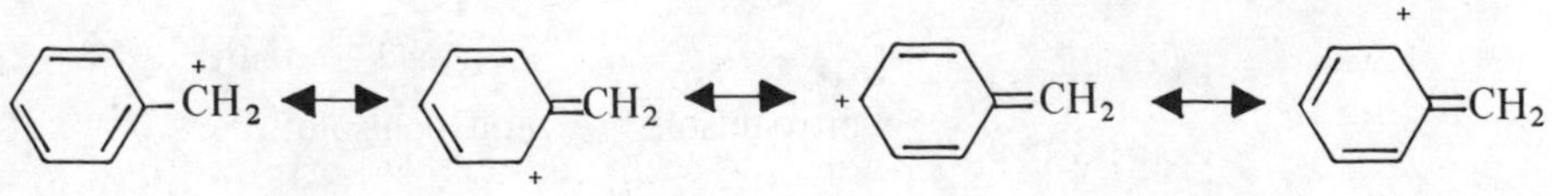

(b) The benzyl cation is stabilized by resonance.

(c) Yes, because the benzyl cation is a hybrid of the structures given in (a).

(d) Over the ortho and para ring carbons and the benzylic carbon.

(e) π_1, π_2, π_3 (see Figure 13.6).

(f) Since π_4 is vacant, molecular orbital theory predicts the same thing as resonance theory—that the positive charge is delocalized over the benzylic carbon and the ortho and para ring carbons.

13.24
(a) Anti addition gives:

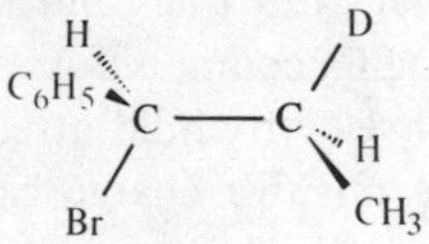

Syn-addition gives:

enantiomer

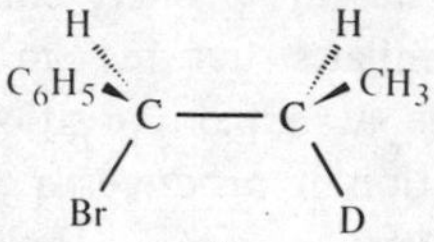

enantiomer

(b) Yes, anti-addition and syn-addition give diastereomeric products.

13.25

(a)

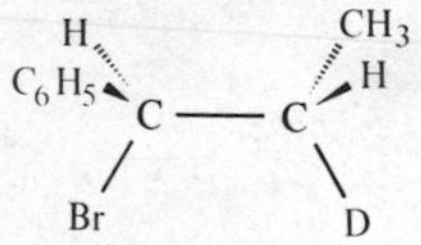

(b)

enantiomer

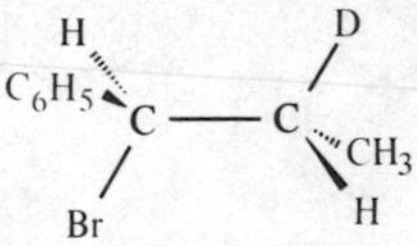

enantiomer

13.26
Chlorinate the ring first. If we were to introduce the side-chain double bond first, chlorination of the ring would result in addition of chlorine to the side-chain double bond.

13.27

(a)

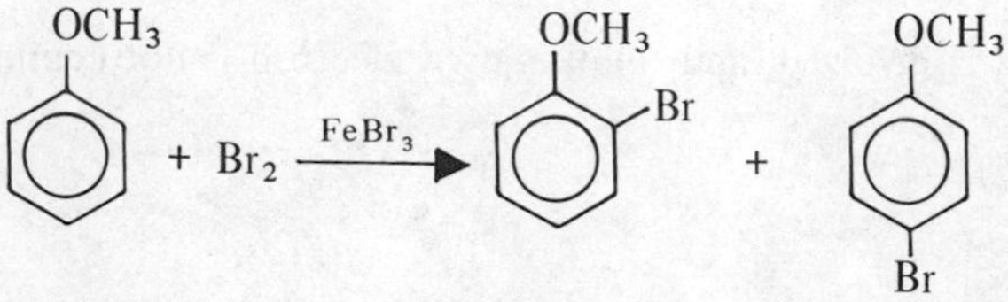

o-Bromoanisole p-Bromoanisole

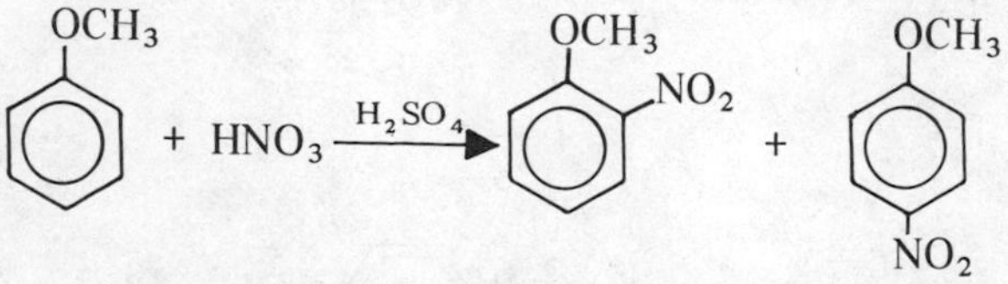

o-nitroanisole p-nitroanisole

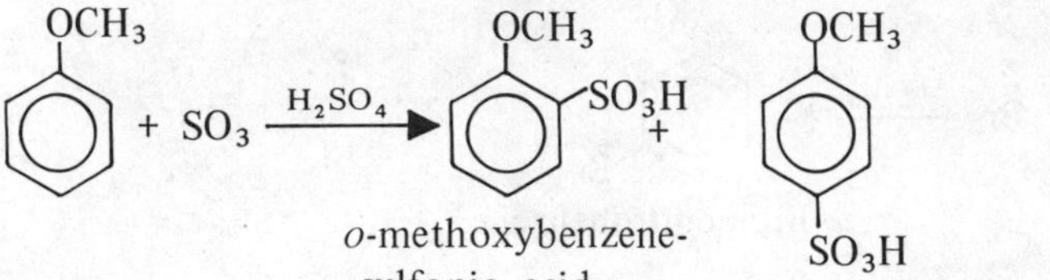

o-methoxybenzene-
sulfonic acid

p-methoxybenzene-
sulfonic acid

Reactions are faster than the corresponding reactions of benzene.

(b)

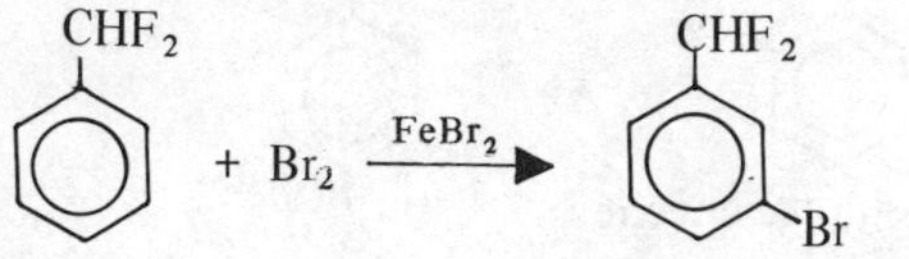

m-Bromobenzal
difluoride

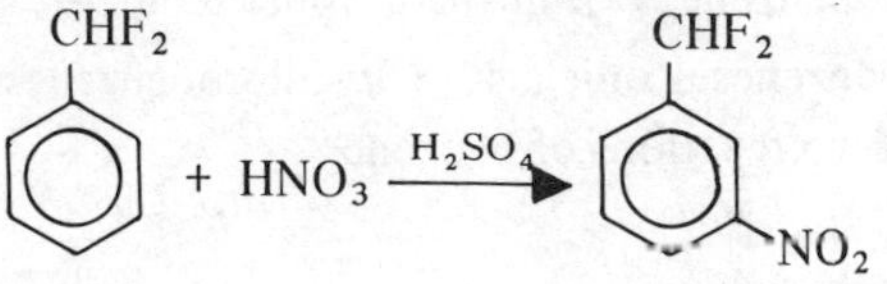

m-Nitrobenzal
difluoride

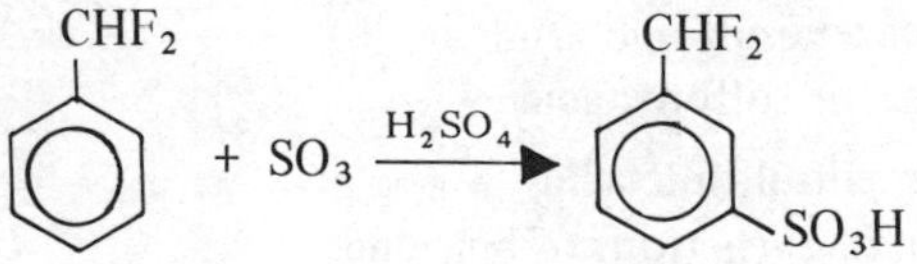

m-Difluoromethyl-
benzenesulfonic acid

Reactions are slower than corresponding reactions of benzene.

(c)

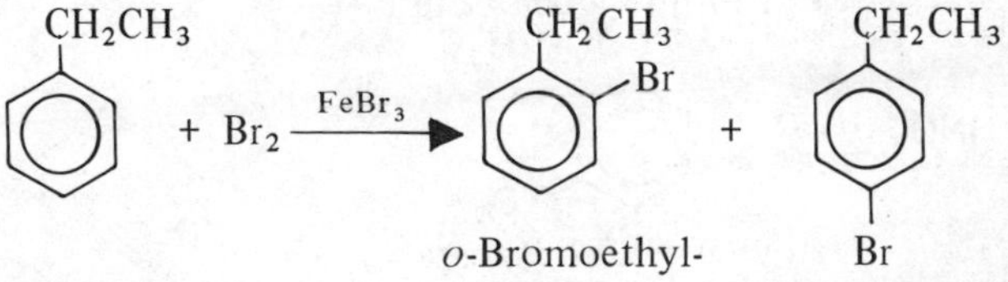

o-Bromoethyl-
benzene

p-Bromoethyl-
benzene

nitration ⟶ *o*-nitroethylbenzene and *p*-nitroethylbenzene

sulfonation ⟶ *o*-ethylbenzenesulfonic acid and *p*-ethylbenzenesulfonic acid.

Reactions are faster than corresponding reactions of benzene.

(d)

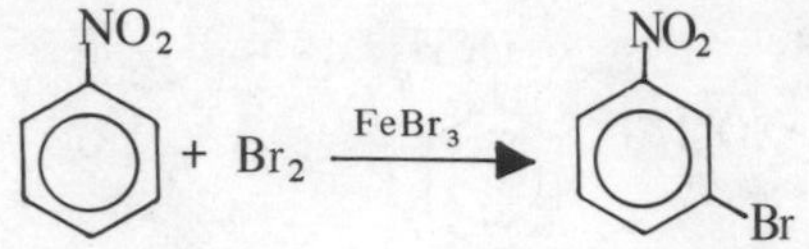

m-Bromonitrobenzene

nitration ⟶ *m*-dinitrobenzene

sulfonation ⟶ *m*-nitrobenzenesulfonic acid

Reactions are slower than corresponding reactions of benzene.

(e)

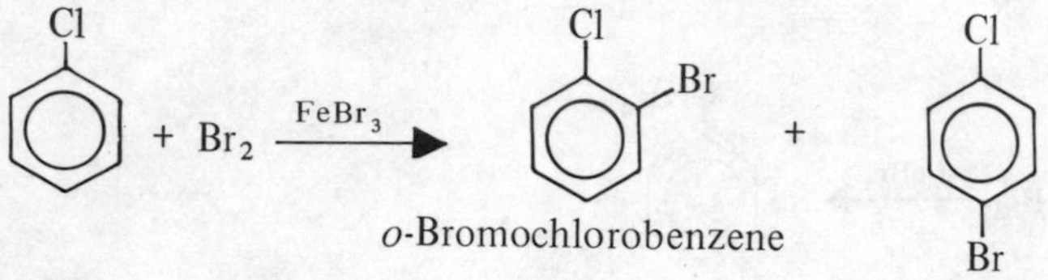

o-Bromochlorobenzene

p-bromochloro-
benzene

nitration ⟶ *o*-nitrochlorobenzene + *p*-nitrochlorobenzene

sulfonation ⟶ *o*-chlorobenzenesulfonic acid + *p*-chlorobenzenesulfonic acid

Reactions are slower than corresponding reactions of benzene.

(f)

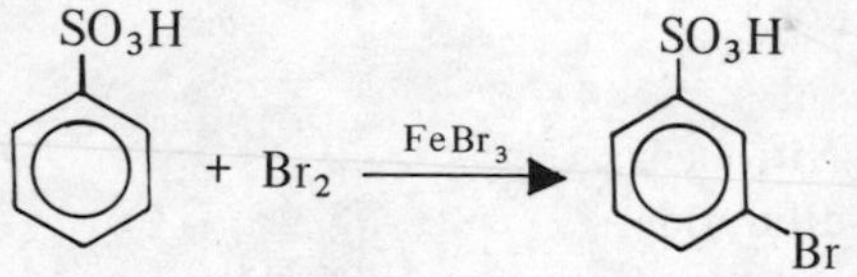

m-Bromobenzenesulfonic acid

nitration ⟶ *m*-nitrobenzene sulfonic acid

sulfonation ⟶ *m*-benzenedisulfonic acid

Reactions are slower than corresponding reactions of benzene.

(g)

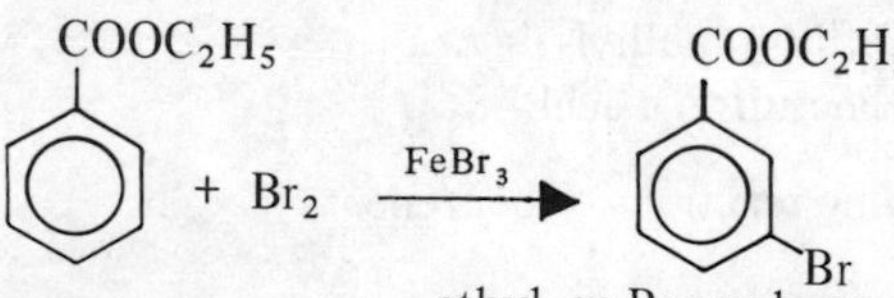

ethyl *m*-Bromobenzoate*

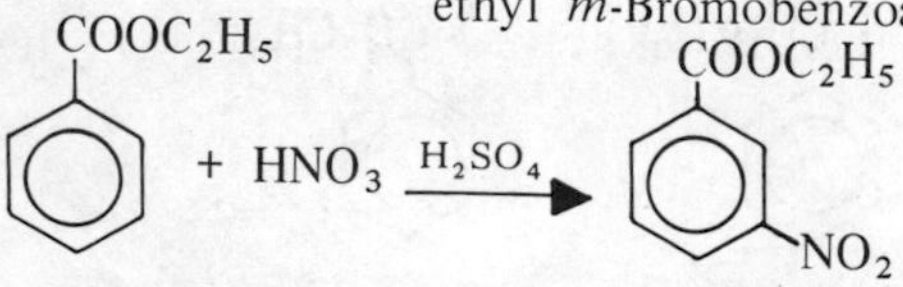

ethyl *m*-Nitrobenzoate*

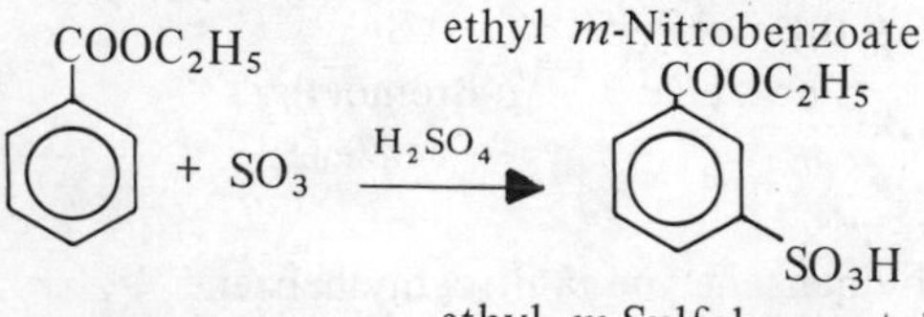

ethyl *m*-Sulfobenzoate*

Reactions are slower than corresponding reactions of benzene.

*Author's apology: Rules for naming the starred compounds in this problem have not been given in the text.

(h)

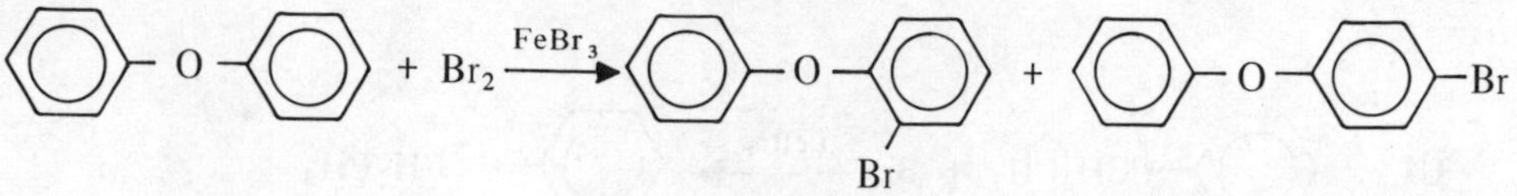

o-bromophenoxybenzene   *p*-bromophenoxybenzene

nitration ⟶ *o*-nitrophenoxybenzene + *p*-nitrophenoxybenzene

sulfonation ⟶ *o*-phenoxybenzenesulfonic acid + *p*-phenoxybenzenesulfonic
acid

Reactions are faster than corresponding reactions of benzene.

(i)

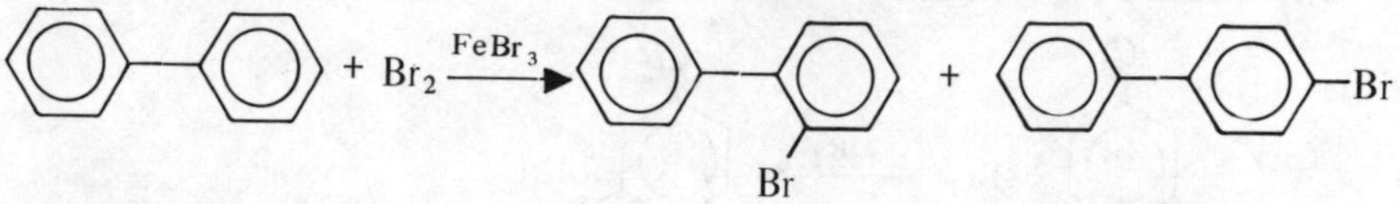

2-Bromobiphenyl*   4-Bromobiphenyl*

nitration ⟶ 2-nitrobiphenyl + 4-nitrobiphenyl

sulfonation ⟶ *o*-phenylbenzenesulfonic acid + *p*-phenylbenzenesulfonic acid

Reactions are faster than corresponding reactions of benzene.

(j)

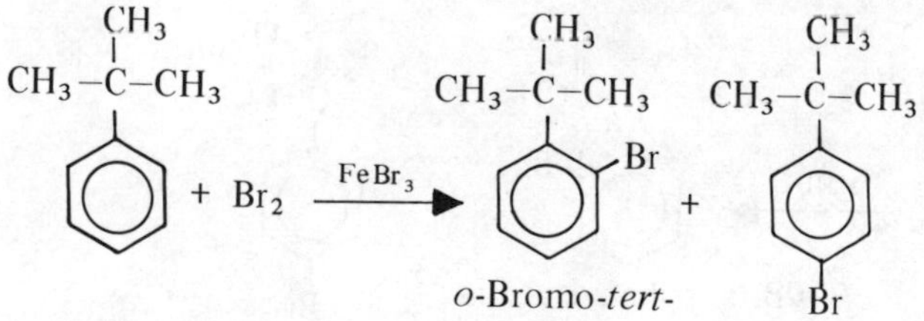

o-Bromo-*tert*-
butylbenzene

p-Bromo-*tert*-
butylbenzene

nitration ⟶ *o*-nitro-*tert*-butylbenzene + *p*-nitro-*tert*-butylbenzene

sulfonation ⟶ *o*-*tert*-butylbenzenesulfonic acid + *p*-*tert*-butylbenzenesulfonic
acid

Reactions are faster than corresponding reactions of benzene.

(k)

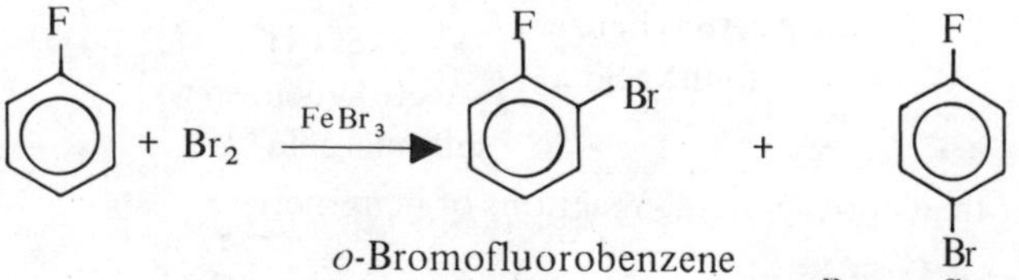

o-Bromofluorobenzene

p-Bromofluorobenzene

nitration ⟶ *o*-nitrofluorobenzene + *p*-nitrofluorobenzene

sulfonation ⟶ *o*-fluorobenzenesulfonic acid + *p*-fluorobenzenesulfonic acid

Reactions are slower than corresponding reactions of benzene.

(l)

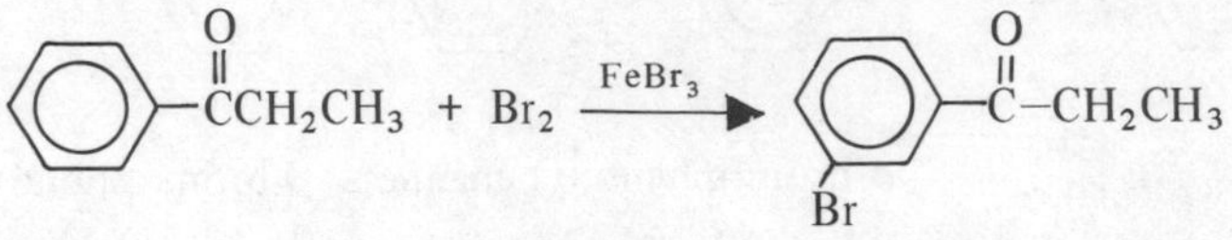

m-Bromopropanoylbenzene

nitration ────▶ *m*-Nitropropanoylbenzene
sulfonation ────▶ *m*-Propanoylbenzenesulfonic acid

Reactions are slower than corresponding reactions of benzene.

(m)

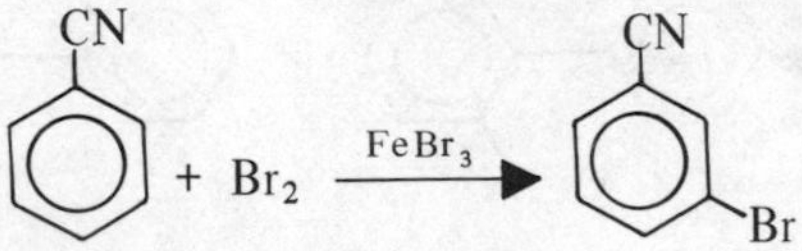

m-Bromobenzonitrile

nitration ────▶ *m*-nitrobenzonitrile
sulfonation ────▶ *m*-cyanobenzenesulfonic acid

Reactions are slower than corresponding reactions of benzene.

(n)

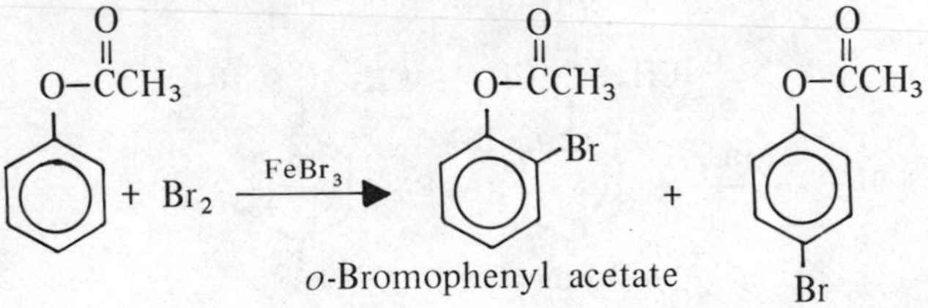

o-Bromophenyl acetate   *p*-Bromophenyl acetate

nitration ────▶ *o*-nitrophenyl acetate + *p*-nitrophenyl acetate

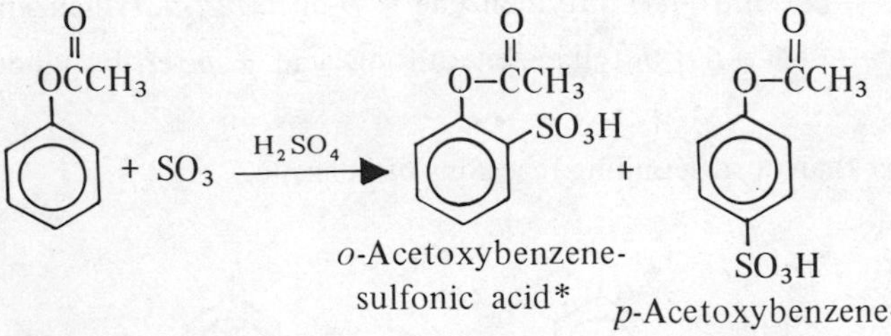

o-Acetoxybenzene-
sulfonic acid*   *p*-Acetoxybenzene-
sulfonic acid*

Reactions are faster than corresponding reactions of benzene.

(o)

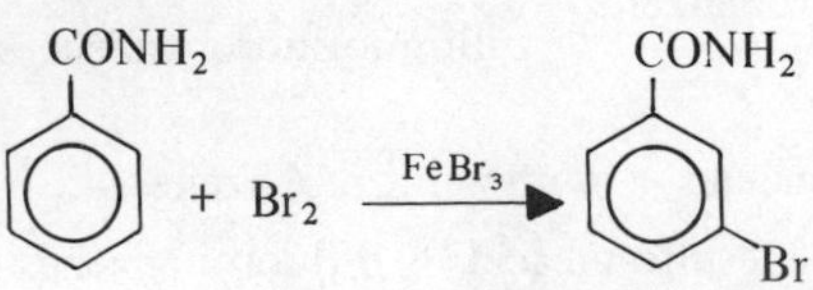

m-Bromobenzamide

nitration ──────▶ *m*-nitrobenzamide

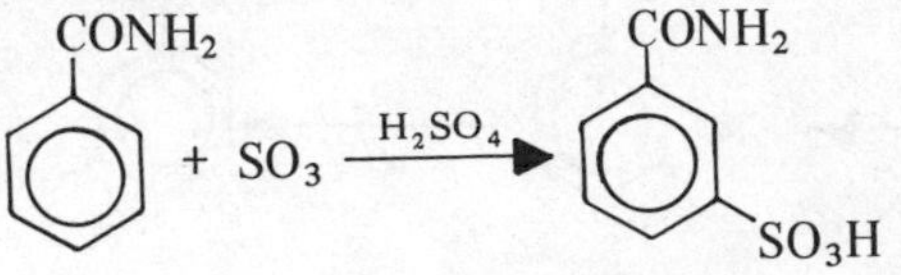

m-Carbamoylbenzene-
sulfonic acid*

Reactions are slower than corresponding reactions of benzene.

(p)

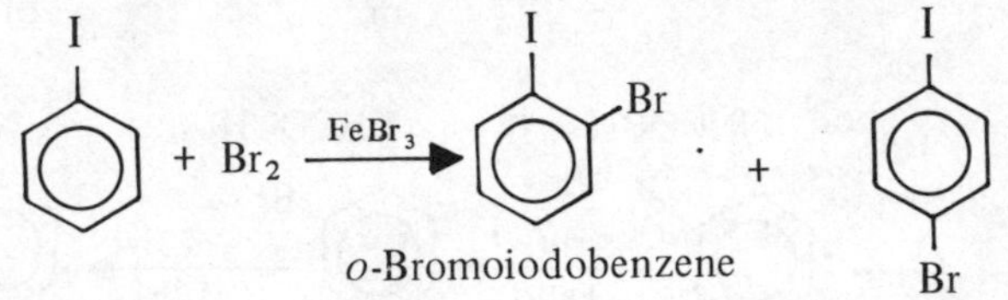

o-Bromoiodobenzene

p-Bromoiodobenzene

nitration ──────▶ *o*-nitroiodobenzene + *p*-nitroiodobenzene

sulfonation ──────▶ *o*-iodobenzenesulfonic acid + *p*-iodobenzenesulfonic acid

Reactions are slower than corresponding reactions of benzene.

13.28

(a)

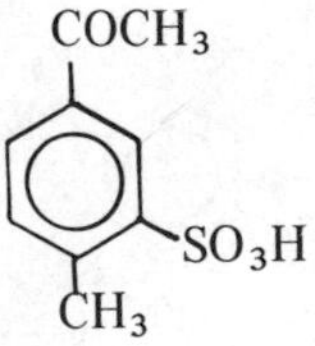

(b)

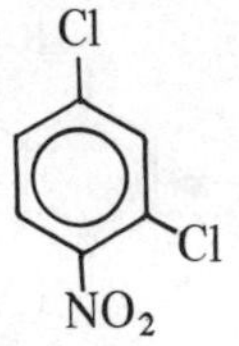

(c)

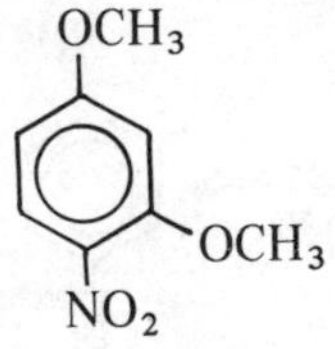

(d)

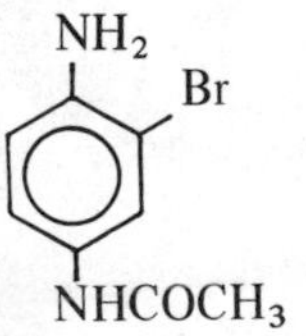

(e)

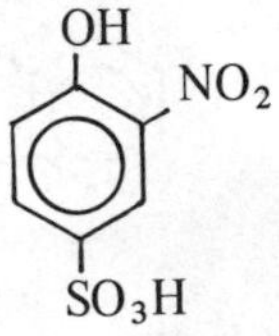

(f)

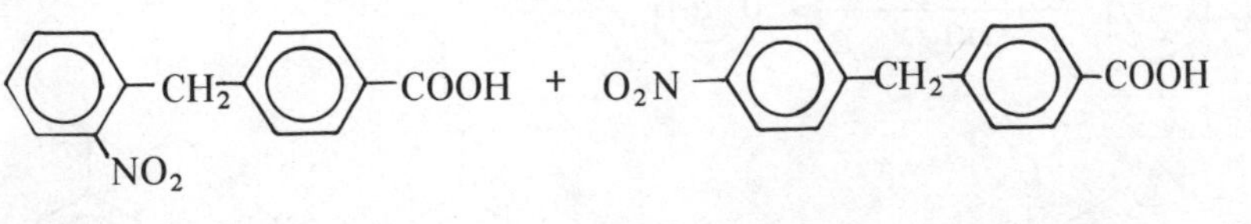

(g)

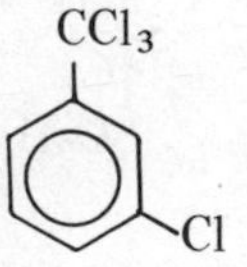

13.29

(a)

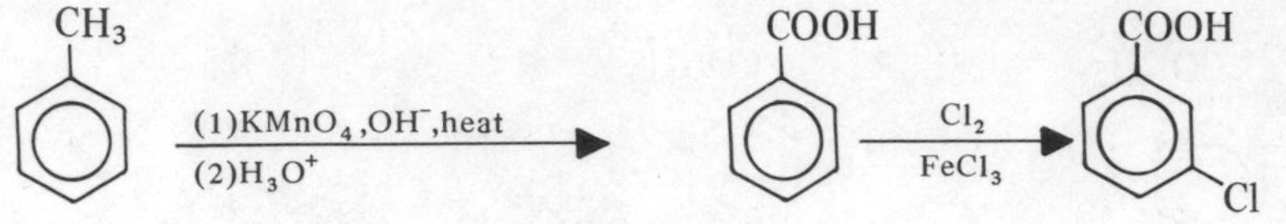

(b)

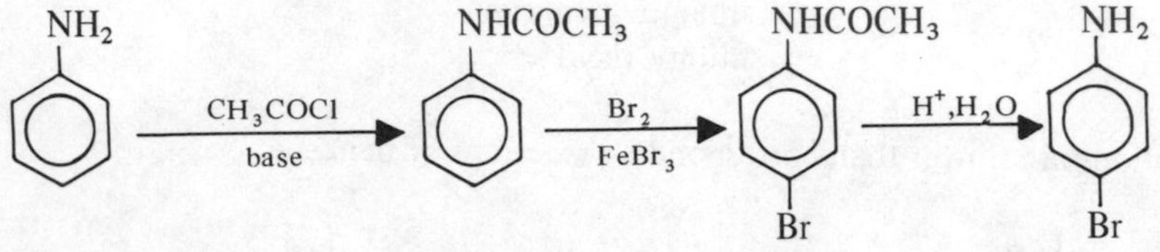

(c)

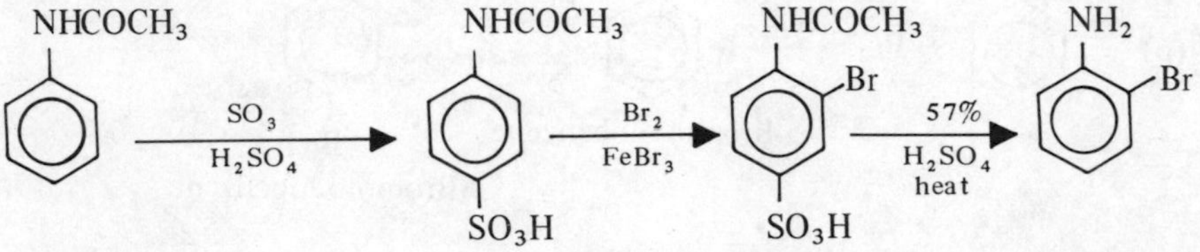

(d)

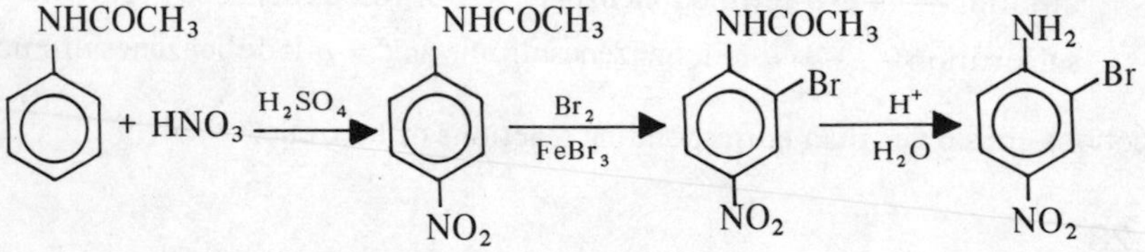

(e)

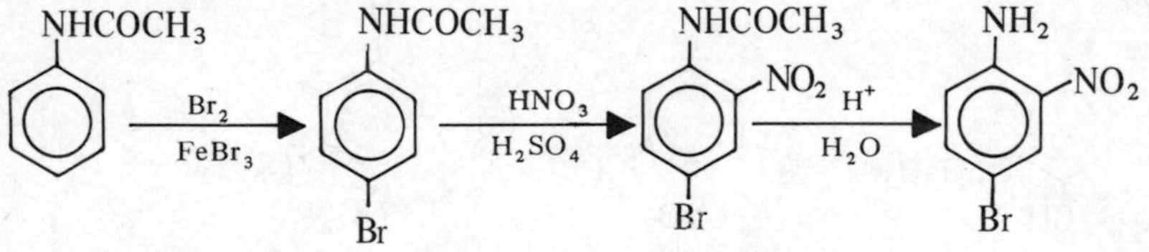

(f)

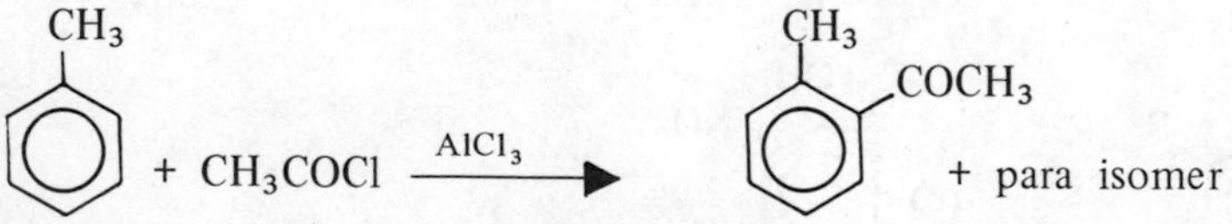

(g)

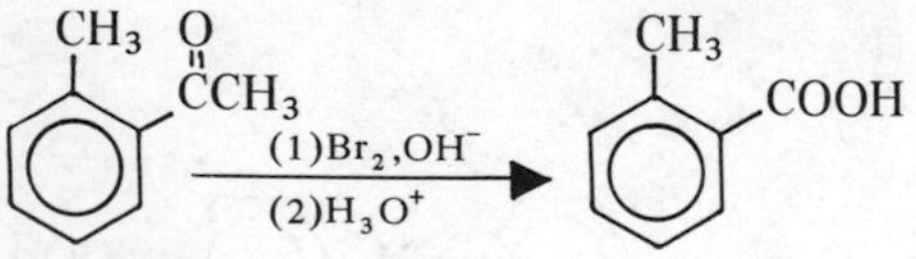

(h)

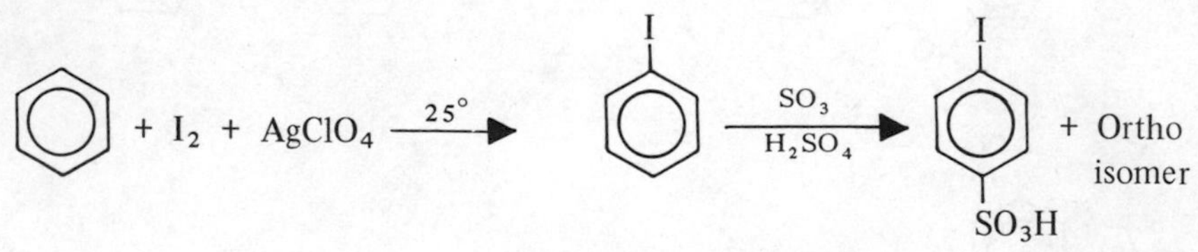

(i)
CH3
+ HNO3
H2SO4
CH3
NO2
Br2
FeBr2
CH3
Br
NO2
+ Ortho isomer

(j)
CH3
(1)KMnO4, OH⁻
(2)H3O⁺
COOH
O
CH3CCl
AlCl3
COOH
CCH3
O
(1)Br2,OH⁻
(2)H3O⁺
COOH
COOH

(k)
O
+ Cl–CCH2CH2CH3
AlCl3
O
CCH2CH2CH3
Zn(Hg)
HCl
CH2CH2CH2CH3

(l)
CH2CH2CH2CH3
Br3
FeBr3
CH2CH2CH2CH3
Br
N-Bromo-
succinimide
Br
CH–CH2CH2CH3
Br
KOH, C2H5OH
CH=CHCH2CH3
Br

(m)
CH3
Cl2 (excess)
light
CCl3
HNO3
H2SO4
CCl3
NO2

(n)
CH3
+ HNO3
(fuming)
H2SO4
heat
CH3
O2N
NO2
NO2

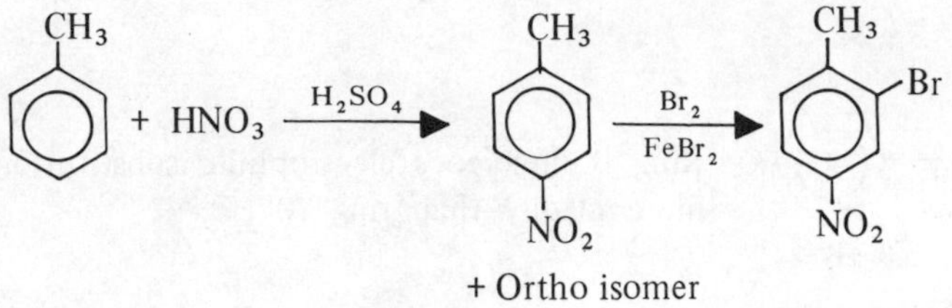

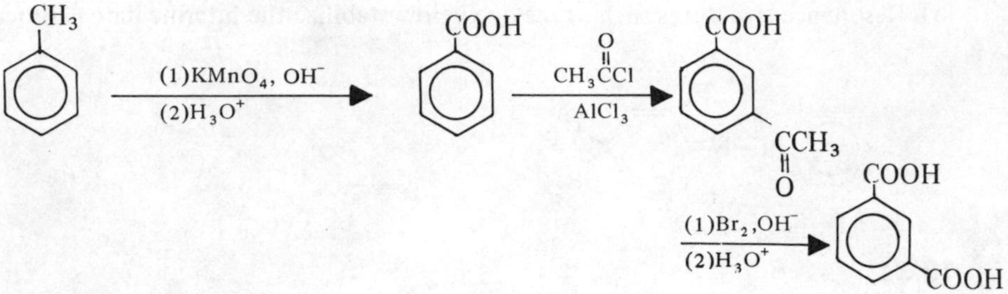

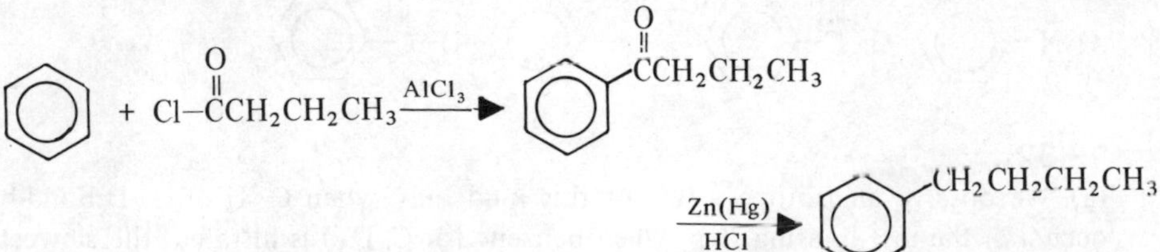

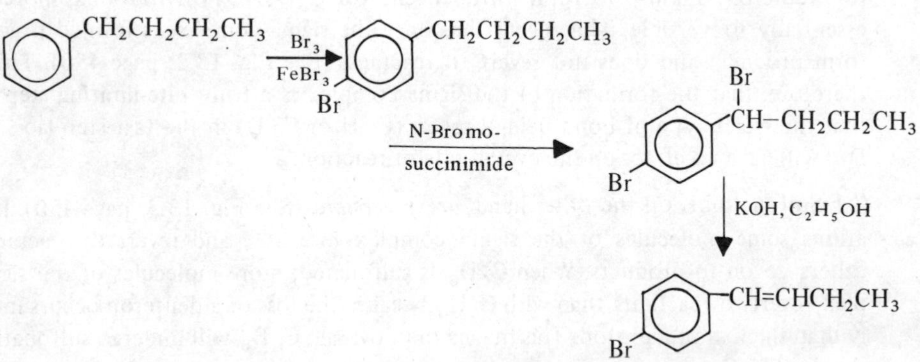

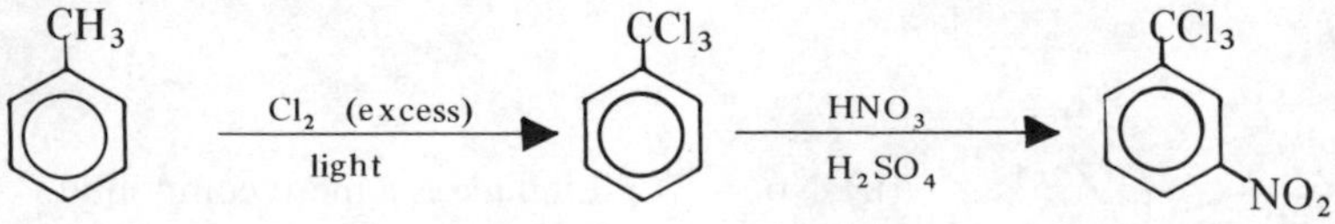

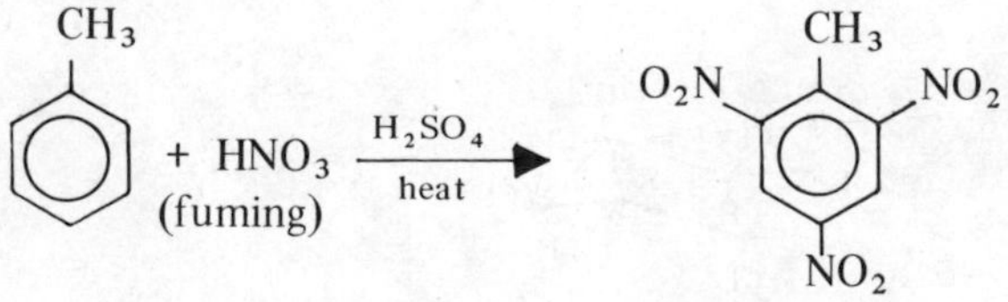

13.30

(a) 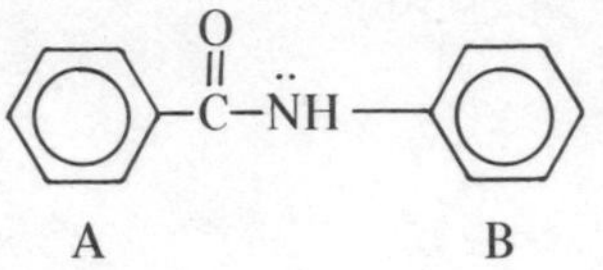 Ring B undergoes electrophilic substitution more readily than ring A.

(b) Resonance structures such as the one below stabilize the intermediate carbocation:

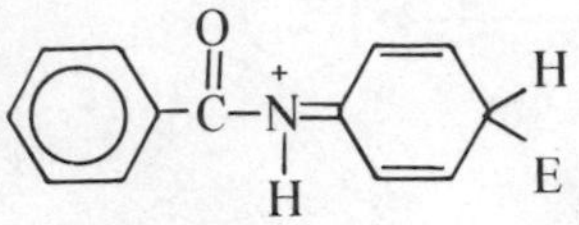

13.31

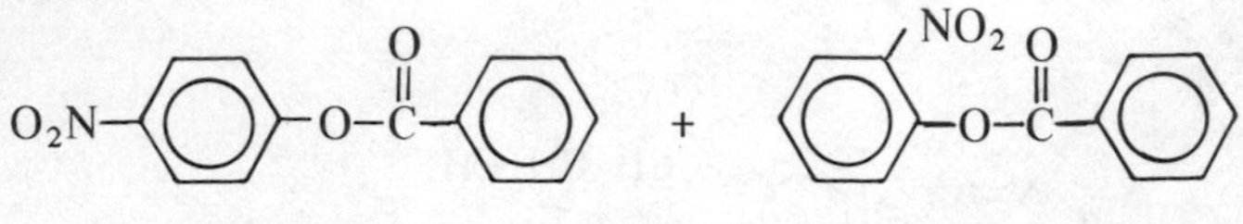

13.32

(a) We observe an isotope effect of this kind only when C—H or C—D bond-breaking occurs in the rate-limiting step. When benzene (or C_6D_6) is nitrated, the slowest step is the formation of the sigma complex. Once the sigma complex is formed it loses a proton (or deuteron) rapidly to form nitrobenzene (or $C_6D_5NO_2$). Nitrations, moreover, are essentially irreversible. This means that once the sigma complex is formed it goes on to form products and does not revert to reactants (See Fig. 13.2, page 450). This means, therefore, that the formation of the sigma complex is a truly rate-limiting step and thus that the exact type of bond being broken (C—H or C—D) in the fast step (loss of H^+ or D^+) will have no effect on the overall rate of reaction.

(b) Sulfonations, on the other hand, *are reversible.* (See Fig. 13.3, page 450). In sulfonations some molecules of the sigma complex lose SO_3 and revert to reactants while others go on to products. When C_6D_6 is sulfonated, more molecules of the sigma complex revert to reactants than with C_6H_6 because the loss of a deuteron occurs more slowly than the loss of a proton. This means that, overall, C_6D_6 will undergo sulfonation more slowly than C_6H_6.

13.33

(a) 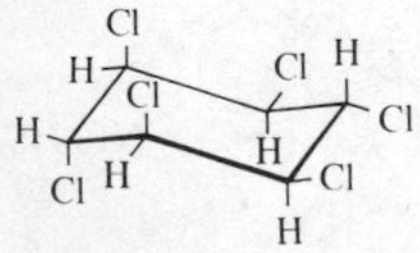 (b) No (c) Lindane is a meso compound.

(d)

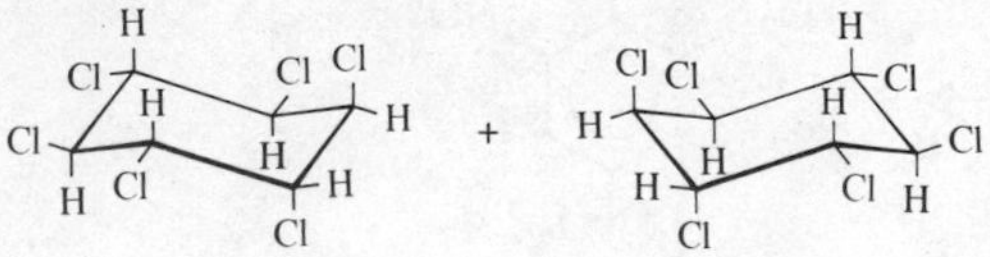

13.34

If we consider resonance structures for the ring that undergoes electrophilic attack, two structures are possible for the sigma complex that forms when attack takes place at the 1-position,

whereas only one is possible when attack takes place at the 2-position,

Attack at the 1-position, therefore, takes place faster.

13.35

13.36

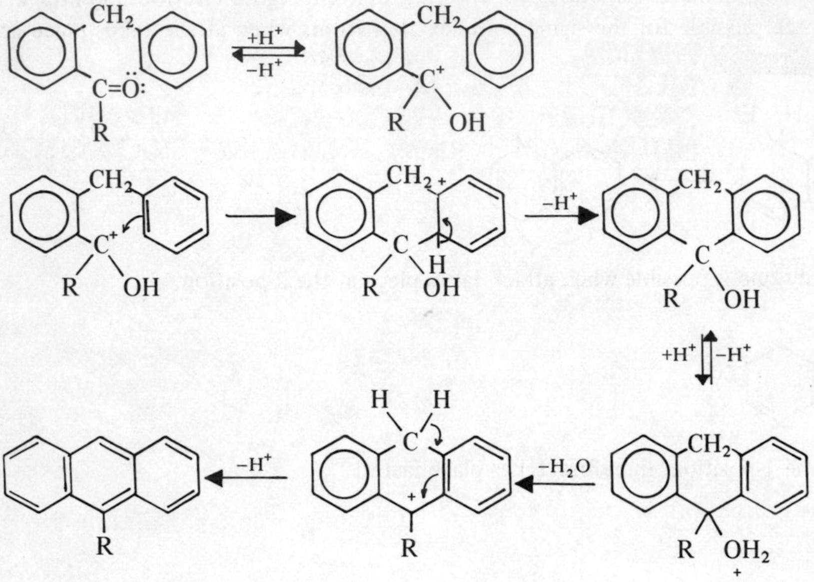

CH_2

$C=\ddot{O}$:

R

$\xrightarrow[-H^+]{+H^+}$

CH_2

C^+

R OH

CH_2

C^+

R OH

$\longrightarrow$

$CH_2 +$

C H

R OH

$\xrightarrow{-H^+}$

CH_2

C

R OH

$+H^+ \updownarrow -H^+$

H H

C

R

$\xleftarrow{-H_2O}$

CH_2

C

R OH_2
        +

R

$\xleftarrow{-H^+}$

PHYSICAL   **NUCLEAR**
METHODS OF   **MAGNETIC**
STRUCTURE   **RESONANCE**   **INFRARED**
DETERMINATION   **SPECTROSCOPY**   **SPECTROSCOPY**

14.1
The methyl protons of 15,16-dimethylpyrene are highly shielded by the induced field in the center of the aromatic system where the induced field opposes the applied field. (p 509).

14.2
(a) The six protons (hydrogens) of ethane are equivalent:

$$\overset{a}{CH_3} - \overset{a}{CH_3}$$

Ethane gives a single signal in its pmr spectrum.

(b) Propane has two different sets of equivalent protons:

$$\overset{a}{CH_3} - \overset{b}{CH_2} - \overset{a}{CH_3}$$

Propane gives two signals.

(c) The six protons of dimethyl ether are equivalent:

$$\overset{a}{CH_3} - O - \overset{a}{CH_3}$$

One signal.

(d) Three different sets of equivalent protons:

Three signals.

(e) Two different sets of equivalent protons:

$$\overset{a}{CH_3} - \overset{\overset{O}{\|}}{C} - O - \overset{b}{CH_3}$$

Two signals.

(f) Three different sets of equivalent protons:

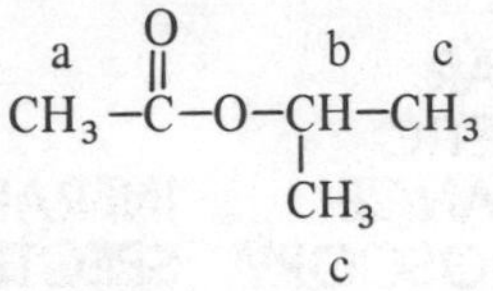

$$\overset{a}{CH_3}-\overset{\overset{\displaystyle O}{\|}}{C}-O-\overset{b}{\underset{\underset{c}{CH_3}}{CH}}-\overset{c}{CH_3}$$

Three signals.

14.3

(a)

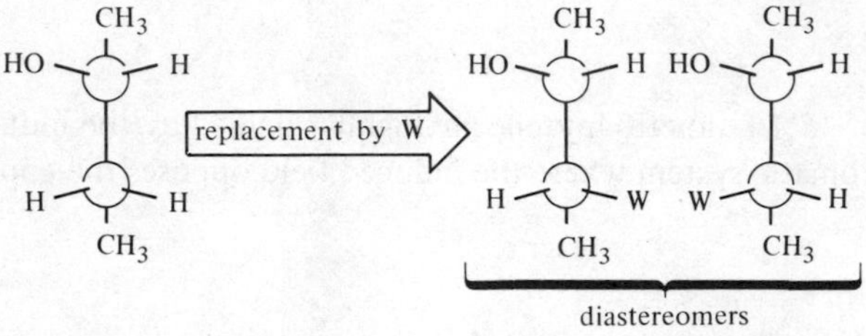

diastereomers

(b) Six,

$$\begin{array}{c} a \\ CH_3 \\ b\ H-\overset{|}{\underset{|}{C}}-OH\ c \\ d\ H-\overset{|}{\underset{|}{C}}-H\ e \\ CH_3 \\ f \end{array}$$

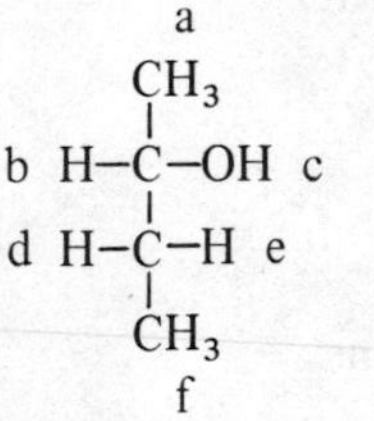

(c) Six signals.

14.4

(a) Two, $\overset{a}{CH_3}-\overset{b}{CH_2}-\overset{b}{CH_2}-\overset{a}{CH_3}$

(b) Three, $\overset{a}{CH_3}-\overset{b}{CH_2}-O-\overset{c}{H}$

(c) Four,

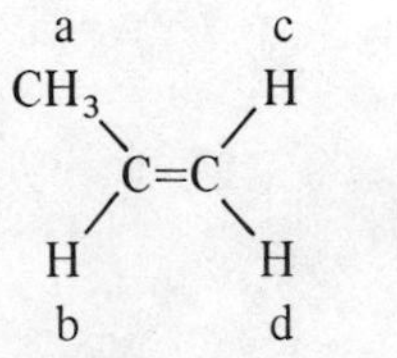

(d) Two,

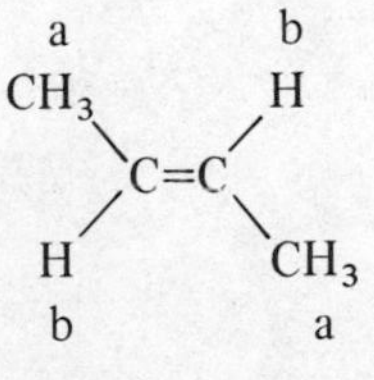

(e) Four, $\overset{a}{CH_3}-\overset{b}{CH}Br-\overset{\overset{\displaystyle H\ c}{|}}{\underset{\underset{H\ d}{|}}{C}}-Br$

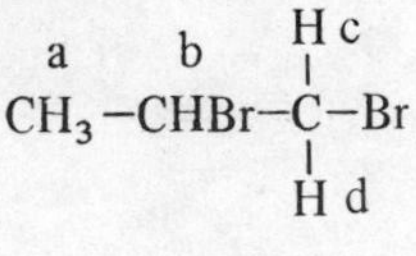

(f) Two,

$$\begin{array}{c} (b)\ H \qquad\qquad H\ (b) \\ (a) \\ (b)\ H \diagdown\ CH_3\diagup H\ (b) \\ \diagdown\diagup \\ CH_3 \\ (a) \end{array}$$

(g) Three,

$$\begin{array}{c} (a) \qquad (b) \\ CH_3\ \ \ H \\ (c) \\ H\diagdown\diagup CH_3 \\ (b) \quad H \quad (a) \\ (c) \end{array}$$

(h) Four,

$$\begin{array}{c} (a) \qquad (a) \\ CH_3 \quad CH_3 \\ (c) \\ H\diagdown\diagup H \\ (b) \quad H \quad (b) \\ (d) \end{array}$$

(i) Six,

$$\begin{array}{c} a \qquad b \qquad c \\ CH_3{-}CH_2{-}CH_2 \diagdown\qquad\diagup H\ e \\ C{=}C \\ \diagup\qquad\diagdown \\ H \qquad\quad H\ f \\ d \end{array}$$

14.5

The pmr spectrum of $CHBr_2CHCl_2$ consists of two doublets. The doublet from the proton of the $-CHCl_2$ group should occur at lowest magnetic field strength because the greater electronegativity of chlorine reduces the electron density in the vicinity of the $-CHCl_2$ proton, and consequently, reduces its shielding relative to $-CHBr_2$.

14.6

The determining factors here are the number of chlorine atoms attached to the carbons bearing protons and the deshielding that results from chlorine's electronegativity. In 1,1,2-trichloroethane the proton that gives rise to the triplet is on a carbon that bears two chlorines, and the signal from this proton is downfield. In 1,1,2,3,3-pentachloropropane the proton that gives rise to the triplet is on a carbon that bears only one chlorine; the signal from this proton is upfield.

14.7

The signal from the three equivalent protons designated **a** should be split into a doublet by the proton **b**. This doublet, because of the electronegativity of the attached chlorines, should occur downfield.

$$\begin{array}{c} a \qquad b \\ (Cl_2CH)_3{-}CH \end{array}$$

The proton designated **b** should be split into a quartet by the three equivalent protons **a**. The quartet should occur upfield.

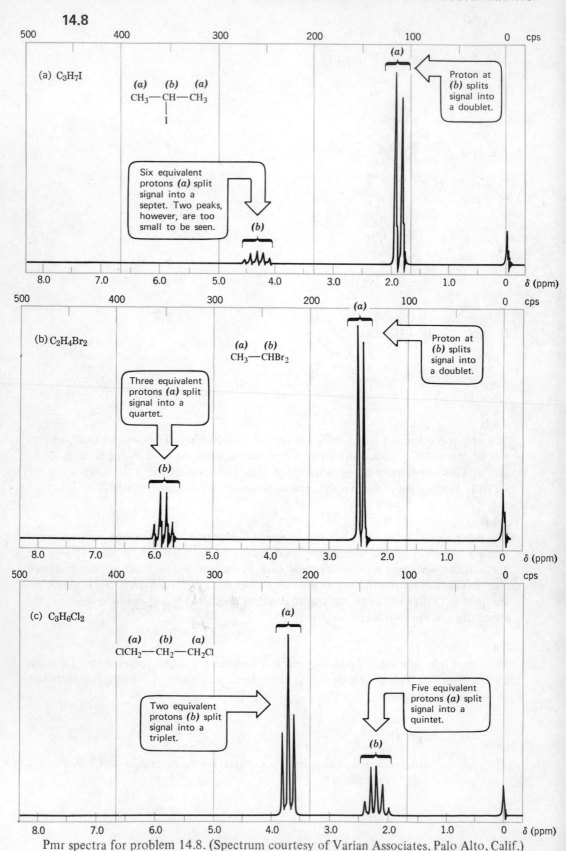

Pmr spectra for problem 14.8. (Spectrum courtesy of Varian Associates, Palo Alto, Calif.)

14.9

(a) $J_{ab} = 2J_{bc}$

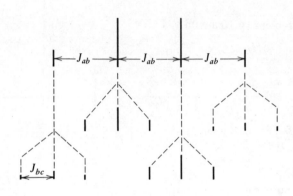

Result: (Nine peaks)

(b) $J_{ab} = J_{bc}$

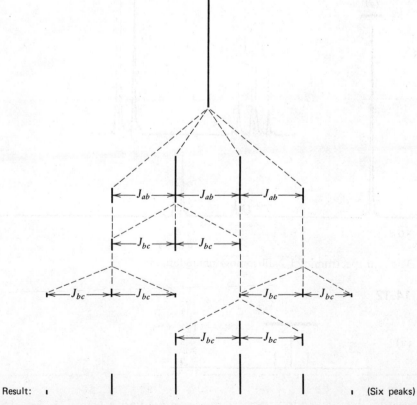

Result: (Six peaks)

14.10

(a) $C_6H_5CH(CH_3)_2$

(b) $C_6H_5\underset{\underset{NH_2}{|}}{C}HCH_3$

(c)

Pmr spectra for problem 14.10 are on p. 133.

14.11

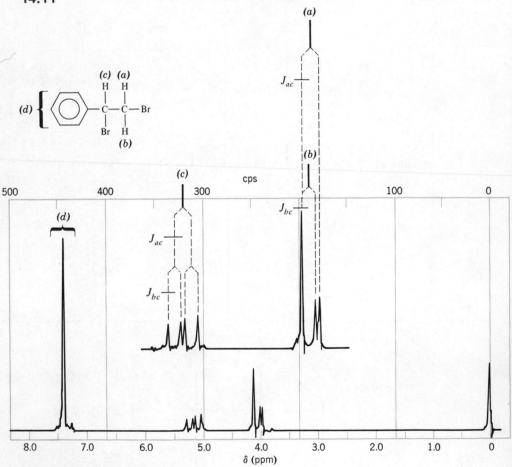

The pmr spectrum of 1,2-dibromo-1-phenylethane.

14.12

(a)

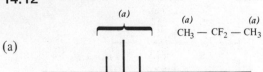

$$CH_3 \underset{(a)}{} - CF_2 - \underset{(a)}{} CH_3$$

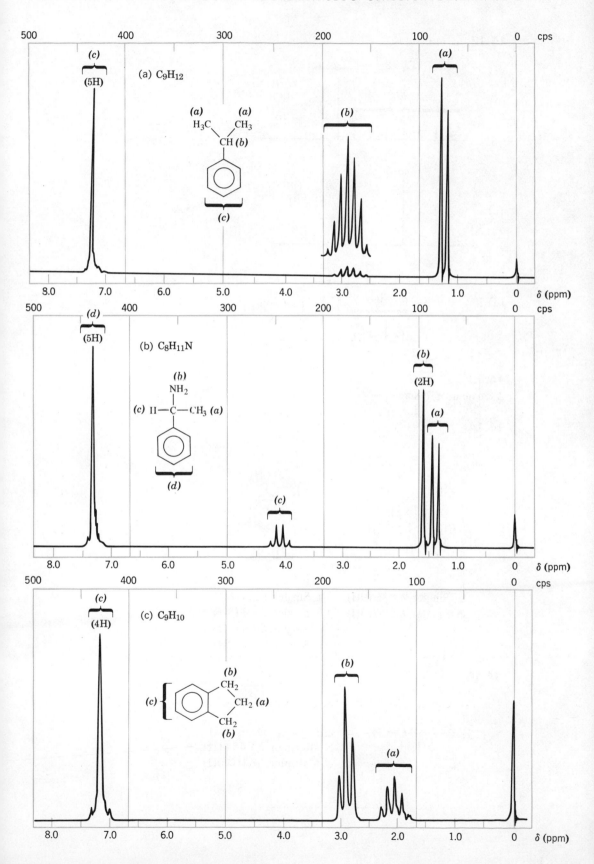

14.12 continued

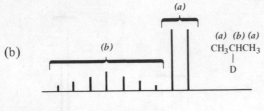

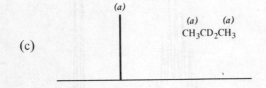

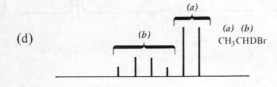

14.13
A single unsplit signal.

14.14

a
CH₃
ᵇH   Hᵇ

CH₃   CH₃
a   a
H
b

a Singlet, δ2.35 (9H)
b Singlet, δ6.70 (3H)

b
CH₃
ᵈH   Hᵈ
        +
CH₃   CH₃
a   a
H   H
c

a Singlet, δ2.8 (6H)
b Singlet, δ2.9 (3H)
c Singlet, δ4.6 (2H)
d Singlet, δ7.7 (2H)

14.15

b
H
|
C₆H₅Ċ———C—CH₃ a
| |
C₆H₅   CH₃
c   a

a Doublet, δ1.48 (6H)
b Multiplet, δ4.45 (1H)
c Multiplet, δ8.0 (10H)

14.16

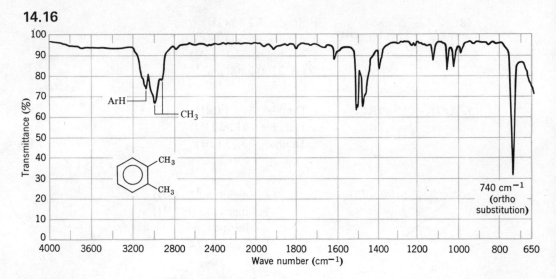

The infrared spectrum of o-xylene.
Similar assignments can be made for m-xylene and p-xylene.

14.17

A   B   C   D

14.18

(a)
$$\underset{a}{CH_3}-\underset{\underset{\underset{a}{CH_3}}{\overset{\overset{a}{CH_3}}{|}}}{C}-OH \ b$$

a Singlet, $\delta 1.28$ (9H)
b Singlet, $\delta 1.35$ (1H)

(b)
$$\underset{a}{CH_3}-\underset{\underset{Br}{|}}{\overset{b}{CH}}-\underset{a}{CH_3}$$

a Doublet, $\delta 1.71$ (6H)
b Septet, $\delta 4.32$ (1H)

(c)
$$\overset{b}{CH_3}-\overset{\overset{O}{||}}{C}-\overset{c}{CH_2}-\overset{a}{CH_3}$$

a Triplet, $\delta 1.05$ (3H)
b Singlet, $\delta 2.13$ (3H)
c Quartet, $\delta 2.47$ (2H)
C=O, 1720 cm^{-1}

(d)

b a
—CH$_2$—OH

c

a Singlet, $\delta 2.43$ (1H)
b Singlet, $\delta 4.58$ (2H)
c Multiplet, $\delta 7.28$ (5H)
O—H, 3200-3600 cm^{-1}

(e)

b   c
CH$_3$—CH—CH$_2$Cl
     |
a   CH$_3$

a Doublet, $\delta 1.04$ (6H)
b Multiplet, $\delta 1.95$ (1H)
c Doublet, $\delta 3.35$ (2H)

(f)

b   O   a
C$_6$H$_5$—CH—C—CH$_3$
     |
     C$_6$H$_5$
c

a Singlet, $\delta 2.20$ (3H)
b Singlet, $\delta 5.08$ (1H)
c Multiplet, $\delta 7.25$ (10H)
C=O, near 1720 cm^{-1}

(g)

a   b   c   d
CH$_3$—CH$_2$—CHCOOH
            |
            Br

a Triplet, $\delta 1.08$ (3H)
b Multiplet, $\delta 2.07$ (2H)
c Triplet, $\delta 4.23$ (1H)
d Singlet, $\delta 10.97$ (1H)
O—H, 2500-3000 cm^{-1}

(h)

b   a
—CH$_2$—CH$_3$

c

a Triplet, $\delta 1.25$ (3H)
b Quartet, $\delta 2.68$ (2H)
c Multiplet, $\delta 7.23$ (5H)

(i)

a   b   c   d
CH$_3$—CH$_2$—O—CH$_2$—COOH

a Triplet, $\delta 1.27$ (3H)
b Quartet, $\delta 3.66$ (2H)
c Singlet, $\delta 4.13$ (2H)
d Singlet, $\delta 10.95$ (1H)
O—H, 2500-3000 cm^{-1}

(j)

a   b   a
CH$_3$—CH—CH$_3$
     |
     NO$_2$

a Doublet, $\delta 1.55$ (6H)
b Septet, $\delta 4.67$ (1H)

(k)

a   b   b   a
CH$_3$O—CH$_2$CH$_2$—OCH$_3$

a Singlet, $\delta 3.25$ (6H)
b Singlet, $\delta 3.45$ (4H)

(l)

b   O   c
CH$_3$—C—CH—CH$_3$
          |
          CH$_3$ a

a Doublet, $\delta 1.10$ (6H)
b Singlet, $\delta 2.10$ (3H)
c Septet, $\delta 2.50$ (1H)
C=O, near 1720 cm^{-1}

(m) ⟨benzene ring⟩–$\overset{b}{C}H$–$\overset{a}{C}H_3$
      |
      Br
   c

a Doublet, $\delta 2.0$ (3H)
b Quartet, $\delta 5.15$ (1H)
c Multiplet, $\delta 7.35$ (5H)

14.19
Compound **E** is phenylacetylene, $C_6H_5C\equiv CH$. We can make the following assignments in the infrared spectrum:

$\sim$ 3300 cm^{-1} , $\equiv C-H$
$\sim$ 3030 cm^{-1} , $Ar-H$
$\sim$ 2100 cm^{-1} (weak) , $-C\equiv C-$

$\sim$ 690 cm^{-1} and $\sim$ 710 cm^{-1} , ⟨benzene ring⟩–

14.20
A pmr signal this far upfield indicates that cyclooctatetraene is a cyclic polyene and is not aromatic.

14.21
Both [14] annulene and dehydro [14] annulene are aromatic as shown by the signals at $\delta 7.78$ (10II) and at $\delta 8.0$ (10H) respectively. [14] Annulene has four "internal" protons (δ -0.61) and dehydro [14] annulene has only two ($\delta 0.0$).

14.22
Compound **F** is *p*-isopropyltoluene.

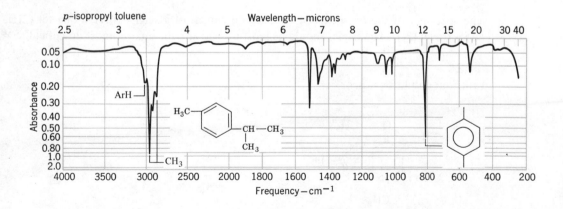

The IR and (on next page) pmr spectra of compound F, problem 14.22. (pmr spectrum adapted from Varian Associates, Palo Alto, Calif. IR spectrum adapted from Sadtler Research Laboratories, Philadelphia, Pa.)

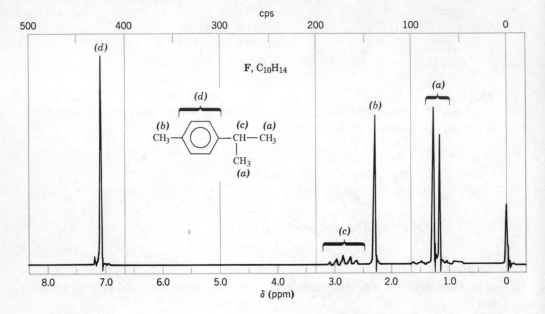

cps

F, $C_{10}H_{14}$

δ (ppm)

14.23

(a) In SbF_5 the carbocations formed initially apparently undergo a complex series of rearrangements to the more stable *tert*-butyl cation.

(b) All of the cations formed intially rearrange to the more stable *tert*-pentyl cation,

$$CH_3CH_2\overset{\overset{\displaystyle CH_3}{|}}{\underset{\underset{\displaystyle CH_3}{|}}{C}}+$$

The spectrum of the *tert*-pentyl cation should consist of a singlet (6H), a quartet (2H), and a triplet (3H). The triplet should be most upfield and the quartet most downfield.

14.24

(a) Four unsplit signals,

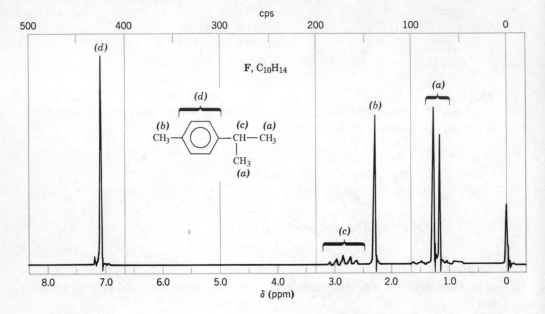

(b) Absorptions arising from: =C—H , CH_3 , and C=O groups.

14.25

Compound **G** is 2-bromobutane.

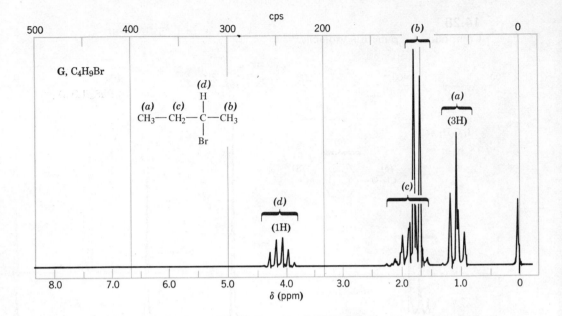

The pmr spectrum of compound **G** (problem 14.25). (Spectrum courtesy of Varian Associates, Palo Alto, Calif.)

Compound **H** is 2,3-dibromopropene.

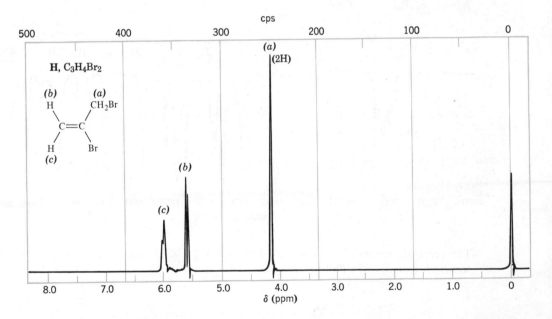

The pmr spectrum of compound **H** (problem 14.25). (Spectrum courtesy of Varian Associates, Palo Alto, Calif.).

14.26

Compound **I** is *p*-methoxytoluene.

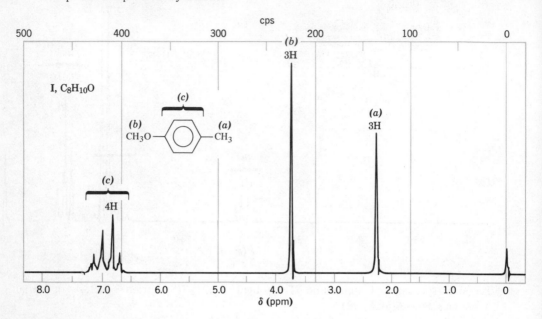

The pmr spectrum of compound **I** (problem 14.26). (Spectrum courtesy of Varian Associates, Palo Alto, Calif.)

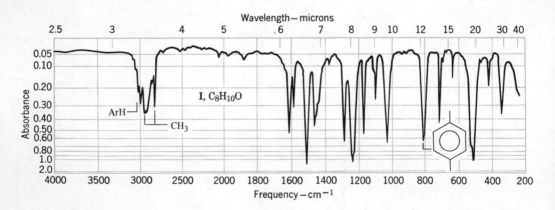

The infrared spectrum of compound **I** (problem 14.26). (Spectrum courtesy of Sadtler Research Laboratories, Philadelphia, Pa.)

15.1

(a) (structure: benzene ring with CH_2–OH, $Tl(OOCCF_3)_3$)

(b) (structure: benzene ring with CH_2–OCH_3, $Tl(OOCCF_3)_3$)

(c) (structure: benzene ring with CH_2–CH_2–O–CH_3, $Tl(OOCCF_3)_3$)

(d) (structure: benzene ring with CH_2–$\overset{|}{C}$–OH, O, $Tl(OOCCF_3)_3$)

(e) (structure: benzene ring with CH_2–$\overset{|}{C}$–OCH_3, O, $Tl(OOCCF_3)_3$)

15.2

In the complex the Tl atom is too far from the ortho position:

(structure: benzene ring with CH_2–CH_2, CH_3–C, O, $Tl(OOCCF_3)_3$)

It must therefore react in the usual way at the para (least crowded) position.

15.3

(a) (structure: benzene ring with $C(CH_3)_3$) $\xrightarrow[Tl(OOCCF_3)_3]{Br_2}$ (structure: benzene ring with $C(CH_3)_3$ and Br para)

(b) (structure: benzene ring with CH_2CH_2COOH) $\xrightarrow[CF_3COOH, 25°]{Tl(OOCCF_3)_3}$ (structure: benzene ring with CH_2CH_2COOH and $Tl(OOCCF_3)_2$) $\xrightarrow[(DMF)]{CuCN}$ (structure: benzene ring with CH_2CH_2COOH and CN)

141

(c) [benzene ring with OCH₃] $\xrightarrow[\substack{(2)\text{Pb(OOCCH}_3)_4,(C_6H_5)_3P \\ (3)\text{OH}^-}]{(1)\text{Tl(OOCCF}_3)_3,\text{CF}_3\text{COOH},25°}$ [benzene ring with OCH₃ (top) and OH (bottom, para)]

(d) [benzene ring with CH₂CH₂CH₃] $\xrightarrow[\substack{(2)\text{KI,H}_2\text{O}}]{(1)\text{Tl(OOCCF}_3)_3,\text{CF}_3\text{COOH},25°}$ [benzene ring with CH₂CH₂CH₃ (top) and I (bottom, para)]

(e) [benzene ring with CH₂CH₂CH₃] $\xrightarrow[\substack{(2)\text{KI,H}_2\text{O}}]{(1)\text{Tl(OOCCF}_3)_3,\text{CF}_3\text{COOH},73°}$ [benzene ring with CH₂CH₂CH₃ and I (meta)]

(f) [benzene ring with two CH₃ groups] $\xrightarrow[\text{above}]{\text{same as (d)}}$ [benzene ring with CH₃, CH₃, and I]

(g) [benzene ring with CH₂CH₃] $\xrightarrow[\substack{(2)\text{CuCN,DMF}}]{(1)\text{Tl(OOCCF}_3)_3,\text{CF}_3\text{COOH},25°}$ [benzene ring with CH₂CH₃ (top) and CN (bottom, para)]

15.4

(a) $\underset{\text{(stronger base)}}{CH_3CH_2CH_2CH_2\!:\!Li} + \underset{\substack{\text{(stronger} \\ \text{acid)}}}{\overset{\delta-\,\curvearrowright\,\delta+}{H:\ddot{O}H}} \longrightarrow \underset{\text{(weaker acid)}}{CH_3CH_2CH_2CH_2-H} + \underset{\substack{\text{(weaker} \\ \text{base)}}}{Li^+\;:\!\ddot{O}H^-}$

(b) $\underset{\text{(stronger base)}}{CH_3CH_2CH_2CH_2\!:\!Li} + \underset{\substack{\text{(stronger} \\ \text{acid)}}}{\overset{\delta-\,\curvearrowright\,\delta+}{H:\ddot{O}CH_2CH_3}} \longrightarrow \underset{\text{(weaker acid}}{CH_3CH_2CH_2CH_2-H} + \underset{\substack{\text{(weaker} \\ \text{base)}}}{Li^+\;\underset{CH_3}{:\!\ddot{O}CH_2}}$

15.5

$$\underset{\underset{CH_3}{|}}{\overset{\overset{CH_3}{|}}{CH_3-C-Br}} + Mg \xrightarrow[35°]{\text{ether}} \underset{\underset{CH_3}{|}}{\overset{\overset{CH_3}{|}}{CH_3-C-MgBr}} \xrightarrow{D_2O} \underset{\underset{CH_3}{|}}{\overset{\overset{CH_3}{|}}{CH_3-C-D}}$$

15.6

(a) $\underset{\underset{OH}{|}}{CH_3CH_2CHCH_3} + PBr_3 \longrightarrow \underset{\underset{Br}{|}}{CH_3CH_2CHCH_3} + P(OH)_3$

(b) $CH_3CH_2CH_2CH_2OH \xrightarrow{PBr_3} CH_3CH_2CH_2CH_2Br \xrightarrow{(CH_3)_3COK}$

$CH_3CH_2CH=CH_2 \xrightarrow[\text{(no peroxides)}]{HBr} CH_3CH_2\underset{Br}{CH}CH_3$

(c) See (b) above.

(d) $CH_3CH_2C \equiv CH \xrightarrow[\text{(no peroxides)}]{HBr} CH_3CH_2\underset{Br}{C}=CH_2 \xrightarrow[Pt]{H_2} CH_3CH_2\underset{Br}{CH}CH_3$

15.7

(a) $CH_3CH_2\underset{OH}{CH}CH_3 \xrightarrow{PBr_3} CH_3CH_2\underset{Br}{CH}CH_3 \xrightarrow{(CH_3)_3COK} CH_3CH_2CH=CH_2$
(+2-butenes)

$\xrightarrow[\text{peroxides}]{HBr} CH_3CH_2CH_2CH_2Br$

(b) $CH_3CH_2CH_2CH_2OH \xrightarrow{PBr_3} CH_3CH_2CH_2CH_2Br + P(OH)_3$

(c) $CH_3CH_2CH=CH_2 \xrightarrow[\text{(peroxides)}]{HBr} CH_3CH_2CH_2CH_2Br$

(d) $CH_3CH_2C \equiv CH \xrightarrow[\text{peroxides}]{HBr} CH_3CH_2CH=CHBr \xrightarrow[H_2]{Pt} CH_3CH_2CH_2CH_2Br$

15.8

(a) 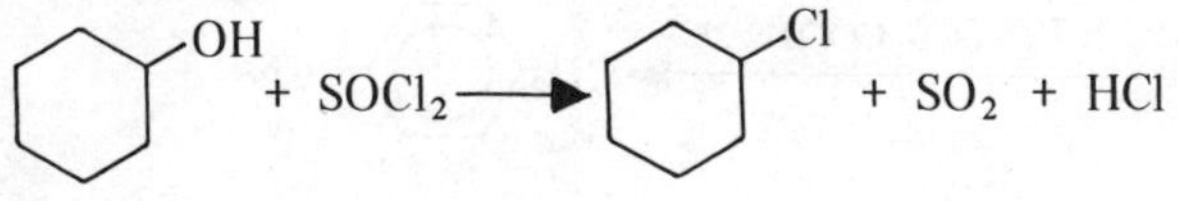 + SOCl₂ ⟶ + SO₂ + HCl

(b) 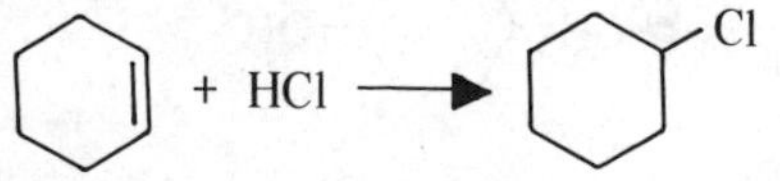 + HCl ⟶

(c) 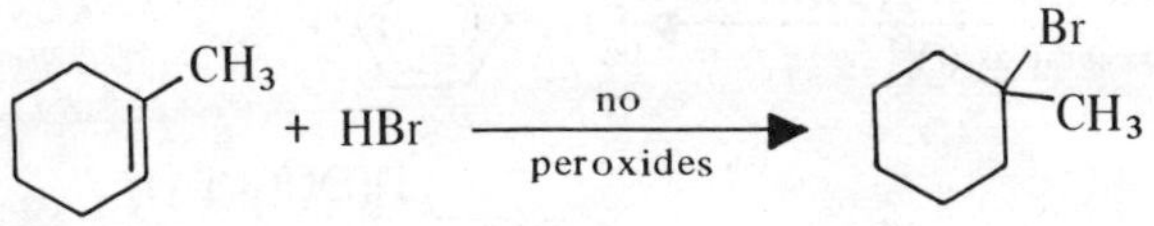 + HBr $\xrightarrow[\text{peroxides}]{\text{no}}$

(d) 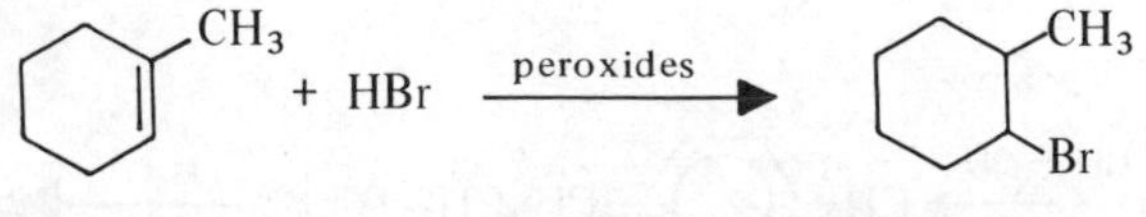 + HBr $\xrightarrow{\text{peroxides}}$

(e) 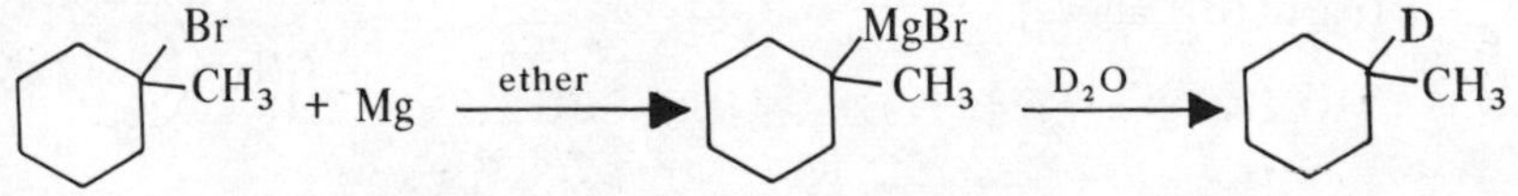 + Mg $\xrightarrow{\text{ether}}$ $\xrightarrow{D_2O}$

15.9

(a)

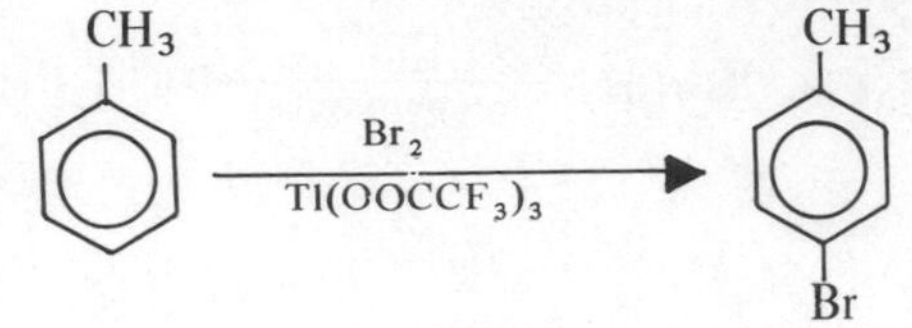

(b)

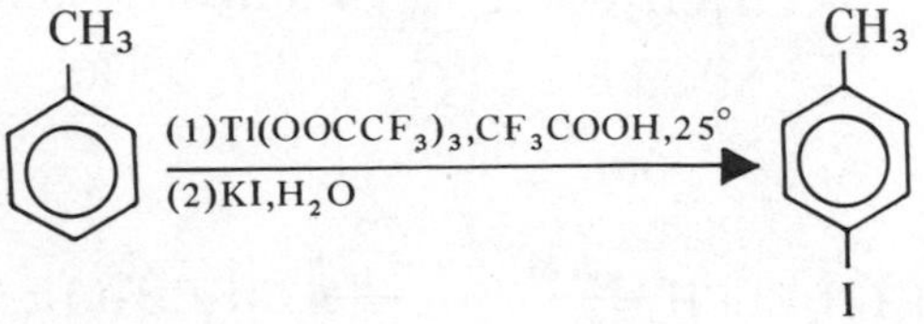

(c)

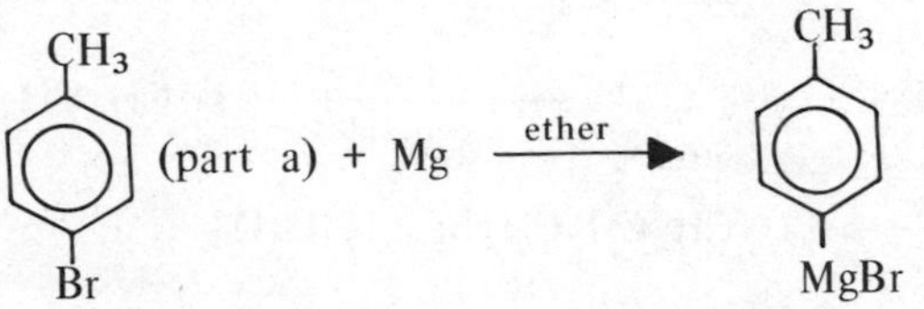

(d) 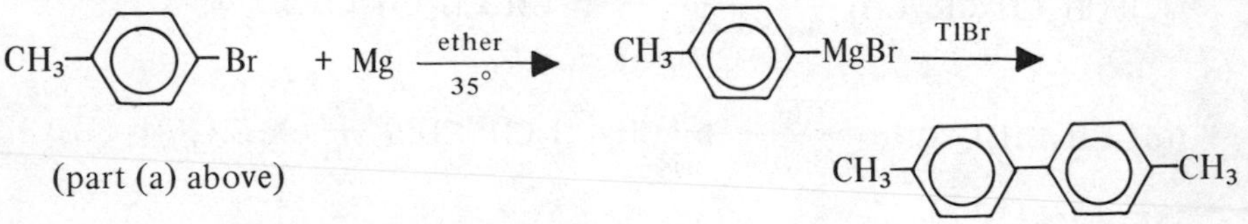 CH_3-⬡$-Br$ + Mg $\xrightarrow[35°]{ether}$ CH_3-⬡$-MgBr$ $\xrightarrow{TlBr}$

(part (a) above)

CH_3-⬡$-$⬡$-CH_3$

(e) 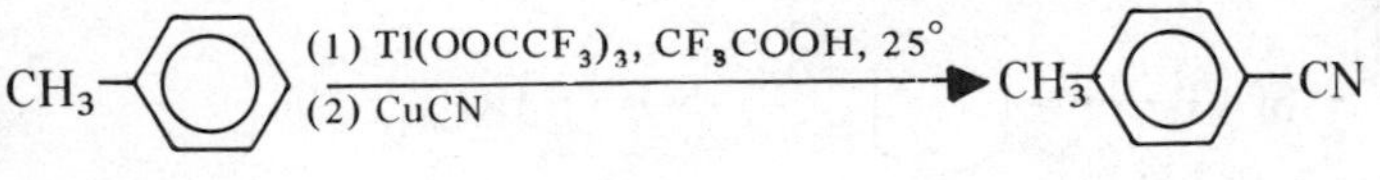 CH_3-⬡ $\xrightarrow[\text{(2) CuCN}]{\text{(1) Tl(OOCCF}_3)_3,\text{ CF}_3\text{COOH, 25°}}$ CH_3-⬡$-CN$

(f) 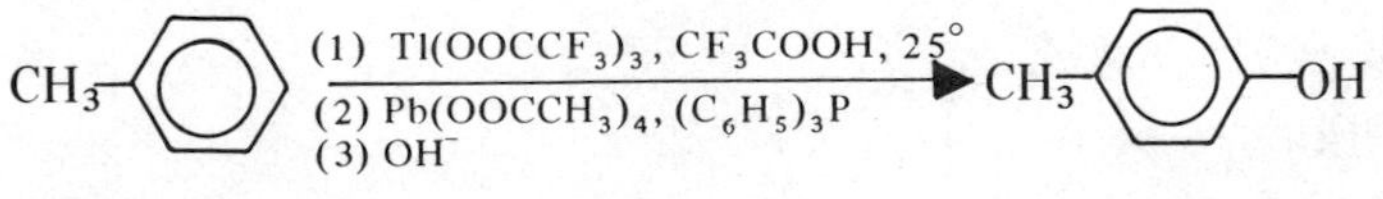 CH_3-⬡ $\xrightarrow[\substack{\text{(2) Pb(OOCCH}_3)_4,\text{ (C}_6\text{H}_5)_3\text{P} \\ \text{(3) OH}^-}]{\text{(1) Tl(OOCCF}_3)_3,\text{ CF}_3\text{COOH, 25°}}$ CH_3-⬡$-OH$

(g) CH_3-⬡ $\xrightarrow[\text{(2) & (3) same as (f)}]{\text{(1) Tl(OOCCF}_3)_3,\text{ CF}_3\text{COOH, 73°}}$ CH_3-⬡$-OH$ (OH in meta position)

(h) CH_3-⬡ $\xrightarrow[73°]{\text{Tl(OOCCF}_3)_3,\text{ CF}_3\text{COOH}}$ CH_3-⬡$-Tl(OOCCF_3)_2$ $\xrightarrow[\text{H}_2\text{O}]{\text{KI}}$ CH_3-⬡$-I$ (I in meta position)

(i) CH_3-⬡$-MgBr$ $\xrightarrow{\overset{O}{\overset{/\backslash}{CH_2-CH_2}}}$ CH_3-⬡$-CH_2CH_2OMgBr$ $\xrightarrow[\text{dilute}]{\text{H}_3\text{O}^+}$

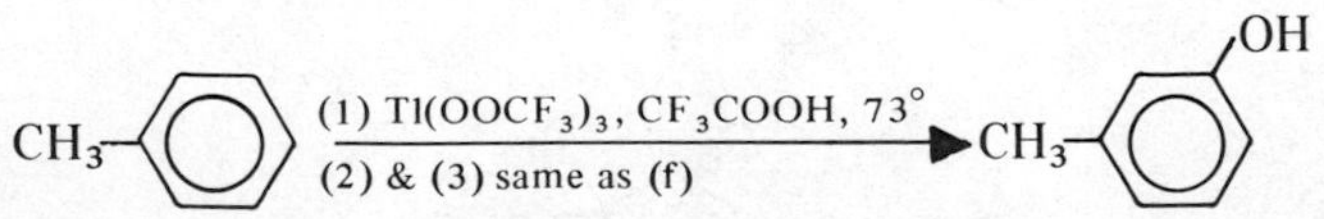

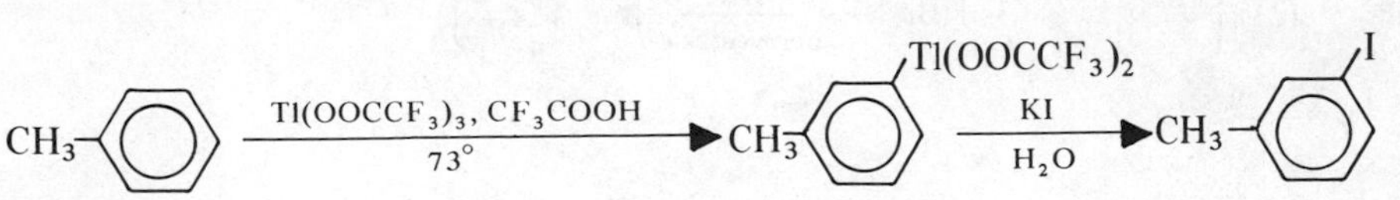

(part (d) above)

CH_3-⬡$-CH_2CH_2OH$

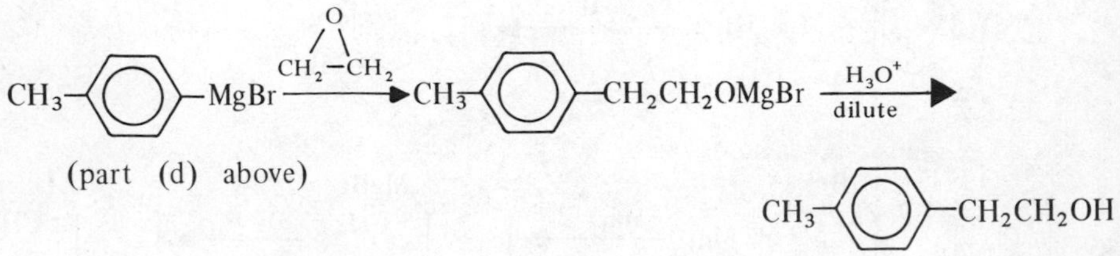

(j) $CH_3-\langle\bigcirc\rangle-MgBr \xrightarrow{D_2O} CH_3-\langle\bigcirc\rangle-D$

15.10
The acetyl group is deactivating, thus the unsubstituted ring of the intermediate, mono-acetylated ferrocene is more reactive than the substituted one.

15.11

(a) $\langle\bigcirc\rangle$-Br $+ (CH_3)_2CuLi \xrightarrow[\text{ether}]{0°} \langle\bigcirc\rangle$-CH$_3$ $+ CH_3Cu + LiBr$

(b) $\langle\bigcirc\rangle$-Br $+ (CH_3)_2CuLi \xrightarrow[\text{ether}]{0°} \langle\bigcirc\rangle$-CH$_3$ $+ CH_3Cu + LiBr$

(c) $CH_2=CH-CH_2Br + (CH_3CH_2)_2CuLi \xrightarrow[\text{ether}]{0°} CH_2=CH-CH_2-CH_2-CH_3 +$
$$CH_3CH_2Cu + LiBr$$

(d) $\begin{matrix} CH_3 & CH_3 \\ \ \ \diagdown & \diagup \\ C=C \\ \diagup & \diagdown \\ H & I \end{matrix}$ $+ (CH_3CH_2CH_2CH_2)_2CuLi \xrightarrow[\text{ether}]{0°}$ $\begin{matrix} CH_3 & CH_3 \\ \diagdown & \diagup \\ C=C \\ \diagup & \diagdown \\ H & CH_2CH_2CH_2CH_3 \end{matrix}$
$$+ CH_3CH_2CH_2CH_2Cu + LiBr$$

15.12

(a) $\langle\bigcirc\rangle$ $+ CH_3COO^- Li^+$ (b) $\langle\bigcirc\rangle$ $+ CH_3O^- Li^+$

(c) $CH_4 + MgBrNH_2$ (d) $(CH_3)_4Si + 4MgBrCl$

(e) $[\langle\bigcirc\rangle]_3$-P $+ 3MgBrCl$ (f) $(CH_3CH_2)_2Cd + 2MgBrCl$

(g) $\langle\bigcirc\rangle$-$\overset{\overset{O}{\|}}{C}$-OH $+ Mg^{++}$ (h) $\langle\bigcirc\rangle$-CH$_2$OH $+ Mg^{++}$

15.13
(a) $CH_2=CH_2$

(b) $BrCH_2-CH_2-Br + Mg \xrightarrow{\text{ether}} \overset{\delta-}{Br}-CH_2-\overset{\delta+}{CH_2}-MgBr \longrightarrow CH_2=CH_2 + MgBr_2$

15.14
(a) Allyl bromide decolorizes Br_2/CCl_4 solution; n-propyl bromide does not.

(b) Benzyl bromide gives a AgBr precipitate with $AgNO_3$ in alcohol; p-bromotoluene does not.

(c) Benzyl chloride gives an AgCl precipitate with $AgNO_3$ in alcohol; or vinyl chloride decolorizes Br_2/CCl_4 solution.

(d) Phenyllithium (a small amount) reacts vigorously with water to give benzene and a strongly basic aqueous solution (LiOH). Diphenylmercury does not react in this way.

(e) Aldrin decolorizes Br_2/CCl_4 solution rapidly. (The double bond of chlordan and the corresponding double bond of aldrin are not reactive.)

15.15

15.16
An elimination reaction.

16 ALCOHOLS, PHENOLS, AND ETHERS

16.1

(a) Alcohols:

$CH_3CH_2CH_2CH_2CH_2OH$ 1-pentanol

$CH_3CH_2CH_2\underset{\underset{OH}{|}}{C}HCH_3$ 2-pentanol

$CH_3CH_2\underset{\underset{OH}{|}}{C}HCH_2CH_3$ 3-pentanol

$CH_3CH_2\overset{\overset{CH_3}{|}}{C}HCH_2OH$ 2-methyl-1-butanol

$CH_3\overset{\overset{CH_3}{|}}{C}HCH_2CH_2OH$ 3-methyl-1-butanol

$CH_3CH_2\overset{\overset{CH_3}{|}}{\underset{\underset{OH}{|}}{C}}CH_3$ 2-methyl-2-butanol

$CH_3\overset{\overset{CH_3}{|}}{C}H\underset{\underset{OH}{|}}{C}HCH_3$ 3-methyl-2-butanol

$CH_3-\overset{\overset{CH_3}{|}}{\underset{\underset{CH_3}{|}}{C}}-CH_2OH$ 2,2-dimethyl-1-propanol (neopentyl alcohol)

Ethers:

$CH_3CH_2CH_2CH_2-O-CH_3$ methyl *n*-butyl ether (1-methoxybutane)

$CH_3\overset{\overset{CH_3}{|}}{C}HCH_2-O-CH_3$ methyl isobutyl ether (2-methyl-1-methoxypropane)

$CH_3-\overset{\overset{CH_3}{|}}{\underset{\underset{CH_3}{|}}{C}}-O-CH_3$ methyl *tert*-butyl ether (2-methyl-2-methoxypropane)

$$\underset{\underset{CH_3}{|}}{CH_3CH_2CH}-O-CH_3$$

methyl *sec*-butyl ether
(2-methoxybutane)

$$CH_3CH_2CH_2-O-CH_2CH_3$$

ethyl *n*-propyl ether
(1-ethoxypropane)

$$\underset{\underset{CH_3}{|}}{CH_3CH}-O-CH_2CH_3$$

ethyl isopropyl ether
(2-ethoxypropane)

(b) Alcohols: $CH_3CH_2CH=CHCH_2OH$

2-penten-1-ol

$CH_3CH=CHCH_2CH_2OH$

3-penten-1-ol

$CH_2=CHCH_2CH_2CH_2OH$

4-penten-1-ol

$$\underset{\underset{CH_3}{|}}{CH_3CH=CHCHOH}$$

3-penten-2-ol

$$\underset{\underset{CH_3}{|}}{CH_2=CHCH_2CHOH}$$

4-penten-2-ol

$$\underset{\underset{OH}{|}}{CH_2=CHCHCH_2CH_2}$$

1-penten-3-ol

$$\underset{\overset{\overset{CH_2}{\|}}{}}{CH_3CH_2CCH_2OH}$$

2-ethyl-2-propen-1-ol

$$\underset{\overset{\overset{CH_2}{\|}}{}}{CH_3CCH_2CH_2OH}$$

3-methyl-3-buten-1-ol

$$\underset{\underset{OH}{|}}{\overset{\overset{CH_2}{\|}}{CH_3CCHCH_3}}$$

3-methyl-3-buten-2-ol

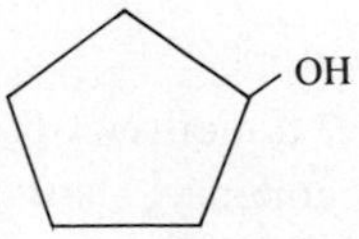

cyclopentanol

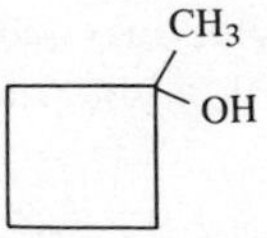

1-methylcyclobutanol

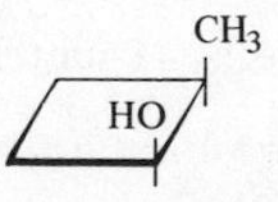

cis-2-methylcyclobutanol

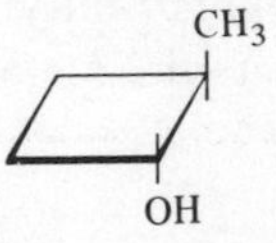

trans-2-methylcyclobutanol

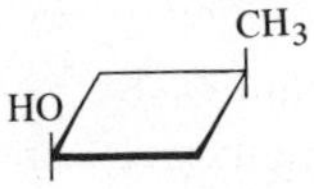

cis-3-methylcyclobutanol

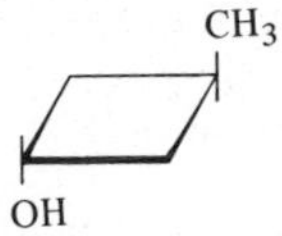

trans-3-methylcyclobutanol

There are also six isomers containing cyclopropane rings.

Ethers:		
$CH_2{=}CHCH_2CH_2{-}O{-}CH_3$	4-methoxy-1-butene	

$$CH_2{=}CH{-}\underset{\underset{CH_3}{|}}{CH}{-}O{-}CH_3 \qquad \text{3-methoxy-1-butene}$$

$$CH_2{=}\underset{\underset{CH_2CH_3}{|}}{C}{-}OCH_3 \qquad \text{2-methoxy-1-butene}$$

$CH_3{-}O{-}CH{=}CHCH_2CH_2$ 1-methoxy-1-butene

$CH_3CH{=}CHCH_2{-}O{-}CH_3$ 1-methoxy-2-butene

$$CH_3CH{=}\underset{\underset{CH_3}{|}}{C}{-}OCH_3 \qquad \text{2-methoxy 2-butene}$$

$CH_2{=}CHCH_2{-}O{-}CH_2CH_3$ 3-ethoxypropene

$$CH_2{=}\underset{\underset{CH_3}{|}}{C}{-}O{-}CH_2CH_3 \qquad \text{2-ethoxypropene}$$

$CH_3CH_2{-}O{-}CH{=}CHCH_2$ 1-ethoxypropene

$CH_3CH_2CH_2{-}O{-}CH{=}CH_2$ *n*-propyl vinyl ether

$$\underset{\underset{}{}}{CH_3}\\ CH_3\overset{\overset{CH_3}{|}}{C}H{-}O{-}CH{=}CH_2 \qquad \text{isopropyl vinyl ether}$$

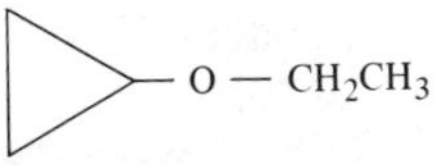

$O - CH_2CH_3$ ethoxycyclopropane

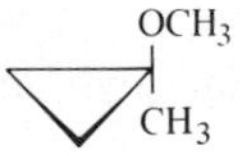

1-methoxy-1-methylcyclopropane

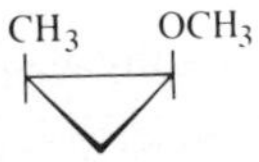

cis-1-methoxy-2-methylcyclopropane

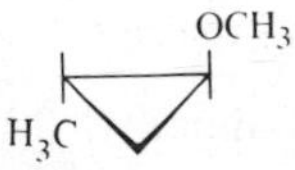

trans-1-methoxy-2-methylcyclopropane

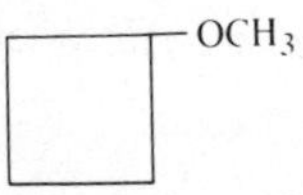

methoxycyclobutane

(c) Alcohols: $CH_3CH_2CH_2CH_2CH_2CH_2OH$ 1-hexanol

$$CH_3CH_2CH_2CH_2\overset{\overset{\displaystyle OH}{|}}{C}HCH_3$$ 2-hexanol

$$CH_3CH_2CH_2\overset{\overset{\displaystyle OH}{|}}{C}HCH_2CH_3$$ 3-hexanol

$$CH_3CH_2CH_2\underset{\underset{\displaystyle CH_3}{|}}{C}HCH_2OH$$ 2-methyl-1-pentanol

$$CH_3CH_2\underset{\underset{\displaystyle CH_3}{|}}{C}HCH_2CH_2OH$$ 3-methyl-1-pentanol

$$CH_3\underset{\underset{\displaystyle CH_3}{|}}{C}HCH_2CH_2CH_2OH$$ 4-methyl-1-pentanol

$$CH_3CH_2CH_2-\overset{\overset{\displaystyle OH}{|}}{\underset{\underset{\displaystyle CH_3}{|}}{C}}-CH_3$$ 2-methyl-2-pentanol

$$CH_3CH_2\underset{\underset{\displaystyle CH_3}{|}}{C}H-\overset{\overset{\displaystyle }{}}{\underset{\underset{\displaystyle OH}{|}}{C}}HCH_3$$ 3-methyl-2-pentanol

$$CH_3\underset{\underset{\displaystyle CH_3}{|}}{C}HCH_2\underset{\underset{\displaystyle OH}{|}}{C}HCH_3$$ 4-methyl-2-pentanol

$$CH_3CH_2\underset{\underset{\displaystyle OH}{|}}{C}H-\underset{\underset{\displaystyle CH_3}{|}}{C}HCH_3$$ 2-methyl-3-pentanol

$$CH_3CH_2-\overset{\overset{\displaystyle OH}{|}}{\underset{\underset{\displaystyle CH_3}{|}}{C}}-CH_2CH_3$$ 3-methyl-3-pentanol

$$CH_3CH_2\overset{\overset{\displaystyle CH_3}{|}}{\underset{\underset{\displaystyle CH_3}{|}}{C}}CH_2OH$$ 2,2-dimethyl-1-butanol

$$CH_3\overset{\overset{\displaystyle CH_3}{|}}{C}H-\overset{\overset{\displaystyle CH_3}{|}}{C}HCH_2OH$$ 2,3-dimethyl-1-butanol

$$CH_3\overset{\overset{\displaystyle CH_3}{|}}{\underset{\underset{\displaystyle CH_3}{|}}{C}}CH_2CH_2OH$$ 3,3-dimethyl-1-butanol

$$CH_3\overset{\overset{\displaystyle CH_3}{|}}{C}H-\overset{\overset{\displaystyle CH_3}{|}}{\underset{\underset{\displaystyle OH}{|}}{C}}-CH_3$$ 2,3-dimethyl- 2-butanol

$$
\begin{array}{c}
\quad\quad\ \ CH_3 \\
\quad\quad\ \ | \\
CH_3-C-CH-CH_3 \\
\quad\quad\ \ | \quad\ \ | \\
\quad\quad\ \ CH_3 \ OH
\end{array}
\qquad \text{3,3-dimethyl-2-butanol}
$$

Ethers:

$CH_3CH_2CH_2CH_2CH_2-O-CH_3$ 1-methoxypentane

$$
\begin{array}{c}
\quad\quad\quad\ \ CH_3 \\
\quad\quad\quad\ \ | \\
CH_3CH_2CH_2CHOCH_3
\end{array}
\qquad \text{2-methoxypentane}
$$

$$
\begin{array}{c}
\quad\quad\ OCH_3 \\
\quad\quad\ | \\
CH_3CH_2CHCH_2CH_3
\end{array}
\qquad \text{3-methoxypentane}
$$

$CH_3CH_2CH_2CH_2-O-CH_2CH_3$ 1-ethoxybutane

$$
\begin{array}{c}
\quad\quad\ \ CH_3 \\
\quad\quad\ \ | \\
CH_3CH_2CH-O-CH_2CH_3
\end{array}
\qquad \text{2-ethoxybutane}
$$

$$
\begin{array}{c}
CH_3CH_2CHCH_2-O-CH_3 \\
\quad\quad\ \ | \\
\quad\quad\ \ CH_3
\end{array}
\qquad \text{1-methoxy-2-methylbutane}
$$

$$
\begin{array}{c}
\quad\quad\ \ CH_3 \\
\quad\quad\ \ | \\
CH_3CH_2C-O-CH_3 \\
\quad\quad\ \ | \\
\quad\quad\ \ CH_3
\end{array}
\qquad \text{2-methoxy-2-methylbutane}
$$

$$
\begin{array}{c}
CH_3 \ OCH_3 \\
| \quad\ \ | \\
CH_3CH-CHCH_3
\end{array}
\qquad \text{2-methoxy-3-methylbutane}
$$

$$
\begin{array}{c}
CH_3CHCH_2CH_2-O-CH_3 \\
\quad | \\
\quad CH_3
\end{array}
\qquad \text{1-methoxy-3-methylbutane}
$$

$$
\begin{array}{c}
CH_3CHCH_2-O-CH_2CH_3 \\
\quad | \\
\quad CH_3
\end{array}
\qquad \text{1-ethoxy-2-methylpropane}
$$

$$
\begin{array}{c}
\quad CH_3 \\
\quad | \\
CH_3C-O-CH_2CH_3 \\
\quad | \\
\quad CH_3
\end{array}
\qquad \text{2-ethoxy-2-methylpropane}
$$

$CH_3CH_2CH_2-O-CH_2CH_2CH_3$ di-*n*-propyl ether

$$
\begin{array}{c}
\quad\quad\quad\quad\ \ CH_3 \\
\quad\quad\quad\quad\ \ | \\
CH_3CH_2CH_2-O-CHCH_2
\end{array}
\qquad \textit{n}\text{-propyl isopropyl ether}
$$

$$
\begin{array}{c}
CH_3 \quad\ CH_3 \\
| \quad\quad\ | \\
CH_3CH-O-CHCH_3
\end{array}
\qquad \text{diisopropyl ether}
$$

16.2

The two hydroxyl groups in ethylene glycol allow the formation of more hydrogen bonds than in the monohydroxy alcohols. Thus a single diol molecule can be associated with many neighboring diol molecules.

16.3

(a) $CH_3\underset{\underset{}{|}}{C}=CH_2 + H_2O \xrightarrow{H^+} CH_3\underset{\underset{OH}{|}}{\overset{\overset{CH_3}{|}}{C}}CH_3$

(b) $CH_3CH_2CH_2CH_2CH=CH_2 + H_2O \xrightarrow{H^+} CH_3CH_2CH_2CH_2\underset{\underset{OH}{|}}{C}HCH_3$

(c) $+ H_2O \xrightarrow{H^+}$

(d) $+ H_2O \xrightarrow{H^+}$

16.4

Rearrangement of the secondary carbocation to the more stable tertiary carbocation,

$$CH_3\underset{\underset{CH_3}{|}}{\overset{\overset{CH_3}{|}}{C}}-CH=CH_2 \underset{}{\overset{H^+}{\rightleftharpoons}} CH_3-\underset{\underset{CH_3}{|}}{\overset{\overset{CH_3}{|}}{\overset{+}{C}}}-CH-CH_3 \longrightarrow CH_3-\underset{\underset{CH_3}{|}}{\overset{\overset{CH_3}{|}}{C}}-\overset{+}{C}H-CH_3$$

followed by reaction of the resulting carbocation with water:

$$CH_3-\underset{\underset{CH_3}{|}}{\overset{\overset{CH_3}{|}}{\overset{+}{C}}}-CHCH_3 + H_2O \rightleftharpoons CH_3-\underset{\underset{CH_3}{|}}{\overset{\overset{+OH_2}{|}}{C}}-\underset{\underset{CH_3}{|}}{C}HCH_3 \rightleftharpoons CH_3-\underset{\underset{CH_3}{|}}{\overset{\overset{OH}{|}}{C}}-\underset{\underset{CH_3}{|}}{C}HCH_3$$

$$+ H^+$$

16.5

(a) $CH_3CH_2CH_2CH_2CH=CH_2 \xrightarrow[\text{THF-H}_2\text{O}]{\text{Hg(OAc)}_2} CH_3CH_2CH_2CH_2\underset{\underset{OH}{|}}{C}HCH_2-HgOAc$

$\xrightarrow[OH^-]{NaBH_4} CH_3CH_2CH_2CH_2\underset{\underset{OH}{|}}{C}HCH_3$

(b) $\xrightarrow[\text{(2) NaBH}_4, \text{OH}^-]{\text{(1) Hg(OAc)}_2, \text{THF}-\text{H}_2\text{O}}$

(c) $CH_3\underset{\underset{CH_3}{|}}{\overset{\overset{CH_3}{|}}{C}}CH_2\underset{}{\overset{\overset{CH_3}{|}}{C}}=CH_2 \xrightarrow{\text{same as (b)}} CH_3\underset{\underset{CH_3}{|}}{\overset{\overset{CH_3}{|}}{C}}CH_2\underset{\underset{OH}{|}}{\overset{\overset{CH_3}{|}}{C}}CH_3$

(d) CH₃C=CH₂ $\xrightarrow{\text{same as (b)}}$ (benzene ring) → CH₃CHCH₃ with OH

16.6

(a) $CH_3-\underset{\underset{CH_3}{|}}{\overset{\overset{CH_3}{|}}{C}}-CH=CH_2$ + (BH₃)₂ $\longrightarrow$ $(CH_3-\underset{\underset{CH_3}{|}}{\overset{\overset{CH_3}{|}}{C}}-CH_2CH_2)_3B$

$\xrightarrow[\text{OH}^-,\text{H}_2\text{O}]{\text{H}_2\text{O}_2}$ $CH_3-\underset{\underset{CH_3}{|}}{\overset{\overset{CH_3}{|}}{C}}-CH_2CH_2OH$

(b) $CH_3CH_2CH_2CH_2CH=CH_2 \xrightarrow[\text{(2)H}_2\text{O}_2,\ \text{OH}^-,\ \text{H}_2\text{O}]{\text{(1)(BH}_3)_2}$ $CH_3CH_2CH_2CH_2CH_2CH_2OH$

(c) (benzene ring)–CH=CH₂ $\xrightarrow{\text{same as (b)}}$ (benzene ring)–CH₂CH₂OH

(d) $\xrightarrow{\text{same as (b)}}$ + enantiomer

16.7

(a) LiAlH₄   (b) NaBH₄   (c) LiAlH₄

16.8

(a) (1)CH_3MgBr + $CH_3\overset{\overset{O}{\|}}{C}CH_3$ $\xrightarrow{\text{(2)H}_3\text{O}^+}$ $CH_3-\underset{\underset{CH_3}{|}}{\overset{\overset{CH_3}{|}}{C}}-OH$

(2) $2CH_3MgBr$ $CH_3\overset{\overset{O}{\|}}{C}-OC_2H_5$ $\xrightarrow{\text{(2)H}_3\text{O}^+}$ $CH_3-\underset{\underset{CH_3}{|}}{\overset{\overset{CH_3}{|}}{C}}-OH$

(b) (1)CH_3MgBr + $CH_3CH_2CH_2\overset{\overset{O}{\|}}{CH}$ $\xrightarrow{\text{(2)H}_3\text{O}^+}$ $CH_3CH_2CH_2\overset{\overset{OH}{|}}{C}HCH_3$

(2)$CH_3CH_2CH_2MgBr$ + $CH_3\overset{\overset{O}{\|}}{CH}$ $\xrightarrow{\text{(2)H}_3\text{O}^+}$ $CH_3CH_2CH_2\overset{\overset{OH}{|}}{C}HCH_3$

(c) (1) $C_6H_5MgBr + CH_3\overset{O}{\overset{\|}{C}}CH_2CH_3 \xrightarrow{(2)H_3O^+} C_6H_5\overset{CH_3}{\underset{OH}{\overset{|}{C}}}CH_2CH_3$

(2) $CH_3MgBr + C_6H_5\overset{O}{\overset{\|}{C}}CH_2CH_3 \xrightarrow{(2)H_3O^+} C_6H_5\overset{CH_3}{\underset{OH}{\overset{|}{C}}}CH_2CH_3$

(3) $CH_3CH_2MgBr + C_6H_5\overset{O}{\overset{\|}{C}}CH_3 \xrightarrow{(2)H_3O^+} C_6H_5\overset{CH_3}{\underset{OH}{\overset{|}{C}}}CH_2CH_3$

(d) (1) $CH_3CH_2CH_2CH_2MgBr + CH_2\overset{O}{-}CH_2 \xrightarrow{(2)H_3O^+} CH_3CH_2CH_2CH_2CH_2CH_2OH$

(2) $CH_3CH_2CH_2CH_2CH_2MgBr + CH_2O \xrightarrow{(2)H_3O^+} CH_3CH_2CH_2CH_2CH_2CH_2OH$

16.9

(a) $CH_3C\equiv\overset{-}{C}: + CH_3CH_2OH \rightleftharpoons CH_3C\equiv CH + CH_3CH_2\overset{-}{O}$

    stronger   stronger   weaker   weaker

    base   acid   acid   base

(b) $CH_3CH_2CH_2\overset{\delta-}{CH_2}: \overset{\delta+}{Li} + CH_3CH_2OH \rightleftharpoons CH_3CH_2CH_2CH_3 + CH_3CH_2\overset{-}{O}\overset{+}{Li}$

      stronger   stronger   weaker   weaker

      base   acid   acid   base

(c) $CH_3\overset{\delta-}{CH_2}: \overset{\delta+}{Mg}Br + CH_3CH_2OH \rightleftharpoons CH_3CH_3 + CH_3CH_2\overset{-}{O}\overset{+}{Mg}Br$

    stronger   stronger   weaker   weaker

    base   acid   acid   base

16.10

(a)

(b)

(c) 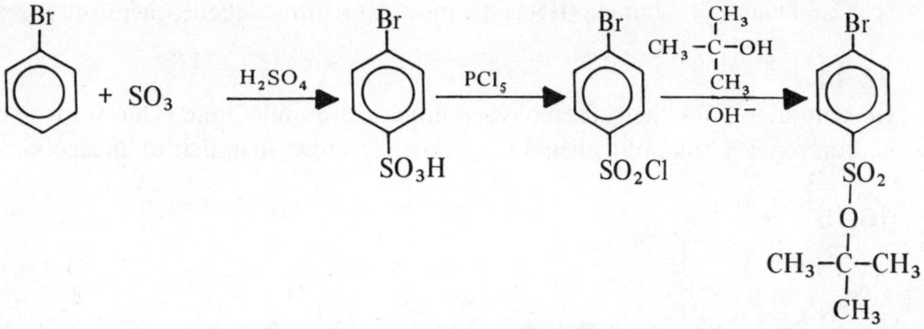 $\quad$ Br–(ring) $+ SO_3 \xrightarrow{H_2SO_4}$ Br–(ring)–SO_3H $\xrightarrow{PCl_5}$ Br–(ring)–SO_2Cl $\xrightarrow[OH^-]{CH_3-\overset{CH_3}{\underset{CH_3}{C}}-OH}$ Br–(ring)–SO_2–O–$\overset{CH_3}{\underset{CH_3}{C}}$–$CH_3$

16.11

(a) $OH^- + CH_3O-\overset{O}{\underset{O}{\overset{\|}{\underset{\|}{S}}}}-CH_3 \longrightarrow CH_3OH + CH_3-\overset{O}{\underset{O}{\overset{\|}{\underset{\|}{S}}}}-O^-$

(b) $OH^- + CH_3-O-\overset{O}{\underset{O}{\overset{\|}{\underset{\|}{S}}}}-O-CH_3 \longrightarrow CH_3OH + {}^-O-SO_2-OCH_3$

$\quad OH^- + CH_3O-SO_2O^- \longrightarrow CH_3OH + {}^-O-SO_2-O^-$

16.12

(a) *Trans*-2-pentene because it is more stable.

(b) A 1-phenylpropene is conjugated and thus is more stable than the unconjugated 3-phenylpropene, and *trans*-3-phenylpropene is more stable than the *cis* isomer.

16.13

(a) 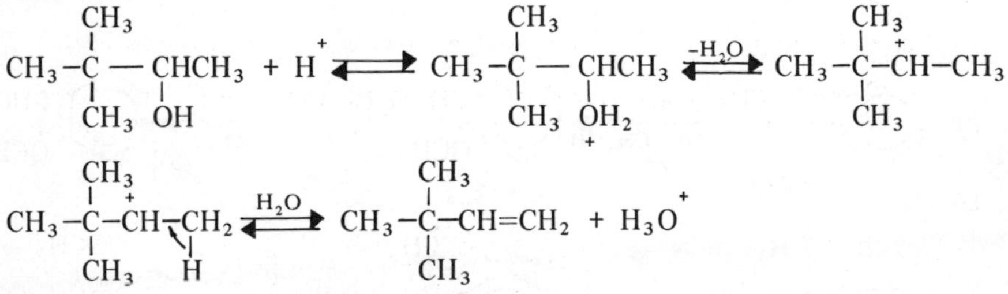 $CH_3-\overset{CH_3}{\underset{CH_3}{C}}-\overset{OH}{\underset{}{CHCH_3}} + H^+ \rightleftharpoons CH_3-\overset{CH_3}{\underset{CH_3}{C}}-\overset{\overset{+}{OH_2}}{\underset{}{CHCH_3}} \overset{-H_2O}{\rightleftharpoons} CH_3-\overset{CH_3}{\underset{CH_3}{C}}-\overset{+}{CH}-CH_3$

$CH_3-\overset{CH_3}{\underset{CH_3}{C}}-\overset{+}{C}H-CH_2-H \overset{H_2O}{\rightleftharpoons} CH_3-\overset{CH_3}{\underset{CH_3}{C}}-CH=CH_2 + H_3O^+$

I

But this secondary carbocation can also rearrange to a tertiary carbocation before losing a proton:

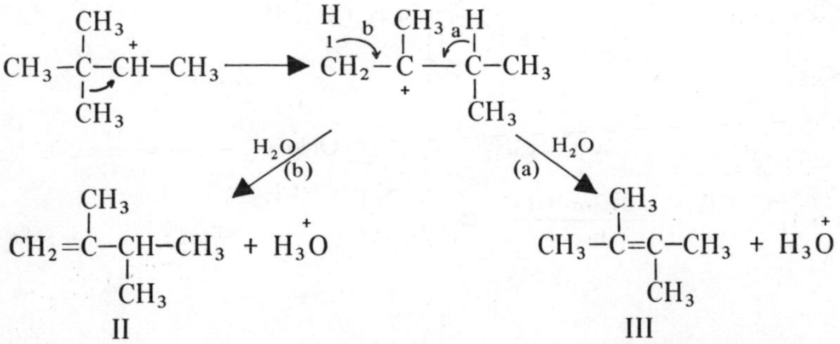

$CH_3-\overset{CH_3}{\underset{CH_3}{C}}-\overset{+}{C}H-CH_3 \longrightarrow \overset{+}{C}H_2-\overset{CH_3}{\underset{CH_3}{C}}-\overset{H}{\underset{}{C}}-CH_3$

$\xrightarrow[(b)]{H_2O}$ $CH_2=\overset{CH_3}{\underset{CH_3}{C}}-CH-CH_3 + H_3\overset{+}{O}$

II

$\xrightarrow[(a)]{H_2O}$ $CH_3-\overset{CH_3}{\underset{CH_3}{C}}=C-CH_3 + H_3\overset{+}{O}$

III

(b) 2,3-Dimethyl-2-butene (III) is the most substituted alkene, therefore it is most stable.

16.14
In structures 2 - 4 the carbon-oxygen bond is a double bond. Thus we would expect the carbon-oxygen bond of a phenol to be much stronger than that of an alcohol.

16.15

16.16

(a) (1)

$$\text{C}_6\text{H}_5\text{—CH=CH}_2 + \text{H}_2\text{O} \xrightarrow{\text{H}^+}$$

(2)

$$\xrightarrow[\text{(2) NaBH}_4, \text{OH}^-]{\text{(1) Hg(OAc)}_2, \text{THF}, \text{H}_2\text{O}}$$

(b)

$$\text{—CH=CH}_2 \xrightarrow{\text{(BH}_3)_2} \left[\text{—CH}_2\text{CH}_2\text{—B} \right]_3 \xrightarrow[\text{OH}^-, \text{H}_2\text{O}]{\text{H}_2\text{O}_2} \text{—CH}_2\text{CH}_2\text{OH}$$

(c)

$$\text{—CH=CH}_2 \xrightarrow[\text{THF—CH}_3\text{OH}]{\text{Hg(OAc)}_2} \underset{\text{OCH}_3}{\text{—CHCH}_2\text{HgOAc}} \xrightarrow[\text{OH}^-]{\text{NaBH}_4} \underset{\text{OCH}_3}{\text{—CHCH}_3}$$

(d)

$$\underset{\text{CH}_3}{\text{—CH—OH}} \xrightarrow{\text{Na}} \underset{\text{CH}_3}{\text{—CH—O}^-} \xrightarrow{\text{CH}_3\text{CH}_2\text{Br}} \underset{\text{CH}_3}{\text{—CH—O—CH}_2\text{CH}_3}$$

(e)

$$\text{—CH}_2\text{COOH} \xrightarrow[\text{ether}]{\text{LiAlH}_4} \text{—CH}_2\text{CH}_2\text{OH}$$

(f)

$$\underset{}{\overset{\text{O}}{\text{—C—CH}_3}} \xrightarrow[\text{ether}]{\text{LiAlH}_4 \text{ or NaBH}_4} \underset{\text{OH}}{\text{—CHCH}_3}$$

(g) $C_6H_5CH_3 \xrightarrow[\substack{CCl_4, \\ light}]{Br_2} C_6H_5CH_2Br \xrightarrow[ether]{Mg} C_6H_5CH_2MgBr$

$C_6H_5CH_2CH_2OH \xleftarrow[(2)\,H_3O^+]{(1)\,CH_2O}$

(h) $C_6H_6 \xrightarrow[FeBr_3]{Br_2} C_6H_5Br \xrightarrow[ether]{Mg} C_6H_5MgBr \xrightarrow[(2)\,H_3O^+]{(1)\,CH_2\overset{O}{\frown}CH_2} C_6H_5CH_2CH_2OH$

(i) $C_6H_5CH_2CO_2CH_3 \xrightarrow[ether]{LiAlH_4} C_6H_5CH_2CH_2OH$

16.17

(a) $CH_3CH_2CH_2CH_2OH \xrightarrow{PBr_3} CH_3CH_2CH_2CH_2Br$

$\xrightarrow{(CH_3)_3COK} CH_3CH_2CH=CH_2$

(b) $CH_3CH_2CH=CH_2 \xrightarrow[(2)\,NaBH_4,\,OH^-]{(1)\,Hg(OAc)_2,\,THF\text{-}H_2O} CH_3CH_2\overset{OH}{\underset{|}{C}}HCH_3$

(c) $CH_3CH_2\overset{OH}{\underset{|}{C}}HCH_3 \xrightarrow[H_2SO_4]{CrO_3} CH_3CH_2\overset{O}{\overset{||}{C}}CH_3$

(d) $CH_3CH_2CH_2CH_2OH \xrightarrow{PBr_3} CH_3CH_2CH_2CH_2Br$

(e) $CH_3CH_2CH=CH_2 + HBr \xrightarrow[(no\ peroxides)]{} CH_3CH_2\overset{Br}{\underset{|}{C}}HCH_3$

(f) $CH_3CH_2CH_2CH_2Br \xrightarrow[ether]{Mg} CH_3CH_2CH_2CH_2MgBr$

$\xrightarrow[(2)\,H_3O^+]{(1)\,CH_2O} CH_3CH_2CH_2CH_2CH_2OH$

(g) $CH_3CH_2CH_2CH_2MgBr \xrightarrow[(2)\,H_3O^+]{(1)\,CH_2\overset{O}{\frown}CH_2} CH_3CH_2CH_2CH_2CH_2CH_2OH$

$\xrightarrow{PBr_3} CH_3CH_2CH_2CH_2CH_2CH_2Br \xrightarrow{(CH_3)_3COK} CH_3CH_2CH_2CH_2CH=CH_2$

(h) $CH_3CH_2CH_2CH_2MgBr + CH_3\overset{O}{\overset{||}{C}}CH_2CH_3 \xrightarrow{(2)\,H_3O^+} CH_3CH_2CH_2CH_2\overset{OH}{\underset{\underset{CH_3}{|}}{C}}CH_2CH_3$

(i) $CH_3CH_2CH_2CH_2OH \xrightarrow{CrO_3 \cdot 2C_5H_5N} CH_3CH_2CH_2\overset{\displaystyle O}{\overset{\|}{C}}H$

(j) $CH_3CH_2CH_2CH_2MgBr + CH_3CH_2CH_2\overset{\displaystyle O}{\overset{\|}{C}}H \xrightarrow{(2)\ H_3O^+}$

$$CH_3CH_2CH_2CH_2\overset{\displaystyle OH}{\overset{|}{C}}HCH_2CH_2CH_3$$

(k) $CH_3CH_2\overset{\displaystyle Br}{\overset{|}{C}}HCH_3 \xrightarrow[\text{ether}]{Mg} CH_3CH_2\overset{\displaystyle CH_3}{\overset{|}{C}}HMgBr$

$\xrightarrow[(2)H_3O^+]{(1)CH_3CH_2CH_2\overset{\displaystyle O}{\overset{\|}{C}}H} CH_3CH_2\overset{\displaystyle CH_3}{\overset{|}{C}}H-\overset{\displaystyle OH}{\overset{|}{C}}HCH_2CH_2CH_3$

(l) $CH_3CH_2CH_2CH_2MgBr + CO_2 \xrightarrow{(2)H_3O^+} CH_3CH_2CH_2CH_2COOH$

(m) (1)$CH_3CH_2\overset{\displaystyle CH_3}{\overset{|}{C}}HOH \xrightarrow{Na} CH_3CH_2\overset{\displaystyle CH_3}{\overset{|}{C}}HONa$

$\xrightarrow{CH_3CH_2CH_2CH_2Br} CH_3CH_2\overset{\displaystyle CH_3}{\overset{|}{C}}H-O-CH_2CH_2CH_2CH_3$

(2)$CH_3CH_2CH_2CH_2OH \xrightarrow{Na} CH_3CH_2CH_2CH_2ONa \xrightarrow{CH_3CH_2\overset{\displaystyle CH_3}{\overset{|}{C}}HBr}$

$CH_3CH_2CH_2CH_2O\overset{\displaystyle CH_3}{\overset{|}{C}}HCH_2CH_3$ (This method would give a lower yield than that of (1) or (3) because of competing elimination.)

(3)$CH_3CH_2CH=CH_2 + Hg(OAc)_2 \xrightarrow[CH_3CH_2CH_2CH_2OH]{THF}$

$CH_3CH_2\overset{\displaystyle}{C}H-CH_2-HgOAc$
$\underset{\displaystyle OCH_2CH_2CH_2CH_3}{|}$

$CH_3CH_2\overset{\displaystyle CH_3}{\overset{|}{C}}H-O-CH_2CH_2CH_2CH_3 \xleftarrow[OH^-]{NaBH_4}$

(n) (1)$2CH_3CH_2CH_2CH_2OH \xrightarrow[140°]{H_2SO_4} (CH_3CH_2CH_2CH_2)_2O$

(2)$CH_3CH_2CH_2CH_2OH + Na \longrightarrow CH_3CH_2CH_2CH_2ONa \xrightarrow{CH_3CH_2CH_2CH_2Br}$

$(CH_3CH_2CH_2CH_2)_2O$

(o) $CH_3CH_2CH_2CH_2Br + 2Li \longrightarrow CH_3CH_2CH_2CH_2Li + LiBr$

(p) $CH_3CH_2CH_2CH_2Li \xrightarrow{CuI} (CH_3CH_2CH_2CH_2)_2CuLi \xrightarrow{CH_3CH_2CH_2CH_2Br}$

$CH_3CH_2CH_2CH_2CH_2CH_2CH_2CH_3$

16.18

(a) $CH_3CH_2CH_2O^- Na^+$ Sodium propoxide

(b) $CH_3CH_2CH_2-O-CH_2CH_2CH_2CH_3$ propyl butyl ether

(c) $CH_3-\overset{\overset{O}{\|}}{\underset{\underset{O}{\|}}{S}}-OCH_2CH_2CH_3$ propyl methanesulfonate

(d) $CH_3-\langle\bigcirc\rangle-SO_2-O-CH_2CH_2CH_3$ propyl p-toluenesulfonate (or propyl tosylate)

(e) $CH_3\overset{\overset{O}{\|}}{C}-O-CH_2CH_2CH_3$ propyl acetate

(f) $CH_3CH_2\overset{\overset{O}{\|}}{C}-O^- K^+$ potassium propanoate

(g) $CH_3CH_2CH_2Cl$ 1-chloropropane

(h) $CH_3CH_2CH_2Cl$ 1-chloropropane

(i) $CH_3CH_2CH_2-O-CH_2CH_2CH_3$ dipropyl ether

(j) $CH_3CH_2CH_2Br$ 1-bromopropane

(k) $CH_3-\overset{\overset{CH_3}{|}}{\underset{\underset{CH_3}{|}}{C}}-O-CH_2CH_2CH_3$ propyl $tert$-butyl ether

(l) $\langle\bigcirc\rangle-CH\overset{CH_3}{\underset{CH_3}{}}$ isopropylbenzene (major) + $\langle\bigcirc\rangle-CH_2CH_2CH_3$ propylbenzene

16.19

(a) $CH_3\overset{\overset{CH_3}{|}}{C}HO^- Na^+$ sodium isopropoxide

(b) $CH_3\overset{\overset{CH_3}{|}}{C}H-O-CH_2CH_2CH_2CH_3$ isopropyl butyl ether

(c) $CH_3SO_2-O-\overset{\overset{CH_3}{|}}{C}HCH_3$ isopropyl methanesulfonate

(d) $CH_3-\langle\bigcirc\rangle-SO_2-O-\overset{\overset{CH_3}{|}}{C}HCH_3$ isopropyl p-toluenesulfonate

(e) $CH_3\overset{\overset{O}{\|}}{C}-O-\overset{\overset{CH_3}{|}}{C}HCH_3$ isopropyl acetate

(f) $CH_3\overset{\overset{O}{\|}}{C}CH_3$ acetone (+ CH_3COOH and CO_2)

(g) $CH_3\overset{\overset{CH_3}{|}}{C}HCl$ 2-chloropropane

(h) Same as (g)

(i) $CH_3\overset{\overset{CH_3}{|}}{C}H-O-\overset{\overset{CH_3}{|}}{C}HCH_3$ diisopropyl ether

(j) $CH_3\overset{\overset{CH_3}{|}}{C}HBr$ 2-bromopropane

(k) $CH_3\underset{\underset{CH_3}{|}}{\overset{\overset{CH_3}{|}}{C}}-O-\overset{\overset{CH_3}{|}}{C}HCH_3$ isopropyl *tert*-butyl ether

(l) ⬡$-\overset{\overset{CH_3}{|}}{C}HCH_3$ isopropyl benzene

16.20

(a) ⬡$-ONa$ + CH_3CH_2OH (e) ⬡$-ONa$ + H_2O,

(b) ⬡ + CH_3CH_2OMgBr (f) ⬡$-OH$ + $NaCl$

(c) CH_3CH_2SNa + H_2O (g) CH_3CH_2OH + $NaOH$

(d) CH_3CH_2ONa + CH_3CH_2SH

16.21

(d), (e), (f). (All are stronger acids than H_2CO_3, see page 612.)

16.22

(a) CH_3Br + CH_3CH_2Br (c) $Br-CH_2CH_2CH_2CH_2-Br$

(b) ⬡$-OH$ + CH_3CH_2Br (d) $Br-CH_2CH_2-Br$ (2 moles)

16.23

(a) CH_3CH_2-⬡ + SO_3 $\xrightarrow{H_2SO_4}$ CH_3CH_2⬡$-SO_3H$ $\xrightarrow[(2)\,dry]{(1)\,NaOH}$

$$CH_3CH_2-\!\!\bigcirc\!\!-SO_3^-\overset{+}{N}a \xrightarrow[(2)H_3O^+]{(1)NaOH-KOH,330^\circ} CH_3CH_2-\!\!\bigcirc\!\!-OH$$

$$CH_3CH_2-\!\!\bigcirc\!\! \xrightarrow[\substack{CF_3COOH \\ 25^\circ}]{Tl(OOCCF_3)_3} CH_3CH_2-\!\!\bigcirc\!\!-Tl(OOCCF_3)_2 \xrightarrow[(2)\,HO^-]{(1)Pb(OOCCH_3)_4 \\ (C_6H_5)_3P}$$

$$CH_3CH_2-\!\!\bigcirc\!\!-OH$$

(b) Thallation at 75° to produce the meta isomer, followed by reaction with $Pb(OOCCH_3)_4 + (C_6H_5)_3P$, then OH^-.

16.24

(a) $CH_3\underset{OH}{CH}\underset{OH}{CH_2}$

(b) $CH_3CH_2-O-CH_2CH_2OH$

(c) $C_6H_5O-CH_2CH_2OH$

(d) $CH_3-\!\!\bigcirc\!\!-O-SO_2-\!\!\bigcirc\!\!-CH_3$

(e) tetrahydropyran with OOCCH_3

(f) phthalic acid phenyl ester; $\begin{array}{c}C-O-C_6H_5 \\ \| \\ O\end{array}$ and $\begin{array}{c}C-OH\\ \|\\ O\end{array}$

(g) 2,6-dibromo-4-methylphenol

(h) $C_6H_5-CH_2-\overset{+}{S}=C\begin{smallmatrix}NH_2\\NH_2\end{smallmatrix}\;Br^-$

(i) $C_6H_5-CH_2SH$

(j) $\bigcirc\!\!-CH_2-S-S-CH_2-\!\!\bigcirc$

(k) $\bigcirc\!\!-CH_2-S^-Na^+$

(l) $\bigcirc\!\!-CH_2-S-CH_2-\!\!\bigcirc$

(m) $\bigcirc\!\!-\overset{O}{\overset{\|}{C}}-O^-K^+$

(n) $\bigcirc\!\!-\overset{O}{\overset{\|}{C}}H$

(o) $\bigcirc\!\!-CH_2O^-Na^+$

(p) $\bigcirc\!\!-O-CH_3$

(q) epoxide with CH_3CH_2 and CH_2CH_3

(r) threo diol + enantiomer

16.25

(a) (1) CH_3CH_2MgBr +

$$\underset{\underset{CH_3}{|}}{\overset{\overset{CH_3}{|}}{C}}=O \xrightarrow{(2)\,H_3O^+} CH_3CH_2\underset{\underset{CH_3}{|}}{\overset{\overset{CH_3}{|}}{C}}-OH$$

(2) CH_3MgBr + $CH_3CH_2\underset{\underset{CH_3}{|}}{C}=O \xrightarrow{(2)\,H_3O^+} CH_3CH_2\underset{\underset{CH_3}{|}}{\overset{\overset{CH_3}{|}}{C}}-OH$

(3) $CH_3CH_2\overset{\overset{O}{||}}{C}-OCH_3$ + $2CH_3MgBr \xrightarrow{(2)\,H_3O^+} CH_3CH_2\underset{\underset{CH_3}{|}}{\overset{\overset{CH_3}{|}}{C}}-OH$

(b) (1) CH_3CH_2MgBr + C₆H₅–$\overset{\overset{O}{||}}{C}$–$CH_2CH_3 \xrightarrow{(2)H_3O^+}$ C₆H₅–$\underset{\underset{CH_2CH_3}{|}}{\overset{\overset{OH}{|}}{C}}$–$CH_2CH_3$

(2) C₆H₅–MgBr + $CH_3CH_2\overset{\overset{O}{||}}{C}CH_2CH_3 \xrightarrow{(2)\,H_3O^+}$ C₆H₅–$\underset{\underset{CH_2CH_3}{|}}{\overset{\overset{OH}{|}}{C}}$–$CH_2CH_3$

(3) C₆H₅–$\overset{\overset{O}{||}}{C}$–$OCH_3$ + $2CH_3CH_2MgBr \xrightarrow{(2)\,H_3O^+}$ C₆H₅–$\underset{\underset{CH_2CH_3}{|}}{\overset{\overset{OH}{|}}{C}}$–$CH_2CH_3$

(c) cyclohexanone (=O) + C_6H_5–MgBr $\xrightarrow{(2)\,H_3O^+}$ 1-phenylcyclohexanol (OH, C_6H_5)

(d) CH_3–C₆H₄–MgBr + CH_2–CH_2 (epoxide, O) $\xrightarrow{(2)\,H_3O^+}$ CH_3–C₆H₄–CH_2CH_2OH

(e) (1) CH_3–C₆H₄–$\overset{\overset{O}{||}}{C}H$ + $CH_3MgBr \xrightarrow{(2)\,H_3O^+}$ CH_3–C₆H₄–$\underset{\underset{}{}}{\overset{\overset{OH}{|}}{C}}HCH_3$

(2) CH_3–C₆H₄–MgBr + $CH_3\overset{\overset{O}{||}}{C}H \xrightarrow{(2)\,H_3O^+}$ CH_3–C₆H₄–$\overset{\overset{OH}{|}}{C}HCH_3$

16.26

(a) *p*-Cresol is soluble in aqueous NaOH; benzyl alcohol is not.

(b) Cyclohexanol is soluble in cold, concentrated H_2SO_4; cyclohexane is not. (Cyclohexanol also gives a positive test with CrO_3 in H_2SO_4, while cyclohexane does not.)

(c) Cyclohexene will decolorize Br_2/CCl_4 solution; cyclohexanol will not.

(d) Allyl propyl ether will decolorize Br_2/CCl_4 solution; dipropyl ether will not.

(e) *p*-Cresol is soluble in aqueous NaOH; anisole is not.

(f) Picric acid is soluble in aqueous $NaHCO_3$: 2,4,6-trimethylphenol is not (cf. Problem 16.21).

16.27

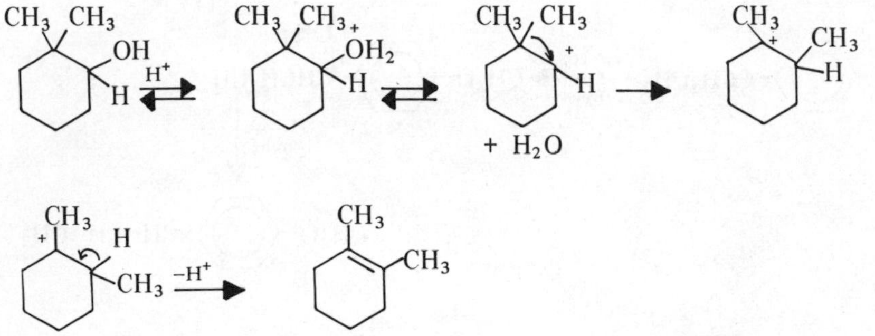

16.28

(a)

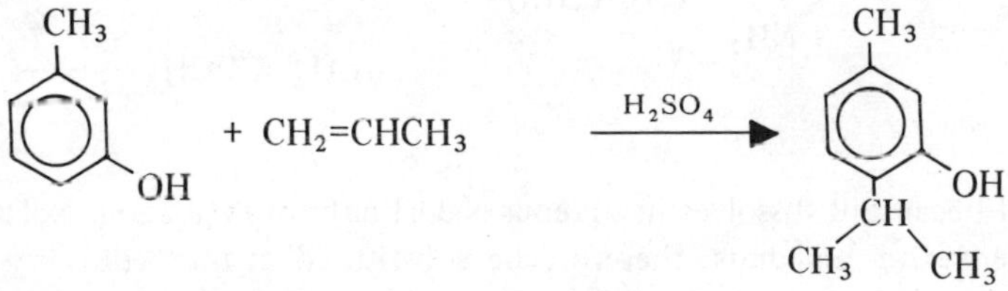

(b)

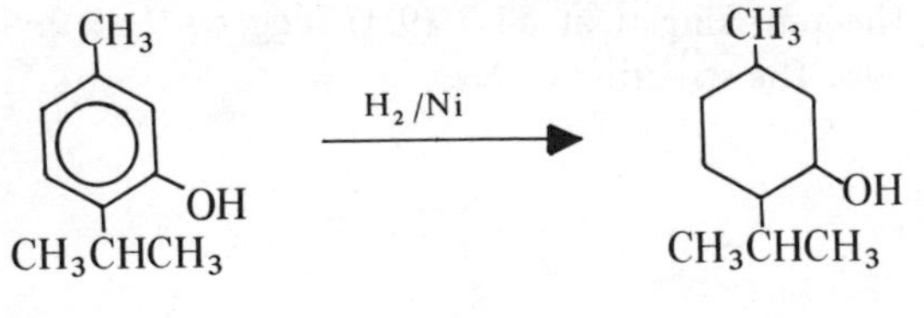

16.29
The position ortho to the isopropyl group is sterically more hindered than the position ortho to the methyl group:

16.30

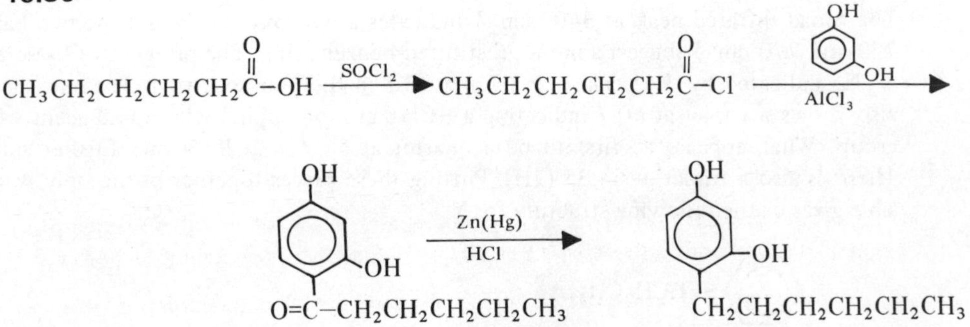

16.31

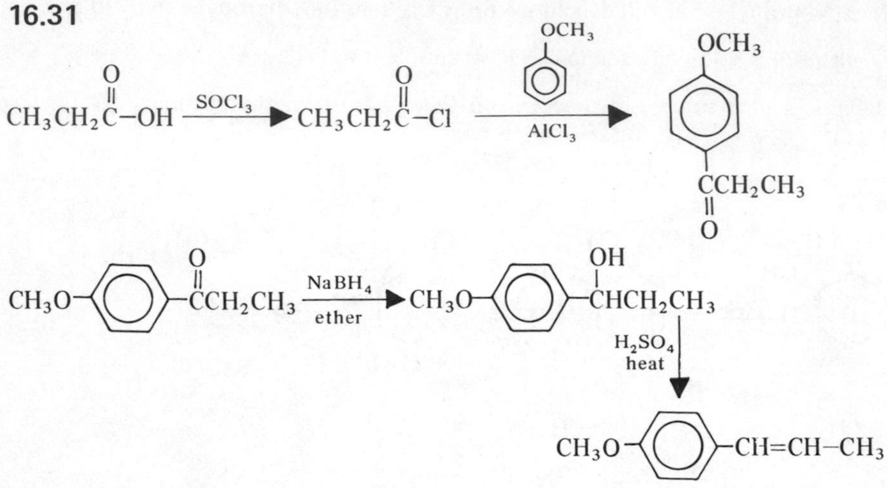

$$CH_3CH_2\overset{O}{\overset{\|}{C}}-OH \xrightarrow{SOCl_3} CH_3CH_2\overset{O}{\overset{\|}{C}}-Cl \xrightarrow[AlCl_3]{\text{OCH}_3}$$

$$CH_3O-\text{⟨○⟩}-\overset{O}{\overset{\|}{C}}CH_2CH_3 \xrightarrow[\text{ether}]{NaBH_4} CH_3O-\text{⟨○⟩}-\overset{OH}{\overset{|}{C}}HCH_2CH_3$$

$$\downarrow \overset{\text{H}_2\text{SO}_4}{\underset{\text{heat}}{}}$$

$$CH_3O-\text{⟨○⟩}-CH{=}CH{-}CH_3$$

16.32

$$CH_2{=}CHCH_2Br + S{=}C\overset{NH_2}{\underset{NH_2}{\diagup\diagdown}} \xrightarrow[(2)\,OH^-,H_2O]{(1)\,CH_3CH_2OH} CH_2{=}CHCH_2SH \xrightarrow{H_2O_2}$$

$$CH_2{=}CHCH_2-S-S-CH_2CH{=}CH_2$$

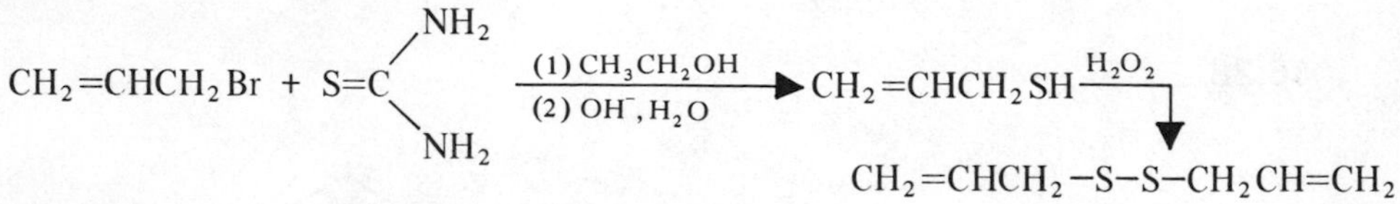

16.33

X is a phenol because it dissolves in aqueous NaOH but not in aqueous $NaHCO_3$. It gives a dibromo derivative, and must therefore be substituted in the ortho or para position. The broad infrared peak at 3250 cm^{-1} also suggests a phenol. The peak at 830 cm^{-1} indicates para substitution. The pmr singlet at $\delta 1.3$ (9H) suggests 9 methyl hydrogens which must be a *tert*-butyl group. The structure of X is:

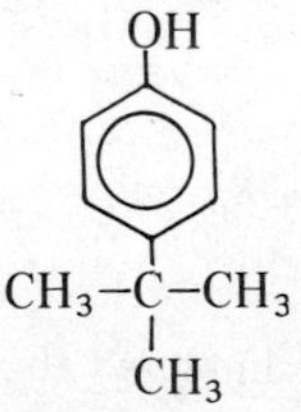

16.34

The broad infrared peak at 3400 cm^{-1} indicates a hydroxy group and the two bands at 720 and 770 cm^{-1} suggest a monosubstituted benzene ring. The presence of these groups is also indicated by the peaks at $\delta 2.7$ and $\delta 7.2$ in the pmr spectrum. The pmr spectrum also shows a triplet at $\delta 0.7$ indicating a $-CH_3$ group coupled with an adjacent $-CH_2-$ group. What appears at first to be a quartet at $\delta 1.9$ actually shows further splitting. There is also a triplet at $\delta 4.35$ (1H). Putting these pieces together in the only way possible gives us the following structure for Y.

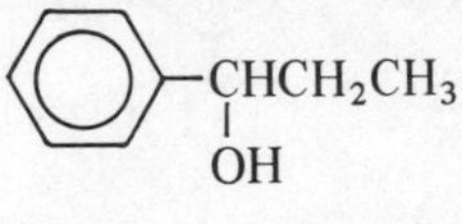

Analysed spectra are shown on p. 165 (Fig. 16.2 in text).

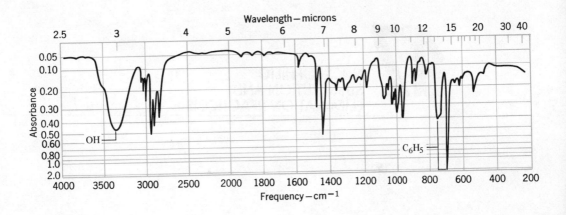

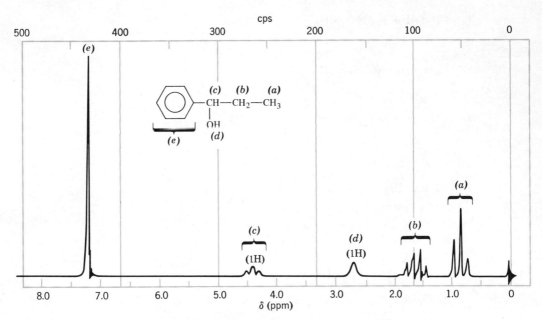

Fig. 16.2. The infrared and pmr spectra of compound Y, problem 16.34 (Spectra courtesy of Sadtler Research Laboratories Inc.)

17

NUCLEOPHILIC SUBSTITUTION AND ELIMINATION REACTIONS

17.1

(a) $CH_3OH + NaO-\overset{\overset{\displaystyle O}{\|}}{\underset{\underset{\displaystyle O}{\|}}{S}}-CH_3$

(b) $C_6H_5CH_2-\overset{+}{N}(CH_3)_3 \ \overset{-}{Br}$

(c) $CH_3CH_2-N_3 + NaO-\overset{\overset{\displaystyle O}{\|}}{\underset{\underset{\displaystyle O}{\|}}{S}}$—⟨benzene ring⟩—$CH_3$

(d) $C_6H_5CH_2-I + NaCl$

(e) $CH_3\underset{\underset{\displaystyle OOCCH_3}{|}}{C}HCH_3 \quad + NaBr$

(f) $\underset{\underset{\underset{\displaystyle O}{\|}}{C_2H_5O\overset{\displaystyle }{C}}}{\overset{\overset{\overset{\displaystyle O}{\|}}{C_2H_5O\overset{\displaystyle }{C}}}{}} CH-CH_2CH_3 + NaBr$

(g) $CH_2{=}CHCH_2CH_2CH_3 + MgXBr$

17.2

$(C_6H_5)_2CH-Cl \underset{}{\overset{slow}{\rightleftharpoons}} (C_6H_5)_2\overset{+}{C}H + \overset{-}{Cl} \quad (step\ 1)$

$(C_6H_5)_2\overset{+}{C}H + \overset{-}{F} \xrightarrow{fast} (C_6H_5)_2CH-F \quad (step\ 2)$

Step 1 is rate-limiting, i.e., much slower than step 2. Only diphenylchloromethane is involved in step 1, therefore the reaction is first order with respect to diphenylchloromethane.

17.3

Compounds that undergo reactions by an S_N1 path must be capable of forming relatively stable carbocations. Primary halides of the type, $ROCH_2X$ form carbocations that are stabilized by resonance:

$$R-\overset{..}{\underset{..}{O}}-\overset{+}{C}H_2 \longleftrightarrow R-\overset{+}{\underset{..}{O}}=CH_2$$

17.4

The relative rates are in the order of the relative stabilities of the carbocations:

$$C_6H_5\overset{+}{C}H_2 < C_6H_5\overset{+}{C}HCH_3 < (C_6H_5)_2\overset{+}{C}H < (C_6H_5)_3\overset{+}{C}$$

The solvolysis reaction involves a carbocation intermediate or a solvent-separated ion-pair.

17.5

(a) Backside attack by the nucleophile is prevented by the cyclic structure. (Notice, too, that the carbon bearing the leaving group is tertiary.)

(b) The bridged cyclic structure prevents the carbon bearing the leaving group from assuming the planar trigonal conformation required of a carbocation.

17.6

(a) NH_2^- (b) Phenoxide ion

17.7

In quinuclidine the three groups attached to nitrogen are "pinned back" in an arrangement that makes approach to the carbon bearing the leaving group relatively unhindered. In triethylamine, on the other hand, the three ethyl groups partially block this approach.

17.8

In strong acid, ethers are protonated:

$$R-O-R + HA \rightleftharpoons R-\overset{+}{\underset{H}{O}}-R + A^-$$

In weak acid or neutral media, protonation does not occur to any appreciable extent. Thus, in strong acid, the leaving group is an alcohol molecule,

$$X^- + R-\overset{+}{\underset{H}{O}}-R \longrightarrow X-R + HO-R$$

whereas in weak acid or neutral media, the leaving group would have to be an alkoxide ion,

$$X^- + R-O-R \longrightarrow X-R + \overset{-}{O}-R$$

R–OH is a good leaving group; $R-O^-$ is a very poor one.

17.9

(a) Decrease   (b) Increase   (c) Decrease   (d) increase   (e)   decrease

(f) increase

17.10

Faster in dimethylsulfoxide. The reaction is S_N2, and dimethylsulfoxide is an aprotic solvent. Aprotic solvents accelerate the rates of all S_N2 reactions because the nucleophiles are unsolvated and, therefore, highly reactive.

17.11

(a) and (b)

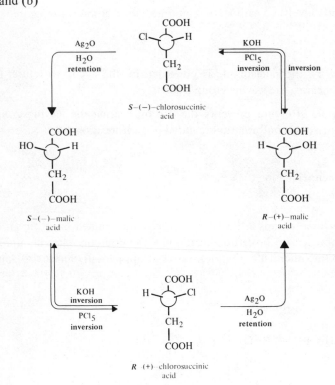

(c) The reaction takes place with retention of configuration.

(d)

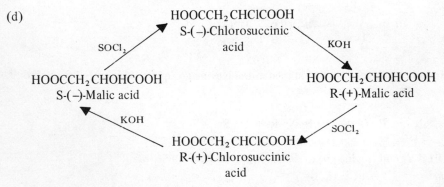

17.12

(a) Reaction of an alkene with halogen in water solution.

(b) $CH_3CH{=}CH_2 + Cl_2 \xrightarrow{H_2O} CH_3CHCH_2Cl \xrightarrow{NaOH} CH_3CH{-}CH_2$

with OH below the first product and an epoxide O below the final product.

(c) The $-\ddot{O}\!:^-$ group must displace the Cl^- from the back side,

$\xrightarrow{OH^-}$ ... $\rightleftarrows$ epoxide

Such displacement is impossible with the *cis*-isomer:

$\xrightarrow{OH^-}$ ... $\rightleftarrows$ No epoxide

17.13
2-(p-hydroxyphenyl)-1-chloropropane > 2-(p-tolyl)-1-chloropropane > 2-phenyl-1-chloropropane > 2-(p-nitrophenyl)-1-chloropropane

17.14
Reaction by a S_N2 pathway is hindered sterically by the large *tert*-butyl group. Reaction by a S_N1 pathway is very slow because a primary carbocation must be formed.

17.15
Only a proton or deuteron *anti* to the bromine can be eliminated. The two conformations of *erythro*-2-bromo-butane-3-*d* in which a proton or deuteron is *anti* to the bromine are I and II below.

Conformation I can undergo loss of HBr to yield *cis*-2-butene-2-*d*. Conformation II can undergo loss of DBr to yield *trans*-2-butene.

17.16
(a) A = $CH_3CH_2-\overset{+}{N}(CH_3)_3\ \overset{-}{Br}$ (b) B = $CH_3CH_2CH_2-\overset{+}{N}(CH_3)_3\ \overset{-}{I}$

(c) C =

17.17

(a) $CH_3\overset{\overset{\displaystyle CH_3}{|}}{C}=CH_2$ + $CH_2=CH_2$ (major product)

(b) $CH_3CH=CH_2$ + $CH_2=CH_2$ (major product)

(c) $(CH_3)_2NCH_2CH_2CH_2CH=CH_2$

(d) $CH_2=CHCH_2CH=CH_2$

17.18

In each case a *syn* elimination takes place from a conformation in which the large phenyl substituents *are not eclipsed*.

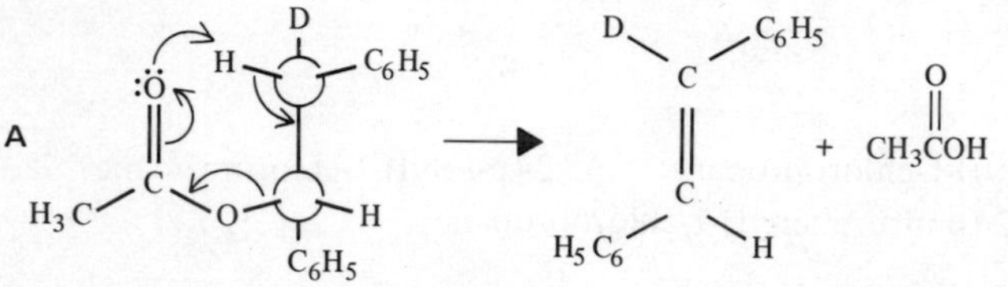

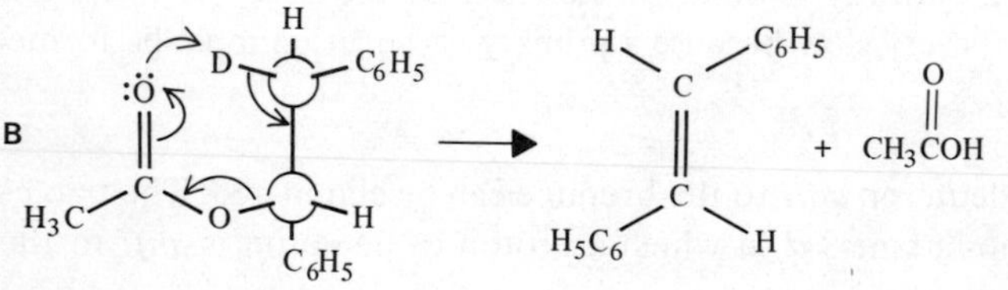

17.19

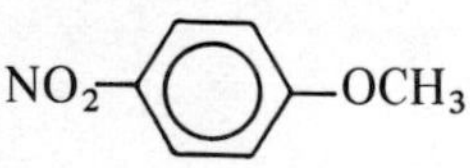

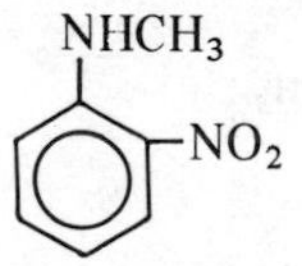

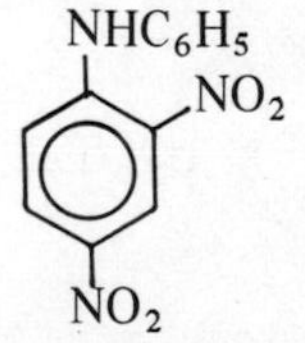

17.20

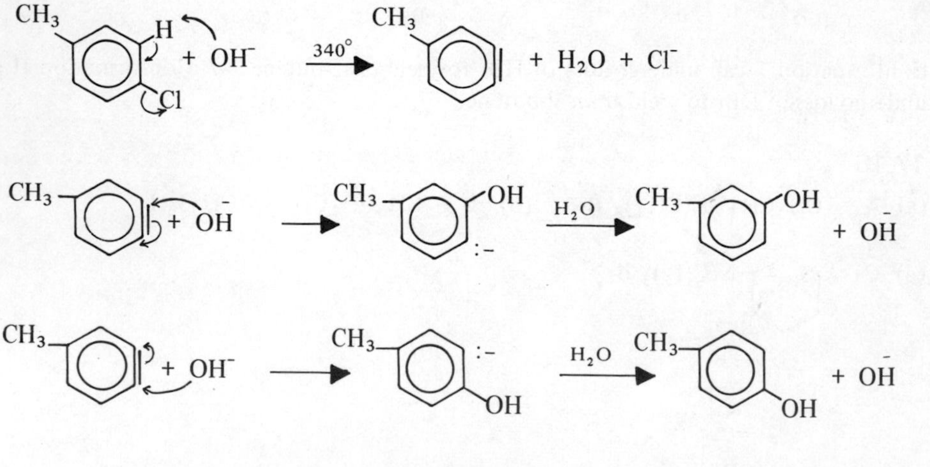

17.21

Since there are no hydrogens *ortho* to the halogen, elimination cannot take place. (Reaction by a bimolecular displacement is not possible either because the substrate lacks strong electron-withdrawing groups.) Thus the absence of a reaction must be due to the inability of 2-bromo-3-methylanisole to form a benzyne intermediate.

17.22

(a) If every substitution *involves an inversion*, then racemization will be complete when only *half* the substrate has incorporated radioactive iodine. (At this point there will be an equimolar mixture of the two enantiomers.) Thus, the rate of racemization *should be twice the rate of incorporation of radioactive iodine*.

(b) If an achiral intermediate such as a carbocation were involved, one would expect the rate of racemization to equal the rate of incorporation of radioactive iodine.

17.23

(a) The carbocation stability is in the order

(b) The methoxy group in the meta position cannot stabilize the carbocation by resonance:

no especially stable structure is possible

The small stabilization resulting from the release of electrons into the ring through resonance is cancelled by oxygen's electron-withdrawing inductive effect.

17.24

Reaction of the alcohol with K and then of the resulting salt with C_2H_5Br does not break bonds to the chiral carbon, and these reactions therefore occur with retention.

Reaction of the tosylate, $C_6H_5CH_2\underset{\underset{\displaystyle OTs}{|}}{C}HCH_3$, with C_2H_5OH in K_2CO_3 solution, however,

is an S_N2 reaction at the chiral carbon and thus it occurs with inversion.

17.25

Elimination is syn; i.e., the acetate group and the hydrogen that is eliminated must be on the same side of the ring. This is true for both ring hydrogens beta to the acetate group in but for only one ring hydrogen in

(In either case hydrogens of the α-CH_3 group can become syn to the acetate group.)

17.26

17.27
The acetate ion is a weaker base than OH^-, and elimination therefore occurs to a smaller extent. (The difference in nucleophilicity of OH^- and CH_3COO^- is not as large as the difference in basicity, however.)

17.28

17.29
(a) Use a strong, hindered base such as $(CH_3)_3COK$ in a solvent of low polarity in order to bring about an E2 reaction.

(b) Here we want an S_N1 reaction. We use ethanol as the solvent *and as the nucleophile*, and we carry out the reaction at a low temperature so that elimination will be minimized.

17.30
(a) Method I succeeds because the S_N2 reaction occurs at a pirmary carbon atom. In the other example (II) the attacking nucleophile, $CH_3CH_2O^-$, would react with a tertiary alkyl halide, and the major reaction would be elimination.

(b) $CH_3-\overset{\displaystyle CH_3}{\underset{\displaystyle CH_3}{\overset{|}{\underset{|}{C}}}}-Br \; + \; \overset{-}{O}C_2H_5 \longrightarrow CH_3-\overset{\displaystyle CH_3}{\overset{|}{C}}=CH_2 \; + \; \overset{-}{Br} \; + \; HOC_2H_5$

18

ALDEHYDES AND KETONES

18.1

(a)

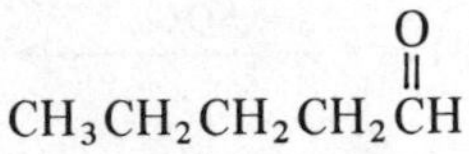

$$CH_3CH_2CH_2CH_2\overset{\overset{\displaystyle O}{\|}}{C}H$$
Pentanal

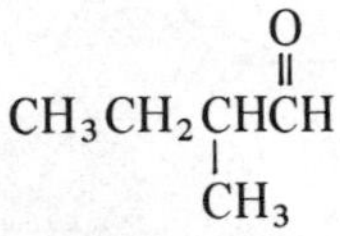

$$CH_3CH_2\overset{\overset{\displaystyle O}{\|}}{C}HCH$$
$$| \atop CH_3$$
2-Methylbutanal

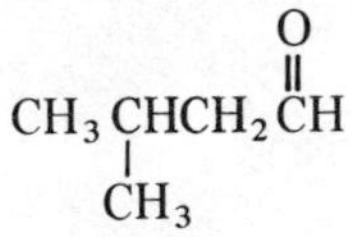

$$CH_3\underset{\underset{\displaystyle CH_3}{|}}{C}HCH_2\overset{\overset{\displaystyle O}{\|}}{C}H$$
3-Methylbutanal

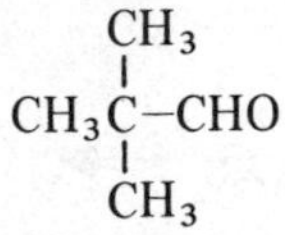

$$CH_3\overset{\overset{\displaystyle CH_3}{|}}{\underset{\underset{\displaystyle CH_3}{|}}{C}}-CHO$$
2,2-Dimethylpropanal

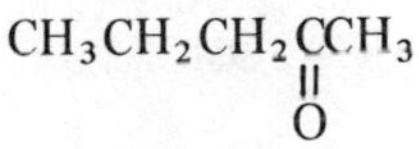

$$CH_3CH_2CH_2\underset{\underset{\displaystyle O}{\|}}{C}CH_3$$
2-Pentanone

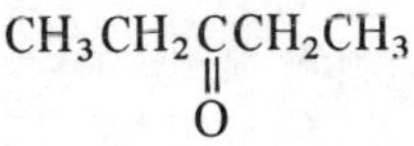

$$CH_3CH_2\underset{\underset{\displaystyle O}{\|}}{C}CH_2CH_3$$
3-Pentanone

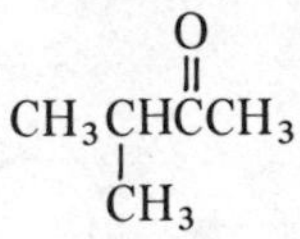

$$CH_3\underset{\underset{\displaystyle CH_3}{|}}{C}H\overset{\overset{\displaystyle O}{\|}}{C}CH_3$$
3-Methyl-2-butanone

(b) and (c)

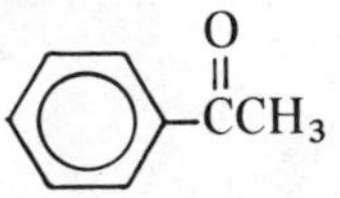

Acetophenone or
phenyl methyl ketone

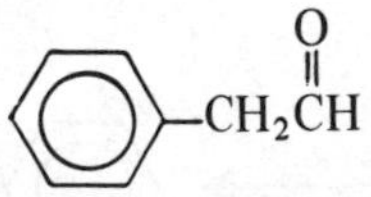

Phenylethanal or
phenylacetaldehyde

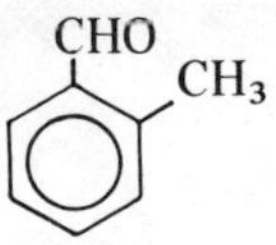

o-Tolualdehyde 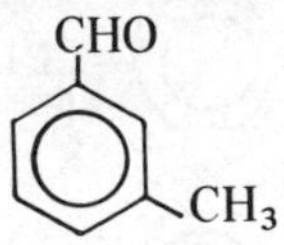   m-Tolualdehyde    p-Tolualdehyde

18.2

(a) 1-Pentanol, because its molecules form hydrogen bonds to each other.

(b) 2-Pentanol, because its molecules form hydrogen bonds to each other.

(c) Pentanal, because its molecules are more polar.

(d) 2-Phenylethanol, because its molecules form hydrogen bonds to each other.

(e) Benzyl alcohol because its molecules form hydrogen bonds to each other.

18.3

(a) $C_6H_6 \xrightarrow{\text{Br}_2, \text{ Fe}} C_6H_5\text{-Br} \xrightarrow[\text{ether}]{\text{Mg}} C_6H_5\text{-MgBr} \xrightarrow[\text{(2) H}^+]{\text{(1) HCHO}}$

$C_6H_5\text{-CH}_2\text{OH} \xrightarrow[\text{CH}_2\text{Cl}_2]{\text{CrO}_3\cdot\text{C}_5\text{H}_5\text{N}} C_6H_5\text{-CHO}$

(b) $C_6H_5\text{-CH}_3 \xrightarrow[\text{(2) H}^+]{\text{(1) KMnO}_4, \text{ OH}^-, \text{ heat}} C_6H_5\text{-COOH} \xrightarrow{\text{SOCl}_2}$

$C_6H_5\text{-COCl} \xrightarrow[\text{ether}]{\text{LiAlH[OC(CH}_3)_3]_3} C_6H_5\text{-CHO}$

$\xrightarrow[\text{Pd(S)}]{\text{H}_2} C_6H_5\text{-CHO}$

(c) $\text{Br-}C_6H_4\text{-CH}_3 \xrightarrow[\text{CS}_2]{\text{CrO}_2\text{Cl}_2} \text{Br-}C_6H_4\text{-CH(OCrOHCl}_2)_2$

$\xrightarrow{\text{H}_2\text{O}} \text{Br-}C_6H_4\text{-CHO}$

(d) $\text{Cl-}C_6H_4\text{-CH}_3 \xrightarrow[\text{H}_2\text{SO}_4]{\substack{\text{CrO}_3 \\ (\text{CH}_3\text{CO})_2\text{O}}} \text{Cl-}C_6H_4\text{-CH(OOCCH}_3)_2 \xrightarrow[\text{H}_2\text{O}]{\text{H}^+} \text{Cl-}C_6H_4\text{-CHO}$

(e) $C_6H_5\text{-CHCH}_3 \xrightarrow[\text{H}_2\text{SO}_4]{\text{CrO}_3} C_6H_5\text{-}\overset{\text{O}}{\underset{||}{\text{C}}}\text{CH}_3$
  (with OH on CHCH₃)

(f) $C_6H_6 \xrightarrow[\text{AlCl}_3]{\text{CH}_3\text{COCl}} C_6H_5\text{-}\overset{\text{O}}{\underset{||}{\text{C}}}\text{CH}_3$

(g) $C_6H_5\text{-}\overset{\text{O}}{\underset{||}{\text{C}}}\text{Cl} \xrightarrow[\substack{\text{or} \\ (\text{CH}_3)_2\text{Cd}}]{(\text{CH}_3)_2\text{CuLi}} C_6H_5\text{-}\overset{\text{O}}{\underset{||}{\text{C}}}\text{CH}_3$

18.4

(a) The nucleophile is the negatively charged carbon of the Grignard reagent *acting as a carbanion*.

(b) The magnesium portion of the Grignard reagent acts as a Lewis acid and accepts an electron pair of the carbonyl oxygen. This makes the carbonyl carbon even more positive and, therefore, even more susceptible to nucleophilic attack.

$$-\overset{\delta+}{\underset{R}{\overset{|}{C}}}=\overset{\delta-}{\overset{..}{O}}\quad \overset{\delta+}{\underset{\delta-}{Mg-X}} \longrightarrow -\overset{|}{\underset{R}{C}}-\overset{..}{\underset{..}{O}}-MgX$$

(c) The product that forms initially (above) is a magnesium halide salt of an alcohol.

(d) On addition of water, the organic product that forms is an alcohol.

18.5
The nucleophile is a hydride ion.

18.6

(a) $CH_3\overset{O}{\overset{||}{C}}H \xrightarrow[H_2O]{NaHSO_3} CH_3\overset{OH}{\underset{|}{C}}HSO_3Na \xrightarrow[H_2O]{NaCN} CH_3\overset{OH}{\underset{|}{C}}HCN$

$$\xrightarrow[\text{reflux}]{HCl} CH_3\overset{OH}{\underset{|}{C}}HCOOH$$
Lactic acid

(b) A racemic modification.

18.7

(a) $\overset{\beta}{CH_3}\,\overset{\alpha}{CH_2}\overset{O}{\overset{||}{C}}H + OH^- \rightleftharpoons CH_3\overset{..}{\overset{-}{C}}H\overset{O}{\overset{||}{C}}H + H_2O$

$CH_3CH_2\overset{O}{\overset{||}{C}}H + {}^-{:}\overset{O}{\underset{\underset{CH_3}{|}}{\overset{||}{C}}}HCH \rightleftharpoons CH_3CH_2\overset{O^-}{\underset{\underset{CH_3}{|}}{\overset{|}{C}}}HCHCH$

$CH_3CH_2\overset{O^-}{\underset{\underset{CH_3}{|}}{\overset{|}{C}}}HCH\overset{O}{\overset{||}{C}}H + HOH \rightleftharpoons CH_3CH_2\overset{OH}{\underset{\underset{CH_3}{|}}{\overset{|}{C}}}HCH\overset{O}{\overset{||}{C}}H$

(b) For $CH_3CH_2\overset{OH}{\underset{|}{C}}HCH_2CH_2\overset{O}{\overset{||}{C}}H$ to form, a hydroxide ion would have to remove a β-proton in the first step. This does not happen because the anion that would be produced, i.e., $^-{:}CH_2CH_2CHO$, cannot be stabilized by resonance.

(c) $CH_3CH_2CH{=}\overset{O}{\underset{\underset{CH_3}{|}}{\overset{||}{C}}}CH$

18.8

(a) $2CH_3CH_2CH_2CHO \xrightarrow[H_2O]{OH^-}$ CH$_3$CH$_2$CH$_2$$\overset{\displaystyle OH}{\overset{|}{C}H}$CHCHO
$\qquad\qquad\qquad\qquad\qquad\qquad\qquad\quad \overset{|}{C}H_2$
$\qquad\qquad\qquad\qquad\qquad\qquad\qquad\quad \overset{|}{C}H_3$

(b) Product of (a) $\xrightarrow[(-H_2O)]{H^+}$ CH$_3$CH$_2$CH$_2$CH=$\overset{}{C}$CHO
$\qquad\qquad\qquad\qquad\qquad\qquad\qquad\quad \overset{|}{C}H_2$
$\qquad\qquad\qquad\qquad\qquad\qquad\qquad\quad \overset{|}{C}H_3$

$\xrightarrow{NaBH_4}$ CH$_3$CH$_2$CH$_2$CH=$\overset{}{C}$CH$_2$OH
$\qquad\qquad\qquad\qquad\qquad\qquad\qquad\quad \overset{|}{C}H_2$
$\qquad\qquad\qquad\qquad\qquad\qquad\qquad\quad \overset{|}{C}H_3$

(c) Product of (b) $\xrightarrow[Ni]{H_2}$ CH$_3$CH$_2$CH$_2$CH$_2$CHCH$_2$OH
$\qquad\qquad\qquad\qquad\qquad\qquad\qquad\quad \overset{|}{C}H_2$
$\qquad\qquad\qquad\qquad\qquad\qquad\qquad\quad \overset{|}{C}H_3$

(d) Product of (a) $\xrightarrow{NaBH_4}$ CH$_3$CH$_2$CH$_2$$\overset{\displaystyle OH}{\overset{|}{C}H}$CHCH$_2$OH
$\qquad\qquad\qquad\qquad\qquad\qquad\qquad\quad \overset{|}{C}H_2$
$\qquad\qquad\qquad\qquad\qquad\qquad\qquad\quad \overset{|}{C}H_3$

18.9

(a) CH$_3$$\overset{\displaystyle O}{\overset{\|}{C}}CH_3$ + OH$^-$ $\rightleftharpoons$ CH$_3$$\overset{\displaystyle O}{\overset{\|}{C}}CH_2$:$^-$ + H$_2$O

CH$_3$$\overset{\displaystyle O}{\overset{\|}{C}}CH_2$:$^-$ + CH$_3$$\overset{\displaystyle O}{\overset{\|}{C}}CH_3$ $\rightleftharpoons$ CH$_3$$\overset{\displaystyle O}{\overset{\|}{C}}CH_2$$\overset{\displaystyle O^-}{\overset{|}{C}}CH_3$
$\qquad\qquad\qquad\qquad\qquad\qquad\qquad\qquad\qquad\quad \overset{|}{C}H_3$

CH$_3$$\overset{\displaystyle O}{\overset{\|}{C}}CH_2$$\overset{\displaystyle O^-}{\overset{|}{C}}CH_3$ + HOH $\rightleftharpoons$ CH$_3$$\overset{\displaystyle O}{\overset{\|}{C}}CH_2$$\overset{\displaystyle OH}{\overset{|}{C}}CH_3$ + OH$^-$
$\qquad\qquad\quad \overset{|}{C}H_3$ $\qquad\qquad\qquad\qquad\qquad\qquad\quad \overset{|}{C}H_3$

(b) CH$_3$$\overset{\displaystyle O}{\overset{\|}{C}}$CH=$\overset{}{C}CH_3$
$\qquad\qquad\qquad \overset{|}{C}H_3$

18.10

CH$_3$CH$_2$$\overset{\displaystyle O}{\overset{\|}{C}}$H + OH$^-$ $\rightleftharpoons$ CH$_3$$\overset{\displaystyle\ddot{}}{C}H\overset{\displaystyle O}{\overset{\|}{C}}$H + H$_2$O

$$CH_3\overset{\overset{O}{\|}}{C}H + CH_3\overset{\cdot\cdot}{C}H\overset{\overset{O}{\|}}{C}H \;\rightleftarrows\; CH_3\overset{\overset{O^-}{|}}{C}H\overset{\overset{O}{\|}}{C}H\overset{}{C}H$$
$$\underset{\overset{|}{CH_3}}{}$$

$$CH_3\overset{\overset{O^-}{|}}{C}H\overset{\overset{O}{\|}}{C}H + H_2O \;\rightleftarrows\; CH_3\overset{\overset{OH}{|}}{C}H\overset{\overset{O}{\|}}{C}H$$
$$\underset{\overset{|}{CH_3}}{}\qquad\qquad\qquad\underset{\overset{|}{CH_3}}{}$$

2-Methyl-3-
hydroxybutanal

$$CH_3\overset{\overset{O}{\|}}{C}H + OH^- \;\rightleftarrows\; {}^-{:}CH_2\overset{\overset{O}{\|}}{C}H + H_2O$$

$$CH_3\,CH_2\overset{\overset{O}{\|}}{C}H + {}^-{:}CH_2\overset{\overset{O}{\|}}{C}H \;\rightleftarrows\; CH_3CH_2\overset{\overset{O^-}{|}}{C}HCH_2\overset{\overset{O}{\|}}{C}H$$

$$CH_3CH_2\overset{\overset{O^-}{|}}{C}HCH_2\overset{\overset{O}{\|}}{C}H + H_2O \;\rightleftarrows\; CH_3CH_2\overset{\overset{OH}{|}}{C}HCH_2\overset{\overset{O}{\|}}{C}H$$
3-Hydroxypentanal

18.11

Three successive aldol additions occur.

First Aldol Addition
$$\begin{cases}
CH_3\overset{\overset{O}{\|}}{C}H + OH^- \;\rightleftarrows\; {}^-{:}CH_2\overset{\overset{O}{\|}}{C}H + H_2O \\[2ex]
HCH \overset{\overset{O}{\|}}{} + {}^-{:}CH_2\overset{\overset{O}{\|}}{C}H \;\rightleftarrows\; {}^-OCH_2CH_2\overset{\overset{O}{\|}}{C}H \\[2ex]
{}^-OCH_2CH\overset{\overset{O}{\|}}{C}H + H_2O \;\rightleftarrows\; HOCH_2CH_2\overset{\overset{O}{\|}}{C}H
\end{cases}$$

Second Aldol Addition
$$\begin{cases}
HOCH_2CH_2\overset{\overset{O}{\|}}{C}H + OH^- \;\rightleftarrows\; HOCH_2\overset{\cdot\cdot}{C}H\overset{\overset{O}{\|}}{C}H + H_2O \\[2ex]
HCH\overset{\overset{O}{\|}}{} + HOCH_2\overset{\cdot\cdot}{C}H\overset{\overset{O}{\|}}{C}H \;\rightleftarrows\; HOCH_2\overset{\overset{CH_2O^-}{|}}{C}HCHO \\[2ex]
HOCH_2\overset{\overset{CH_2O^-}{|}}{C}HCHO + H_2O \;\rightleftarrows\; HOCH_2\overset{\overset{CH_2OH}{|}}{C}HCHO + OH^-
\end{cases}$$

Third
Aldol
Addition

$$\text{HOCH}_2\overset{\displaystyle \text{CH}_2\text{OH}}{\underset{\displaystyle |}{\text{CH}}}-\text{CHO} + \text{OH}^- \rightleftharpoons \text{HOCH}_2\overset{\displaystyle \text{CH}_2\text{OH}}{\underset{\displaystyle |}{\overset{\displaystyle ..}{\text{C}}}}-\text{CHO}$$

$$\text{H}\overset{\displaystyle \text{O}}{\overset{\displaystyle \|}{\text{C}}}\text{H} + \text{HOCH}_2\overset{\displaystyle \text{CH}_2\text{OH}}{\underset{\displaystyle \underset{-}{|}}{\text{C}}}-\text{CHO} \rightleftharpoons \text{HOCH}_2-\overset{\displaystyle \text{CH}_2\text{OH}}{\underset{\displaystyle \underset{\displaystyle \text{CH}_2\text{O}^-}{|}}{\text{C}}}-\text{CHO}$$

$$\text{HOCH}_2-\overset{\displaystyle \text{CH}_2\text{OH}}{\underset{\displaystyle \underset{\displaystyle \text{CH}_2\text{O}^-}{|}}{\text{C}}}-\text{CHO} + \text{H}_2\text{O} \rightleftharpoons \text{HOCH}_2-\overset{\displaystyle \text{CH}_2\text{OH}}{\underset{\displaystyle \underset{\displaystyle \text{CH}_2\text{OH}}{|}}{\text{C}}}-\text{CHO} + \text{OH}^-$$

18.12

(a) $\text{CH}_3\text{COOH} + \text{BF}_3 \rightleftharpoons \text{CH}_3\text{COOBF}_3^- + \text{H}^+$

Pseudoionone

α-Ionone

β-Ionone

(b) In β-ionone the double bonds and the carbonyl group are conjugated, thus it is more stable.

(c) β-Ionone, because it is a conjugated unsaturated system.

18.13

(from Prob. 18.11)

$$H_2O + H-\overset{\displaystyle ||}{\underset{\displaystyle O}{C}}-O^- + H-\overset{\displaystyle O^-}{\underset{\displaystyle H}{\underset{|}{C}}}\overset{|}{\underset{|}{\overset{\displaystyle CH_2OH}{\underset{|}{C}}}}-CH_2OH \rightleftharpoons$$

$$HOCH_2-\overset{\displaystyle CH_2OH}{\underset{\displaystyle CH_2OH}{\underset{|}{\overset{|}{C}}}}-CH_2OH + OH^- + H\overset{\displaystyle ||}{\underset{\displaystyle O}{C}}-O^-$$

Pentaerythritol

18.14

(a) $CH_3(CH_2)_4CH_2Br \xrightarrow{(C_6H_5)_3P} CH_3(CH_2)_4CH_2-P(C_6H_5)_3{}^+ Br^-$

$\xrightarrow{RLi} CH_3(CH_2)_4CH=P(C_6H_5)_3$

(b) $CH_3I \xrightarrow{(C_6H_5)_3P} CH_3-P(C_6H_5)_3{}^+ I^- \xrightarrow{RLi} CH_2=P(C_6H_5)_3$

(c) $BrCH_2CH_2CH_2CH_2Br + 2(C_6H_5)_3P \longrightarrow$

$(C_6H_5)_3\overset{+}{P}-CH_2CH_2CH_2CH_2-\overset{+}{P}(C_6H_5)_3 \xrightarrow{2\ RLi}$
$Br^- \qquad\qquad\qquad\qquad Br^-$

$(C_6H_5)_3P=CHCH_2CH_2CH=P(C_6H_5)_3$

18.15

(a) $C_6H_5CH_2Br \xrightarrow[\text{(2) RLi}]{\text{(1) }(C_6H_5)_3P} C_6H_5CH=P(C_6H_5)_3 \xrightarrow{\overset{\displaystyle O}{\overset{||}{CH_3CCH_3}}}$

$C_6H_5CH=\underset{\displaystyle CH_3}{\underset{|}{C}}CH_3$

(b) $CH_3I \xrightarrow[\text{(2) RLi}]{\text{(1) }(C_6H_5)_3P} CH_2=P(C_6H_5)_3 \xrightarrow{\overset{\displaystyle O}{\overset{||}{C_6H_5CCH_3}}} C_6H_5\underset{\displaystyle CH_3}{\underset{|}{C}}=CH_2$

(c) $CH_3CH_2Br \xrightarrow[\text{(2) RLi}]{\text{(1) }(C_6H_5)_3P} CH_3CH=P(C_6H_5)_3 \xrightarrow{\overset{\displaystyle O}{\overset{||}{C_6H_5CCH_3}}}$

$C_6H_5\underset{\displaystyle CH_3}{\underset{|}{C}}=CHCH_3$

(d) $CH_2=P(C_6H_5)_3 \xrightarrow{\overset{\displaystyle O}{\overset{||}{CH_3CCH_3}}} \underset{\displaystyle CH_3}{\overset{\displaystyle CH_3}{C}}=CH_2$
(from part b)

(e) $CH_2=P(C_6H_5)_3 \longrightarrow$
(from part b)

(f) $CH_3CH_2CH_2Br \xrightarrow[\text{(2) RLi}]{\text{(1) }(C_6H_5)_3P} CH_3CH_2CH=P(C_6H_5)_3$

$$\xrightarrow[\text{CH}_3\overset{\overset{\displaystyle O}{\|}}{C}CH_2CH_3]{} CH_3CH_2CH=\overset{\overset{\displaystyle CH_3}{|}}{C}CH_2CH_3$$

(g) $CH_2=CHCH_2Br \xrightarrow[\text{(2) RLi}]{\text{(1) }(C_6H_5)_3P} CH_2=CHCH=P(C_6H_5)_3$

$$\xrightarrow[\text{C}_6\text{H}_5\overset{\overset{\displaystyle O}{\|}}{C}H]{} C_6H_5CH=CHCH=CH_2$$

(h) $C_6H_5CH=P(C_6H_5)_3 \xrightarrow[]{C_6H_5\overset{\overset{\displaystyle O}{\|}}{C}H} C_6H_5CH=CHC_6H_5$
(from part (a)

18.16

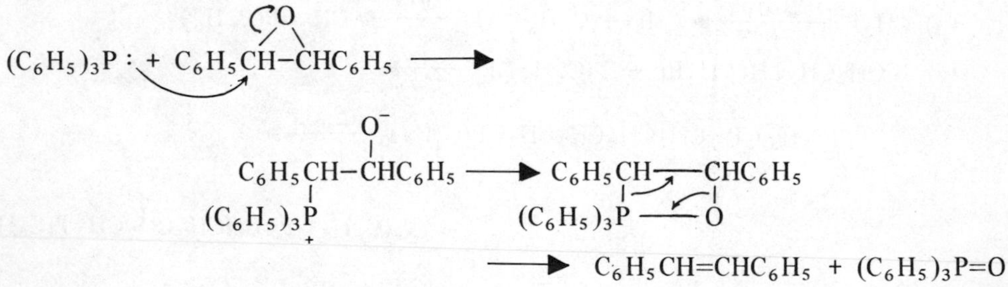

$(C_6H_5)_3P: + C_6H_5\overset{\curvearrowright}{C}H-\overset{\overset{\displaystyle O}{\diagup}}{C}HC_6H_5 \longrightarrow$

$\underset{\underset{\displaystyle (C_6H_5)_3\overset{+}{P}}{|}}{C_6H_5CH}-\overset{\overset{\displaystyle O^-}{|}}{C}HC_6H_5 \longrightarrow \underset{\underset{\displaystyle (C_6H_5)_3P \;\diagdown\; O}{|}}{C_6H_5CH}\;\diagup\!\!\diagdown\;CHC_6H_5$

$\longrightarrow C_6H_5CH=CHC_6H_5 + (C_6H_5)_3P=O$

18.17

(a) $CH_3OCH_2Br + (C_6H_5)_3P \xrightarrow{\text{(2) base}} CH_3OCH=P(C_6H_5)_3$

(b) Hydrolysis of the ether yields a hemiacetal (see the example in part (c) below) that then goes on to form an aldehyde.

(c)

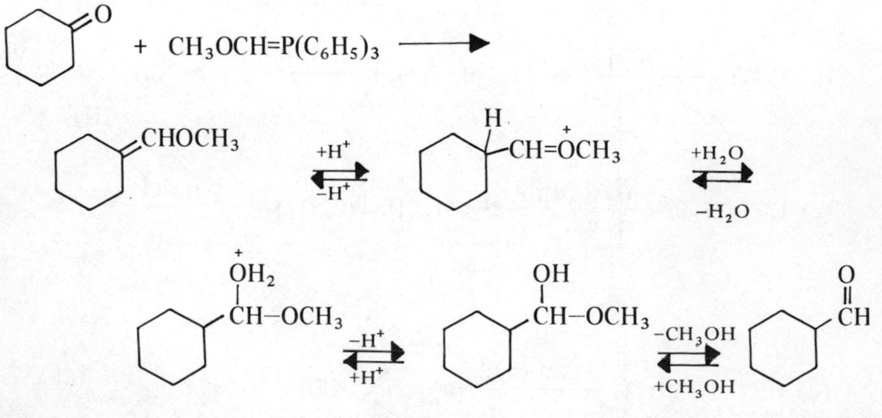

18.18

(a)

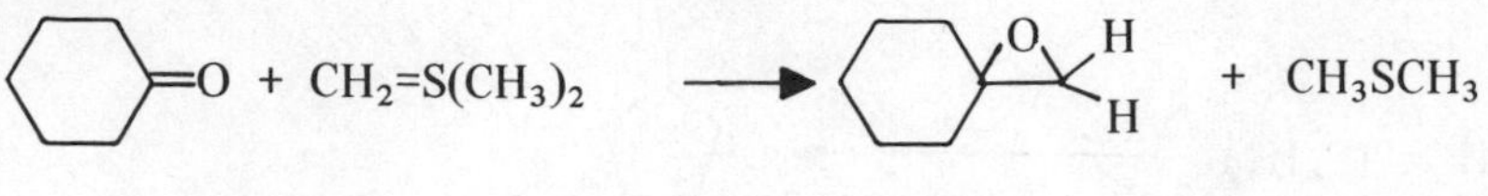

(b)
$$\begin{array}{c} CH_3 \\ CH_3 \end{array}\!\!C{=}O \ + \ CH_2{=}S(CH_3)_2 \ \longrightarrow \ \begin{array}{c} CH_3 \\ CH_3 \end{array}\!\!C\!\!-\!\!CH_2 \ + \ CH_3SCH_3$$

18.19

(a) $RCH{=}\overset{+}{\underset{\cdot\cdot}{O}}{-}R \ \longleftrightarrow \ RCH{-}\underset{\cdot\cdot}{\overset{\cdot\cdot}{O}}{-}R$
 I II

(b) and (c) Structure I should make a greater contribution because in it both the carbon atom and the oxygen atom have an octet of electrons and because it has one more bond.

18.20

18.21

(a)

(b) Addition would take place at the ketone group as well as at the ester group. The product (after hydrolysis) would be,

18.22

(a) $=O$ + $HSCH_2CH_2SH$ $\xrightarrow{BF_3}$

$\xrightarrow[\text{Ni}]{\text{Raney}}$ + CH_3CH_3 + NiS

(b) $CH_3\overset{O}{\overset{\|}{C}}CH_2CH_2CO_2C_2H_5$ + $HSCH_2CH_2SH$ $\xrightarrow{BF_3}$

$$CH_3-\overset{\overset{\displaystyle CH_2-CH_2}{\underset{\displaystyle S \qquad S}{\diagup\diagdown}}}{C}-CH_2CH_2CO_2C_2H_5 \xrightarrow[\text{Ni}]{\text{Raney}} CH_3CH_2CH_2CH_2CO_2C_2H_5$$
$$+ CH_3CH_3$$
$$+ NiS$$

18.23

The reaction is said to be "base promoted" because base is consumed as the reaction takes place. A catalyst is, by definition, not consumed.

18.24

(a) The slow step in base-catalyzed racemization is the same as that in base-promoted halogenation—*the formation of an enolate ion*. (Formation of an enolate ion from phenyl *sec*-butyl ketone leads to racemization because the enolate ion is achiral. When it accepts a proton it yields a racemic modification.) The slow step in acid-catalyzed racemization is also the same as that in acid-catalyzed halogenation—*the formation of an enol*. (The enol, like the enolate ion, is achiral and tautomerizes to yield a racemic modification of the ketone.)

(b) According to the mechanism given, the slow step for acid-catalyzed iodination (formation of the enol) is the same as that for acid-catalyzed bromination. Thus we would expect both reactions to occur at the same rate.

(c) Again, the slow step for both reactions (formation of the enolate ion) is the same, and consequently, both reactions take place at the same rate.

18.25

(a) Acetone, $CH_3\overset{O}{\overset{\|}{C}}CH_3$

(b) Acetophenone, $C_6H_5\overset{O}{\overset{\|}{C}}CH_3$

(d) 2-Pentanone, $CH_3CH_2CH_2\overset{O}{\overset{\|}{C}}CH_3$

(f) 1-Phenylethanol, $C_6H_5\overset{OH}{\overset{|}{C}H}CH_3$

(h) 2-Butanol, $CH_3CH_2\overset{\overset{\displaystyle OH}{|}}{C}HCH_3$

(i) 1-Acetylnaphthalene,

18.26

(b) 2-Methyl-1,3-cyclohexanedione is more acidic because its enolate ion is stabilized by an additional resonance structure.

18.27

(a) $C_6H_5\overset{\overset{\displaystyle O}{\|}}{C}CH_3 \underset{+H^+}{\overset{-H^+}{\rightleftharpoons}} C_6H_5\overset{\overset{\displaystyle O}{\|}}{C}CH_2{:}^-$

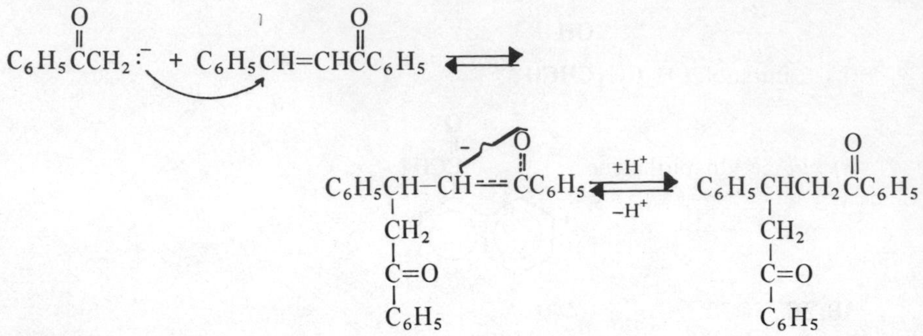

$$C_6H_5\overset{O}{\overset{\|}{C}}CH_2{:}^- \;+\; C_6H_5CH=CH\overset{O}{\overset{\|}{C}}C_6H_5 \;\rightleftharpoons$$

$$C_6H_5CH-CH{=\!=\!=}\overset{-O}{\overset{\|}{C}}C_6H_5 \quad \underset{-H^+}{\overset{+H^+}{\rightleftharpoons}} \quad C_6H_5CHCH_2\overset{O}{\overset{\|}{C}}C_6H_5$$

$$\underset{\underset{C_6H_5}{\overset{|}{\underset{\|}{C=O}}}}{\overset{|}{CH_2}} \qquad\qquad \underset{\underset{C_6H_5}{\overset{|}{\underset{\|}{C=O}}}}{\overset{|}{CH_2}}$$

(b)

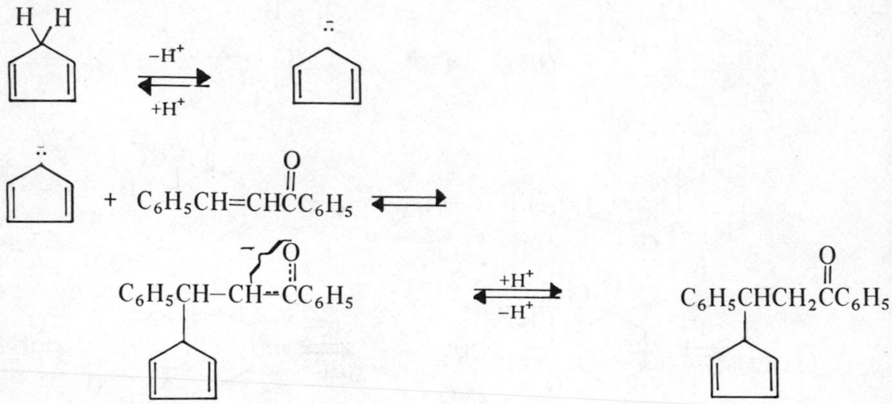

$$+ \; C_6H_5CH=CH\overset{O}{\overset{\|}{C}}C_6H_5 \;\rightleftharpoons$$

$$C_6H_5CH-CH{\cdots}\overset{-O}{\overset{\|}{C}}C_6H_5 \qquad \underset{-H^+}{\overset{+H^+}{\longrightarrow}} \qquad C_6H_5CHCH_2\overset{O}{\overset{\|}{C}}C_6H_5$$

18.28

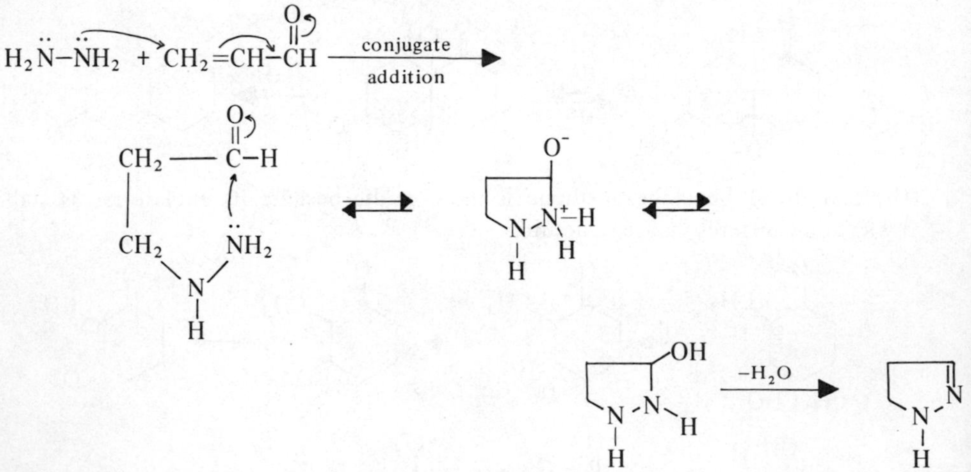

$$H_2\ddot{N}-\ddot{N}H_2 \;+\; CH_2=CH-\overset{O}{\overset{\|}{C}}H \xrightarrow[\text{addition}]{\text{conjugate}}$$

$$\xrightarrow{-H_2O}$$

18.29

(a) HCHO   (d) CH_3COCH_3

(b) CH_3CHO (e) $CH_3COCH_2CH_3$

(c) $C_6H_5CH_2CHO$ (f) $CH_3COC_6H_5$

(g) $C_6H_5CH=CHCOCH_3$

(j)

(h) $C_6H_5CH=CHCOC_6H_5$

(k)

(i) $C_6H_5COC_6H_5$

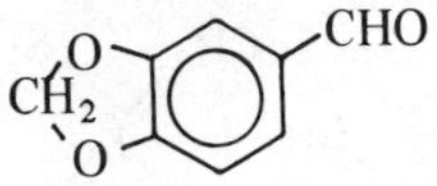

(l)

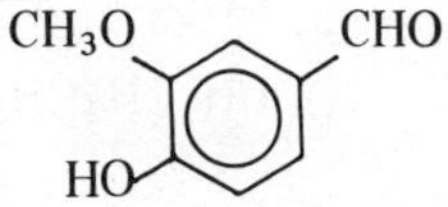

18.30

(a) $CH_3CH_2CH_2OH$

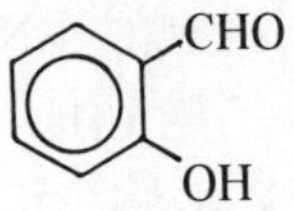

(b) $CH_3CH_2CHOHC_6H_5$

(c) $CH_3CH_2CH_2OH$

(d) $CH_3CH_2CHOHSO_3Na$

(e) $CH_3CH_2CHOHCN$

(f) $CH_3CH_2CHOHCH(CH_3)CHO$

(g) $CH_3CH_2CH=C(CH_3)CHO$

(h) $CH_3CH_2CH_2OH$

(i) 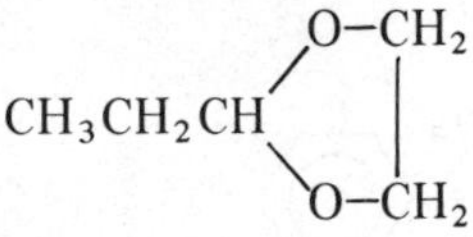 $CH_3CH_2CH\begin{smallmatrix}O-CH_2\\ | \\ O-CH_2\end{smallmatrix}$

(j) $CH_3CH_2CH=CHCH_3$

(k) $CH_3CHBrCHO$

(l) $CH_3CH_2COO^- + Ag\downarrow$

(m) $CH_3CH_2CH=NOH$

(n) $CH_3CH_2CH=NNHCONH_2$

(o) $CH_3CH_2CH=NNHC_6H_5$

(p) CH_3CH_2COOH

(q) 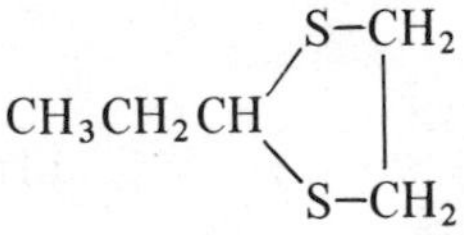 $CH_3CH_2CH\begin{smallmatrix}S-CH_2\\ | \\ S-CH_2\end{smallmatrix}$

(r) $CH_3CH_2CH_3 + CH_3CH_3 + NiS$

18.31

(a) $CH_3CHOHCH_3$

(b) $C_6H_5\underset{\underset{CH_3}{|}}{C}OHCH_3$

(c) $CH_3CHOHCH_3$

(d) $CH_3\underset{\underset{CH_3}{|}}{\overset{\overset{OH}{|}}{C}}SO_3Na$

(e) $CH_3\underset{\underset{CH_3}{|}}{\overset{\overset{OH}{|}}{C}}CN$

(f) $CH_3COCH_2\underset{\underset{CH_3}{|}}{\overset{\overset{OH}{|}}{C}}CH_3$

(g) $CH_3COCH=\underset{\underset{CH_3}{|}}{C}CH_3$

(h) $CH_3CHOHCH_3$

(i)

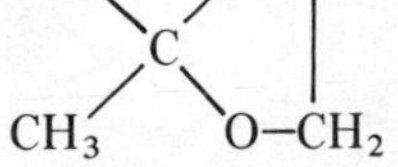

(j) $CH_3CH=C(CH_3)_2$

(k) CH_3COCH_2Br

(o) $CH_3C{=}NNHC_6H_5$
　　　　　|
　　　　CH_3

(l) No reaction

(p) No reaction

(m) $CH_3C{=}NOH$
　　　　|
　　　CH_3

(q)

$$\underset{CH_3}{\overset{CH_3}{{>}}}C\underset{S-CH_2}{\overset{S-CH_2}{{<}}}$$

(n) $CH_3C{=}NNHCONH_2$
　　　　|
　　　CH_3

(r) $CH_3CH_2CH_3 + CH_3CH_3 + NiS$

18.32

(a) $CH_3-\langle\bigcirc\rangle-CH{=}CHCHO$

(d) $CH_3-\langle\bigcirc\rangle-COOH$

(b) $CH_3-\langle\bigcirc\rangle-COO^- + CH_3-\langle\bigcirc\rangle-CH_2OH$

(e) $HOOC-\langle\bigcirc\rangle-COOH$

(c) $CH_3-\langle\bigcirc\rangle-CH_2OH + HCOO^-$

(f) $CH_3-\langle\bigcirc\rangle-CH{=}CH_2$

18.33

(a)

$$\langle\bigcirc\rangle\overset{\overset{O}{\parallel}}{\underset{NO_2}{C-CH_3}}$$

(d)

$$\langle\bigcirc\rangle-\underset{OH}{\overset{}{CHCH_3}}$$

(b) $\langle\bigcirc\rangle-COO^- + CHCl_3$

(e)

$$\langle\bigcirc\rangle-\underset{OH}{\overset{CH_3}{\underset{|}{C}}}-\langle\bigcirc\rangle$$

(c)

$$\langle\bigcirc\rangle-C\underset{CH_3}{\overset{CH_2}{{<}}}$$

18.34

(a) $\langle\bigcirc\rangle$ + $CH_3CH_2CH_2COCl$ $\xrightarrow{AlCl_3}$ $\langle\bigcirc\rangle-\overset{\overset{O}{\parallel}}{C}CH_2CH_2CH_3$

$\langle\bigcirc\rangle$ + $(CH_3CH_2CH_2CO)_2O$ $\xrightarrow{AlCl_3}$ $\langle\bigcirc\rangle-\overset{\overset{O}{\parallel}}{C}CH_2CH_2CH_3$

$\langle\bigcirc\rangle$ $\xrightarrow[Fe]{Br_2}$ $\langle\bigcirc\rangle-Br$ $\xrightarrow[(2)\ CdCl_2]{(1)\ Mg,\ ether}$ $\left(\langle\bigcirc\rangle-\right)_2 Cd$

$$CH_3CH_2CH_2\overset{\overset{\displaystyle O}{\|}}{C}Cl \longrightarrow \langle\!\bigcirc\!\rangle-\overset{\overset{\displaystyle O}{\|}}{C}CH_2CH_2CH_3$$

(b)

$$\langle\!\bigcirc\!\rangle-\overset{\overset{\displaystyle O}{\|}}{C}CH_2CH_2CH_3$$

$$\xrightarrow[\text{HCl}]{\text{Zn(Hg)}} \langle\!\bigcirc\!\rangle-CH_2CH_2CH_2CH_3$$

$$\xrightarrow[\text{OH}^-]{\text{NH}_2\text{NH}_2} \langle\!\bigcirc\!\rangle-CH_2CH_2CH_2CH_3$$

$$\xrightarrow[\text{H}^+]{\text{HSCH}_2\text{CH}_2\text{H}} \langle\!\bigcirc\!\rangle-C\overset{S-CH_2}{\underset{S-CH_2}{<}}$$

with CH_2, CH_2, CH_3 chain

$$\xrightarrow[\text{Ni}]{\text{Raney}} \langle\!\bigcirc\!\rangle-CH_2CH_2CH_2CH_3$$

18.35

(a) $\langle\!\bigcirc\!\rangle-CH_2OD$ + $\langle\!\bigcirc\!\rangle-COO^-$

(b) $\langle\!\bigcirc\!\rangle-CH_2OH$ + $\langle\!\bigcirc\!\rangle-COO^-$

(c) Yes. In both reactions a hydride ion (rather than a deuteride ion) is transferred to benzaldehyde. This shows that the hydride ion is transferred from one benzaldehyde molecule to another (as shown on page 705) and not from the solvent to benzaldehyde.

18.36

(a) $Ag(NH_3)_2{}^+OH^-$ (positive test with benzaldehyde)

(b) $Ag(NH_3)_2{}^+OH^-$ (positive test with hexanal)

(c) I_2 in NaOH (Iodoform, from 2-hexanone)

(d) CrO_3 in H_2SO_4 (positive test with 2-hexanol)

(e) I_2 in NaOH (Iodoform from 2-hexanol)

(f) CrO_3 in H_2SO_4 (positive test with 3-hexanol)

(g) Br_2 in CCl_4 (decolorization with benzalacetophenone)

(h) I_2 in NaOH (Iodoform from 1-phenylethanol)

(i) $Ag(NH_3)_2{}^+OH^-$ (positive test with pentanal)

(j) Br$_2$ in CCl$_4$ (immediate decolorization occurs with enol form)

(k) Ag(NH$_3$)$_2$$^+OH^-$ (positive test with cyclic hemiacetal)

18.37
(a) In simple addition the carbonyl peak (1665-1780 cm^{-1} region) does not appear in the product; in conjugate addition it does.

(b) As the reaction takes place the long wavelength absorption arising from the conjugated system should disappear. One could follow the rate of the reaction by following the rate at which this absorption peak disappears.

18.38
(a) The conjugate base is a hybrid of the following structures:

$$^-:CH_2-CH=CH-\overset{\overset{O}{\|}}{C}H \longleftrightarrow CH_2=CH-\overset{\overset{O}{\|}}{C}H-\overset{..}{C}H \longleftrightarrow CH_2=CH-CH=\overset{\overset{O^-}{|}}{C}H$$

This structure is especially stable because the negative charge is on oxygen

(b) $CH_3CH=CHCHO \underset{+H^+}{\overset{-H^+}{\rightleftharpoons}} :^-CH_2CH=CHCHO$

$$C_6H_5CH=CH\overset{\overset{O}{\|}}{C}H + {}^-:CH_2CH=CHCHO \rightleftharpoons$$

$$C_6H_5\overset{\overset{O^-}{|}}{C}H=CHCH-CH_2CH=CHCHO \underset{-H^+}{\overset{+H^+}{\rightleftharpoons}} C_6H_5\overset{\overset{OH}{|}}{C}H=CHCH-CH_2CH=CHCHO$$

$$\overset{-H_2O}{\longrightarrow} C_6H_5CH=CHCH=CHCH=CHCHO$$

18.39

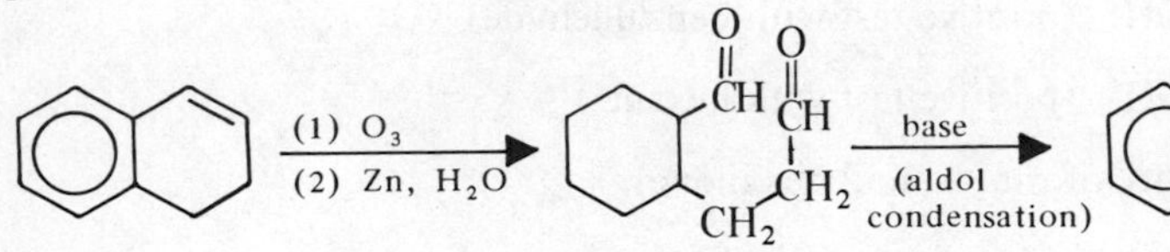

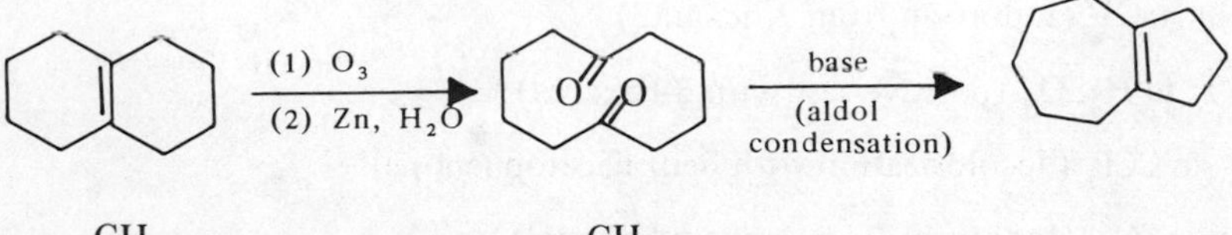

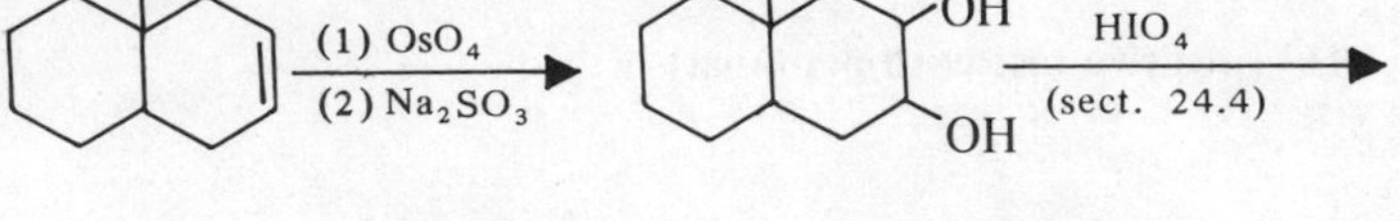

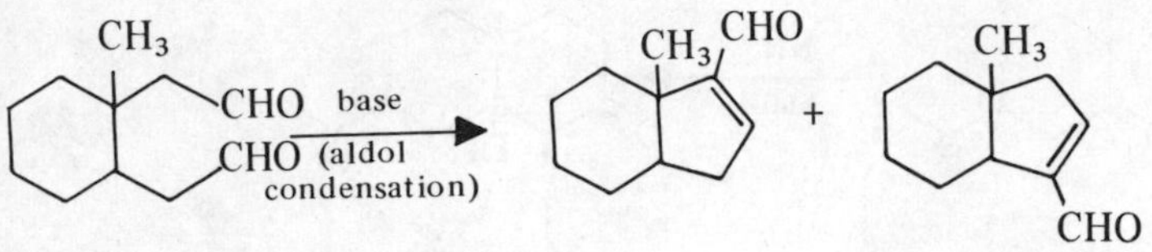

(d)

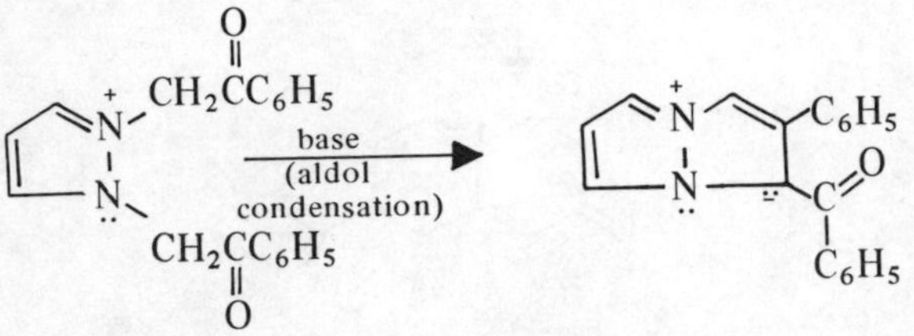

18.40

(a)

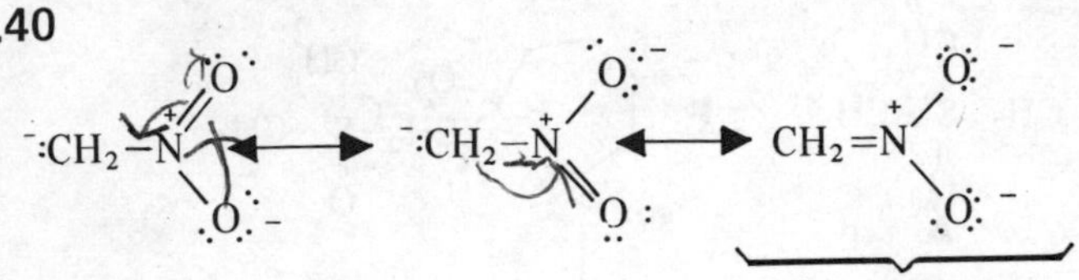

This structure is especially stable because both negative charges are on oxygen

(b) All of these syntheses are variations of a crossed aldol addition or condensation.

$$\underset{HCH}{\overset{O}{\parallel}} + CH_3NO_2 \xrightarrow{base} HOCH_2CH_2NO_2$$

$$C_6H_5CHO + CH_3NO_2 \xrightarrow[(-H_2O)]{base} C_6H_5CH=CHNO_2$$

$$C_6H_5CHO + CH_3CH_2NO_2 \xrightarrow[(-H_2O)]{base} C_6H_5CH=\underset{\underset{CH_3}{|}}{C}NO_2$$

18.41

First an elimination takes place,

$$R_3\overset{+}{N}CH_2CH_2\overset{O}{\overset{\parallel}{C}}CH_2CH_3 + NH_2^- \longrightarrow CH_2=CH\overset{O}{\overset{\parallel}{C}}CH_2CH_3 + R_3N + NH_3$$

then a conjugate addition occurs, followed by an aldol addition:

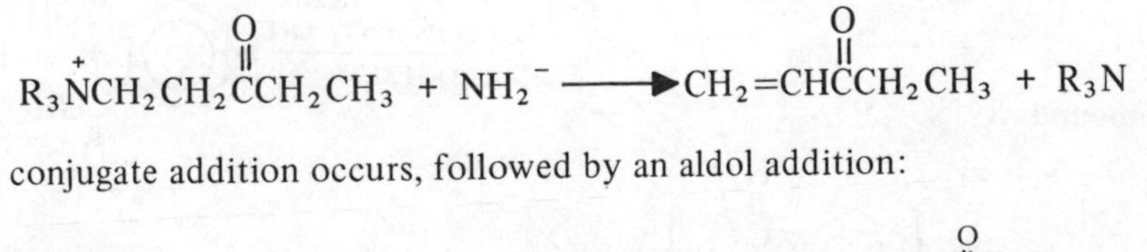

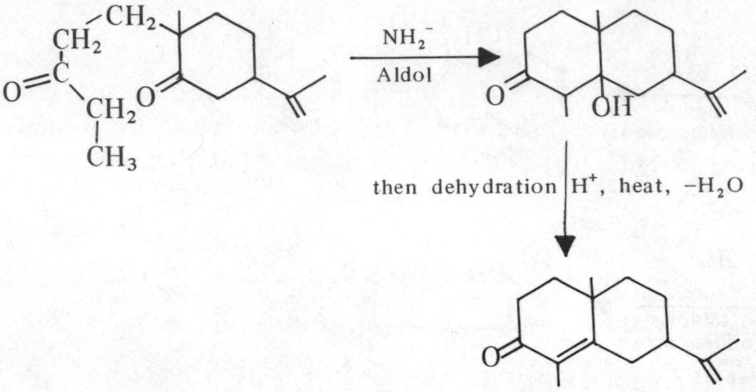

18.42

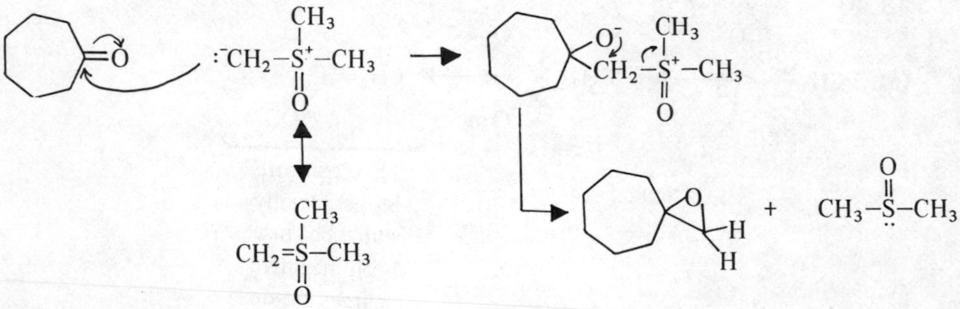

18.43

(a) Compound U is phenyl ethyl ketone:

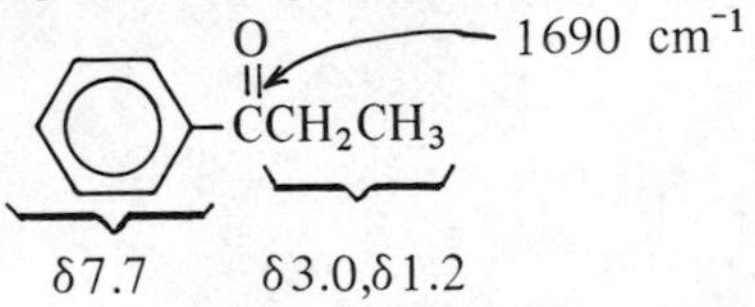

$\delta 7.7 \quad \delta 3.0, \delta 1.2$

(b) Compound V is benzyl methyl ketone:

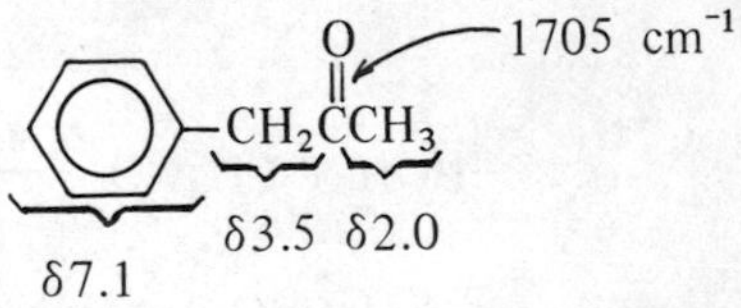

$\delta 7.1$

$\delta 3.5 \quad \delta 2.0$

18.44

Compound W is:

multiplet, δ 7.3

singlet δ 3.4

infrared peak near 1715 cm^{-1}

$$\xrightarrow[\text{(2) } H_3O^+]{\substack{\text{heat} \\ \text{(1) KMnO}_4, \text{ OH}^-}}$$

Phthalic aicd

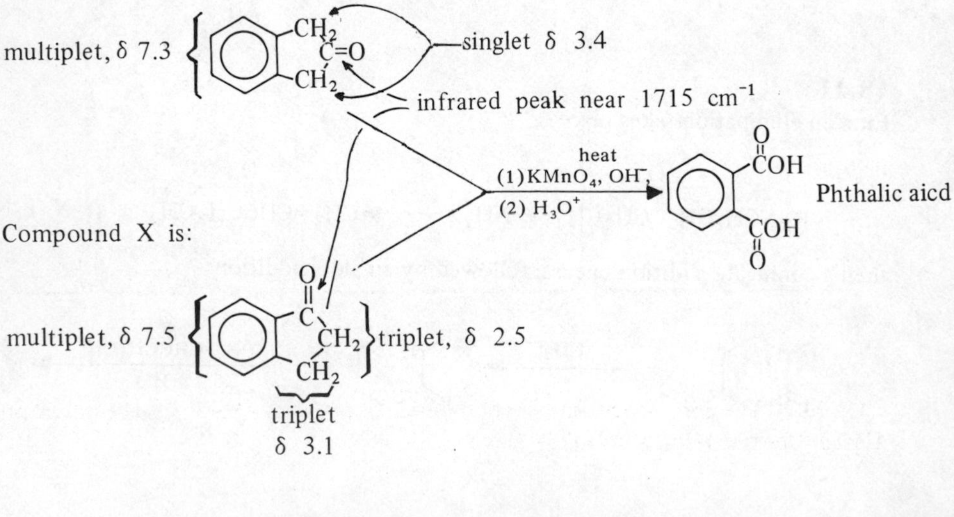

Compound X is:

multiplet, δ 7.5

triplet, δ 2.5

triplet
δ 3.1

18.45

The pmr spectra (Figures 18.2 and 18.3) each have a five hydrogen peak near $\delta 7.1$, suggesting that Y and Z each have a C_6H_5- group. The infrared spectrum of each compound shows a strong peak near 1705 cm^{-1}. This indicates that each compound has a C=O group. We have, therefore, the following pieces,

$$\langle\bigcirc\rangle- \quad \text{and} \quad -\overset{\overset{\text{O}}{\|}}{\text{C}}-$$

If we subtract the atoms of these pieces from the molecular formula,

$$\begin{array}{l} C_{10}H_{12}O \\ -\ \underline{C_7\ H_5\ O}\ (C_6H_5\ +\ C{=}O) \end{array}$$

We are left with, $\overline{C_3\ H_7}$

In the pmr spectrum of Y we see an ethyl group [triplet, $\delta 1.0$ (3H) and quartet, $\delta 2.3$ (2H)] and an unsplit $-CH_2-$ group [singlet, $\delta 3.7$ (2H)]. This means that Y must be,

$$\langle\bigcirc\rangle-CH_2\overset{\overset{\text{O}}{\|}}{C}CH_2CH_3$$

1-Phenyl-2-butanone

In the pmr spectrum of Z, we see an unsplit $-CH_3$ group [singlet, $\delta 2.0$ (3H)] and a multiplet (actually two superimposed triplets) at $\delta 2.8$. This means Z must be,

$$\langle\bigcirc\rangle-CH_2CH_2\overset{\overset{\text{O}}{\|}}{C}CH_3$$

4-Phenyl-2-butanone

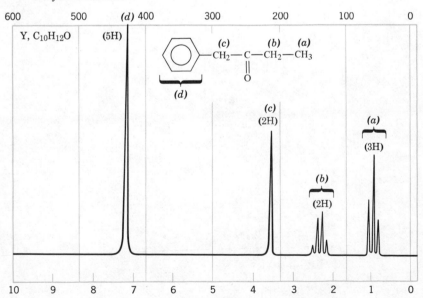

FIG. 18.2. The pmr spectrum of compound Y, problem 18.45. (Spectrum courtesy of Aldrich Chemical Co., Milwaukee, Wis.)

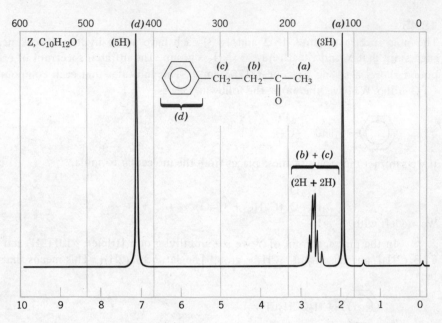

FIG. 18.3. The pmr spectrum of compound Z, problem 18.45. (Spectrum courtesy of Aldrich Chemical Co., Milwaukee, Wis.)

19

CARBOXYLIC ACIDS
AND THEIR DERIVATIVES:
NUCLEOPHILIC SUBSTITUTION
AT ACYL CARBON

19.1

(a) Carbon dioxide is an acid; it converts an aqueous solution of the strong base, NaOH, into an aqueous solution of the weaker base, $NaHCO_3$.

$$NaOH_{(aq)} + CO_2 \longrightarrow NaHCO_{3\,(aq)}$$

In this new solution, the more strongly basic p-cresoxide ion accepts a proton and becomes p-cresol,

$$p\text{-}CH_3C_6H_4O^- + HCO_3^- \rightleftharpoons p\text{-}CH_3C_6H_4OH + CO_3^=$$
$$\text{water-insoluble}$$

The more weakly basic benzoate ion remains in solution.

(b) Dissolve all three compounds in an organic solvent such as CH_2Cl_2, then extract with aqueous NaOH. The organic layer will contain cyclohexanol, which can be separated by distillation. The aqueous layer will contain the benzoic acid, as sodium benzoate, and the p-cresol, as sodium p-cresoxide.

Now pass CO_2 into the aqueous layer; this will cause p-cresol to separate (it can then be extracted into an organic solvent and purified by distillation). After separation of the p-cresol, the aqueous phase can be acidified with aqueous HCl to yield benzoic acid as a precipitate.

19.2

(a) CH_2FCOOH (F— is more electronegative than H—)

(b) CH_2FCOOH (F— is more electronegative than Cl—)

(c) $CH_2ClCOOH$ (Cl— is more electronegative than Br—)

(d) $CH_3CHClCH_2COOH$ (Cl— is closer to —COOH)

(e) $CH_3CH_2CHClCOOH$ (Cl— is closer to —COOH)

(f) $(CH_3)_3\overset{+}{N}$—⬡—COOH [$(CH_3)_3\overset{+}{N}$— is more electronegative than H—]

(g) CF_3—⬡—COOH (CF_3— is more electronegative than CH_3—)

193

19.3

(a) The carboxyl group is an electron-withdrawing group; thus in a dicarboxylic acid such as those in Table 19.3, one carboxyl group increases the acidity of the other.

(b) As the distance between the carboxyl groups increases the acid-strengthening, inductive effect decreases.

19.4

These syntheses are easy to see if we work backward.

(a) $C_6H_5CH_2COOH \xleftarrow[\text{(2) H}^+]{\text{(1) CO}_2} C_6H_5CH_2MgBr$

$\uparrow$ Mg, ether

$C_6H_5CH_2Br$

(b) $CH_3CH_2CH_2\overset{\overset{\displaystyle CH_3}{|}}{\underset{\underset{\displaystyle CH_3}{|}}{C}}COOH \xleftarrow[\text{(2) H}^+]{\text{(1) CO}_2} CH_3CH_2CH_2\overset{\overset{\displaystyle CH_3}{|}}{\underset{\underset{\displaystyle CH_3}{|}}{C}}MgBr$

$\uparrow$ Mg, ether

$CH_3CH_2CH_2\overset{\overset{\displaystyle CH_3}{|}}{\underset{\underset{\displaystyle CH_3}{|}}{C}}Br$

(c) $CH_2{=}CHCH_2COOH \xleftarrow[\text{(2) H}^+]{\text{(1) CO}_2} CH_2{=}CHCH_2MgBr$

$\uparrow$ Mg, ether

$CH_2{=}CHCH_2Br$

(d) $CH_3-\!\!\bigcirc\!\!-COOH \xleftarrow[\text{(2) H}^+]{\text{(1) CO}_2} CH_3-\!\!\bigcirc\!\!-MgBr$

$\uparrow$ Mg, ether

$CH_3-\!\!\bigcirc\!\!-Br$

(e) $CH_3CH_2CH_2CH_2CH_2COOH \xleftarrow[\text{(2) H}_2\text{O}]{\text{(1) CO}_2} CH_3CH_2CH_2CH_2CH_2MgBr$

$\uparrow$ Mg, ether

$CH_3CH_2CH_2CH_2CH_2Br$

19.5

(a) $C_6H_5CH_2COOH \xleftarrow[\text{(2) H}^+,\ \text{H}_2\text{O, heat}]{\text{(1) CN}^-} C_6H_5CH_2Br$

$$CH_2=CHCH_2COOH \xleftarrow[\text{(2) H}^+\text{, H}_2\text{O, heat}]{\text{(1) CN}^-} CH_2=CHCH_2Br$$

$$CH_3CH_2CH_2CH_2COOH \xleftarrow[\text{(2) H}^+\text{, H}_2\text{O, heat}]{\text{(1) CN}^-} CH_3CH_2CH_2CH_2Br$$

(b) A nitrile synthesis. Preparation of a Grignard reagent from $HOCH_2CH_2CH_2CH_2Br$ would not be possible because of the presence of the acidic hydroxyl group.

19.6

(a) $CH_3COOH + C_6H_5COCl \xrightarrow{\text{pyridine}}$ $CH_3\overset{\overset{\displaystyle O}{\|}}{C}O\overset{\overset{\displaystyle O}{\|}}{C}C_6H_5$

(b) $CH_3(CH_2)_4COOH + (CH_3CO)_2O \xrightarrow{\text{heat}}$

$$CH_3(CH_2)_4\overset{\overset{\displaystyle O}{\|}}{C}O\overset{\overset{\displaystyle O}{\|}}{C}(CH_2)_4CH_3 + 2CH_3COOH$$
(remove by distillation)

(c)

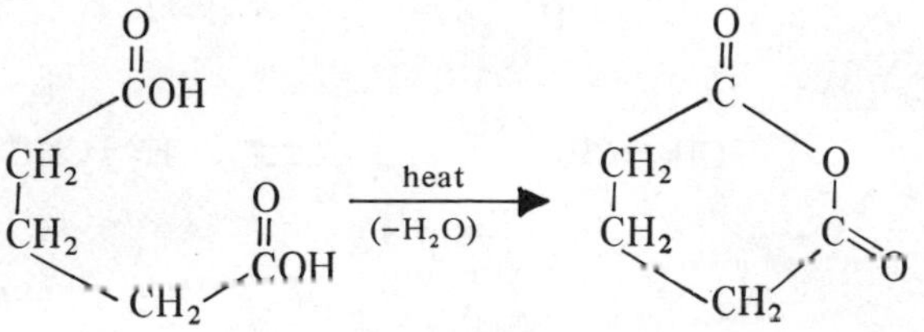

19.7

Since maleic acid is a *cis*-dicarboxylic acid, dehydration occurs readily:

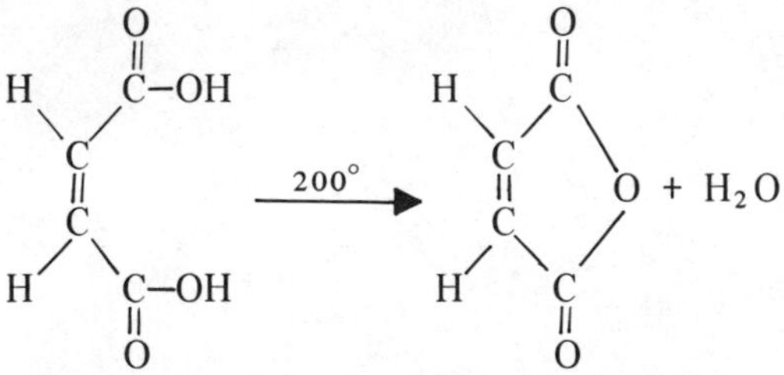

Maleic acid   Maleic anhydride

Being a *trans*-dicarboxylic acid, fumaric acid must undergo isomerization to maleic acid first. This requires a higher temperature.

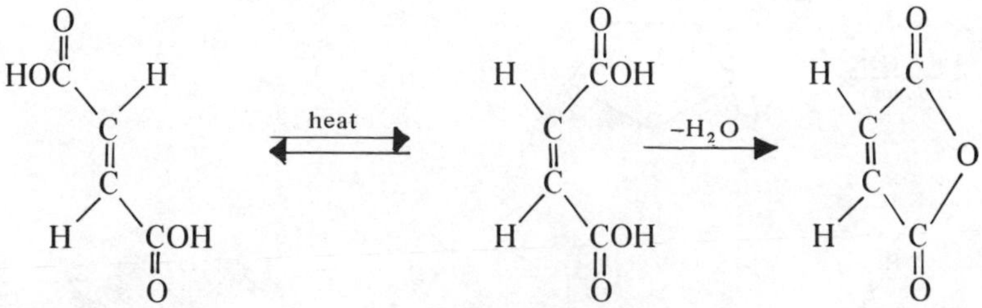

Fumaric acid

19.8

The labeled oxygen should appear in the carboxyl group of the acid. (Follow the reverse steps of the mechanism on page 756 of the text using $H_2{}^{18}O$.)

19.9

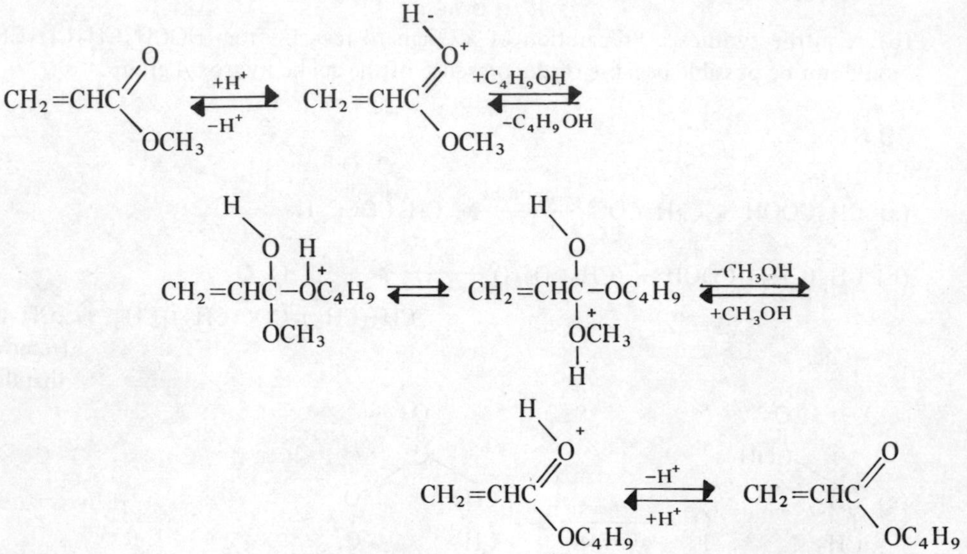

$$CH_2=CHC\underset{OCH_3}{\overset{O}{\diagup}} \underset{-H^+}{\overset{+H^+}{\rightleftharpoons}} CH_2=CHC\underset{OCH_3}{\overset{\overset{H-}{\overset{|}{O^+}}}{\diagup}} \underset{-C_4H_9OH}{\overset{+C_4H_9OH}{\rightleftharpoons}}$$

$$CH_2=CHC\overset{\overset{H}{\underset{|}{O}}\ \overset{}{\overset{H}{\underset{+}{|}}}}{\underset{\overset{|}{OCH_3}}{-OC_4H_9}} \rightleftharpoons CH_2=CHC\overset{\overset{H}{\underset{|}{O}}}{\underset{\overset{\overset{|}{OCH_3}}{\overset{+}{|}}}{-OC_4H_9}}\overset{}{\underset{H}{}} \underset{+CH_3OH}{\overset{-CH_3OH}{\rightleftharpoons}}$$

$$CH_2=CHC\underset{OC_4H_9}{\overset{\overset{H}{\underset{|}{O^+}}}{\diagup}} \underset{+H^+}{\overset{-H^+}{\rightleftharpoons}} CH_2=CHC\underset{OC_4H_9}{\overset{O}{\diagup}}$$

19.10

(a)

(1) $+\ C_6H_5SO_2Cl \longrightarrow$ (A) $\xrightarrow[\text{(inversion)}]{OH^-,\ heat}$

(B) $+\ C_6H_5SO_3^-$

(2) $+\ C_6H_5\overset{O}{\overset{||}{C}}-Cl \longrightarrow$ (C)

$\xrightarrow[\text{(retention)}]{OH^-,\ heat}$ (D) $+\ C_6H_5CO_2^-$

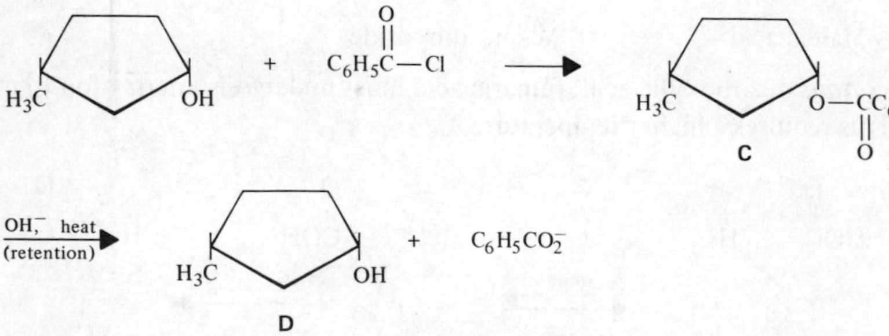

(3) $+\ CH_3\overset{O}{\overset{||}{C}}O^-Na^+ \xrightarrow[\text{(inversion)}]{}$ (E)

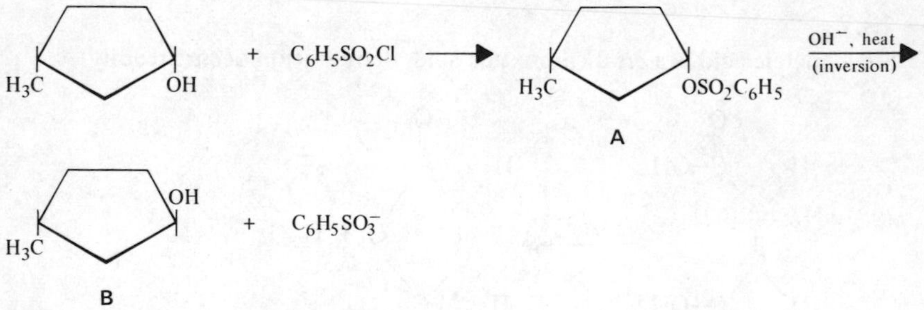

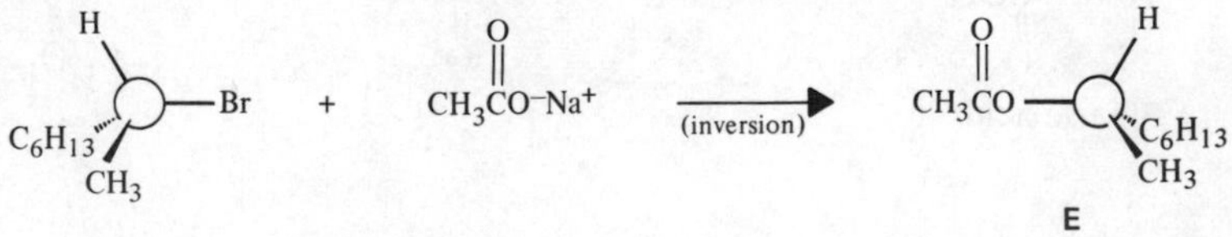

$$\xrightarrow[\text{(retention)}]{\text{OH}^-,\ \text{heat}}$$

HO—C(H)("C_6H_{13})(CH_3)

F

(4) (H)(C_6H_{13})(CH_3)C—Br $\xrightarrow[\text{(inversion)}]{\text{OH}^-,\ \text{heat}}$ HO—C(H)("C_6H_{13})(CH_3)

(b) Method (3) should give a higher yield of **F** than method (4). Since the hydroxide ion is a strong base and since the alkyl halide is secondary, method (4) is likely to be accompanied by considerable elimination. Method (3), on the other hand, employs a weaker base, acetate ion, in the S_N2 step and is less likely to be complicated by elimination. Hydrolysis of the ester **E** that results should also proceed in high yield.

19.11

(a) Steric hindrance presented by the di-*ortho* methyl groups of methyl mesitoate prevents formation of the tetrahedral intermediate that must accompany attack at the acyl carbon.

(b) Carry out hydrolysis with labeled OH^- in labeled H_2O. The label should appear in the methanol.

19.12

(a) $C_6H_5\overset{O}{\underset{\|}{C}}N(CH_2CH_3)_2$

$\xrightarrow[H_2O]{OH^-}$ $C_6H_5COO^-$ + $(CH_3CH_2)_2NH$

$\xrightarrow[H_2O]{H^+}$ C_6H_5COOH + $(CH_3CH_2)_2\overset{+}{N}H_2$

(b) [cyclic amide structure]

$\xrightarrow[H_2O]{OH^-}$ $^-O\overset{O}{\underset{\|}{C}}CH_2CH_2CH_2CH_2NH_2$

$\xrightarrow[H_2O]{H^+}$ $HO\overset{O}{\underset{\|}{C}}CH_2CH_2CH_2CH_2\overset{+}{N}H_3$

(c) $HOOC\underset{\underset{CH_3}{|}}{C}H-NH\overset{O}{\underset{\|}{C}}\underset{\underset{\underset{C_6H_5}{|}}{CH_2}}{C}HNH_2$

$\xrightarrow[H_2O]{OH^-}$ $^-OOCC\underset{\underset{CH_3}{|}}{H}NH_2$ + $^-OOCC\underset{\underset{\underset{C_6H_5}{|}}{CH_2}}{H}NH_2$

$\xrightarrow[H_2O]{H^+}$ $HOOCC\underset{\underset{CH_3}{|}}{H}\overset{+}{N}H_3$ + $HOOCC\underset{\underset{\underset{C_6H_5}{|}}{CH_2}}{H}\overset{+}{N}H_3$

19.13

(a) $(CH_3)_3CCOOH \xrightarrow{SOCl_2} (CH_3)_3CCOCl$

$\xrightarrow{NH_3} (CH_3)_3CCONH_2 \xrightarrow[\text{heat}]{P_2O_5} (CH_3)_3CC\equiv N$

(b) An elimination reaction would take place.

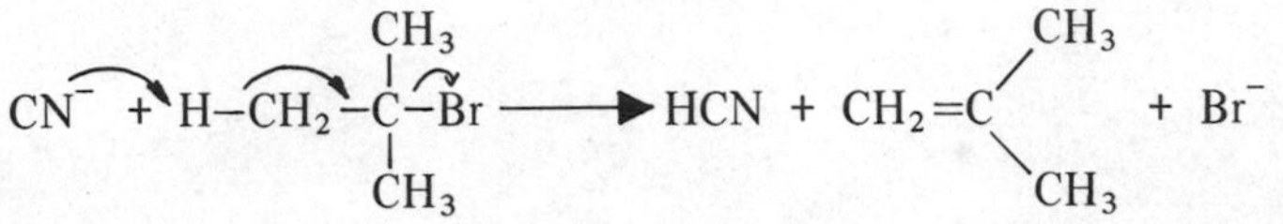

$$CN^- + H-CH_2-\underset{\underset{CH_3}{|}}{\overset{\overset{CH_3}{|}}{C}}-Br \longrightarrow HCN + CH_2=C\overset{CH_3}{\underset{CH_3}{}} + Br^-$$

19.14

(a) $CH_3(CH_2)_4COOH$

(b) $CH_3(CH_2)_4CONH_2$

(c) $CH_3(CH_2)_4CONHC_2H_5$

(d) $CH_3(CH_2)_4CON(C_2H_5)_2$

(e) $CH_3CH_2CH=CHCH_2COOH$

(f) $CH_3CH=CHCH_2\underset{\underset{CH_3}{|}}{C}HCOOH$

(g) $HOOCCH_2CH_2CH_2CH_2COOH$

(h)

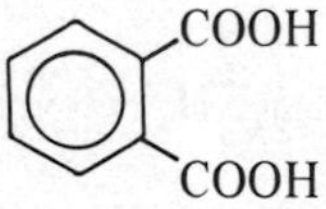

(i)

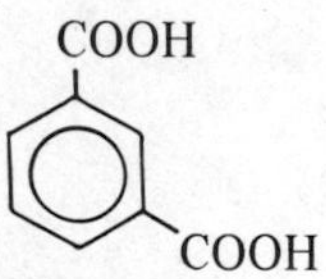

(j)

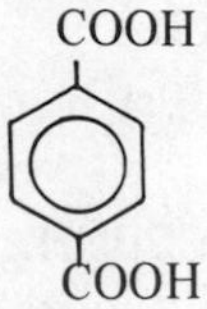

(k) $C_2H_5OOC-COOC_2H_5$

(l) $C_2H_5OOC(CH_2)_4COOC_2H_5$

(m) $CH_3CH_2COOCH_2CH(CH_3)_2$

(n)

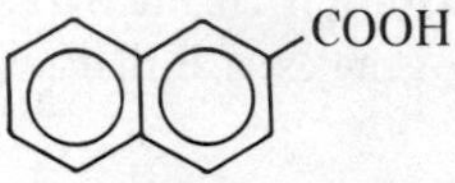

(o)

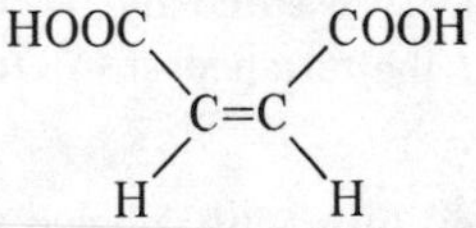

(p) $HOOCCHOHCH_2COOH$

(q)

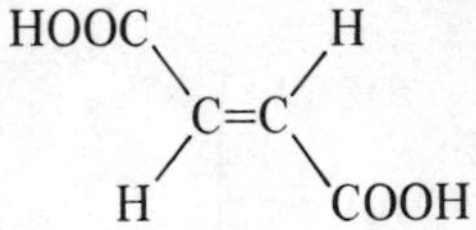

(r) $HOOCCH_2CH_2COOH$

(s)

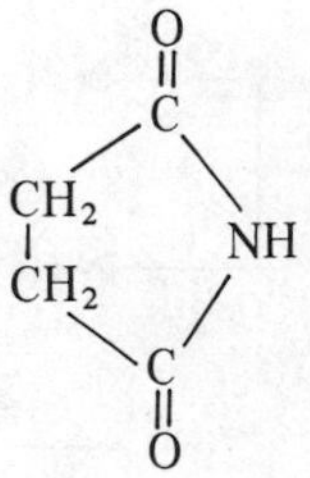

(t) $HOOCCH_2COOH$

(u) $C_2H_5OOCCH_2COOC_2H_5$

19.15

(a) Benzoic acid

(b) Benzoyl chloride

(c) Benzamide

(d) Benzoic anhydride

(e) Benzyl benzoate

(f) Phenyl benzoate

(g) Isopropyl acetate

(h) N,N-Dimethylacetamide

(i) Acetonitrile

(j) Maleic anhydride

(k) Phthalic anhydride

(l) Phthalimide

(m) Glyceryl tripalmitate

(n) α-Ketosuccinic acid (oxaloacetic acid)

(o) Methyl salicylate

19.16

(a) $\text{C}_6\text{H}_5\text{—Br} \xrightarrow{\text{Mg, ether}} \text{C}_6\text{H}_5\text{—MgBr} \xrightarrow[\text{(2) } H_3O^+]{\text{(1) } CO_2} \text{C}_6\text{H}_5\text{—COOH}$

(b) $\text{C}_6\text{H}_5\text{—CH}_3 \xrightarrow[\text{(2) } H_3O^+]{\text{(1) } KMnO_4, \ OH^-, \ heat} \text{C}_6\text{H}_5\text{—COOH}$

(c) $\text{C}_6\text{H}_5\text{—CN} \xrightarrow[\text{heat}]{H_3O^+, \ H_2O} \text{C}_6\text{H}_5\text{—COOH} + NH_4^+$

(d) $\text{C}_6\text{H}_5\overset{\overset{\text{O}}{\|}}{\text{—CCH}_3} \xrightarrow[(-CHCl_3)]{Cl_2, \ OH^-} \text{C}_6\text{H}_5\text{—COO}^- \xrightarrow{H_3O^+} \text{C}_6\text{H}_5\text{—COOH}$

(e) $\text{C}_6\text{H}_5\text{—CHO} \xrightarrow{Ag(NH_3)_2^+ \ OH^-} \text{C}_6\text{H}_5\text{—COO}^- \xrightarrow{H_3O^+} \text{C}_6\text{H}_5\text{—COOH}$

(f) $\text{C}_6\text{H}_5\text{—CH=CH}_2 \xrightarrow[\text{(2) } H_3O^+]{\text{(1) } KMnO_4, \ OH^-, \ heat} \text{C}_6\text{H}_5\text{—COOH}$

(g) $\text{C}_6\text{H}_5\text{—CH}_2\text{OH} \xrightarrow[\text{(2) } H_3O^+]{\text{(1) } KMnO_4, \ OH^-, \ heat} \text{C}_6\text{H}_5\text{—COOH}$

19.17

(a) $\text{C}_6\text{H}_5\text{—CH}_2\text{CHO} \xrightarrow{Ag(NH_3)_2^+ \ OH^-} \text{C}_6\text{H}_5\text{—CH}_2\text{COO}^- \xrightarrow{H_3O^+}$

$\text{C}_6\text{H}_5\text{—CH}_2\text{COOH}$

(b) $\bigcirc$—CH$_2$Br $\xrightarrow[\text{(2) CO}_2]{\text{(1) Mg, ether}}$ $\bigcirc$—CH$_2$COOMgBr $\xrightarrow{\text{H}_3\text{O}^+}$

$\bigcirc$—CH$_2$COOH

$\bigcirc$—CH$_2$Br $\xrightarrow{\text{CN}^-}$ $\bigcirc$—CH$_2$CN $\xrightarrow[\text{heat}]{\text{H}_3\text{O}^+, \text{ H}_2\text{O}}$

$\bigcirc$—CH$_2$COOH

19.18

(a) CH$_3$CH$_2$CH$_2$CH$_2$CH$_2$OH $\xrightarrow[\text{(2) H}_3\text{O}^+]{\text{(1) KMnO}_4\text{, OH}^-\text{, heat}}$ CH$_3$CH$_2$CH$_2$CH$_2$COOH

(b) CH$_3$CH$_2$CH$_2$CH$_2$Br $\xrightarrow[\text{(2) CO}_2]{\text{(1) Mg, ether}}$ CH$_3$CH$_2$CH$_2$CH$_2$COOMgBr $\xrightarrow{\text{H}_3\text{O}^+}$

CH$_3$CH$_2$CH$_2$CH$_2$COOH

CH$_3$CH$_2$CH$_2$CH$_2$Br $\xrightarrow{\text{CN}^-}$ CH$_3$CH$_2$CH$_2$CH$_2$CN $\xrightarrow[\text{heat}]{\text{H}_3\text{O}^+, \text{ H}_2\text{O}}$

CH$_3$CH$_2$CH$_2$CH$_2$COOH

(c) CH$_3$CH$_2$CH$_2$CH$_2$CCH$_3$ $\xrightarrow[\text{(−CHCl}_3\text{)}]{\text{Cl}_2\text{, OH}^-}$ CH$_3$CH$_2$CH$_2$CH$_2$COO$^-$ $\xrightarrow{\text{H}_3\text{O}^+}$
$\quad\quad\quad\quad\quad\;\; \| $
$\quad\quad\quad\quad\quad\; \text{O}$

CH$_3$CH$_2$CH$_2$CH$_2$COOH

(d) CH$_3$(CH$_2$)$_3$CH=CH(CH$_2$)$_3$CH$_3$ $\xrightarrow[\text{(2) H}_3\text{O}^+]{\text{(1) KMnO}_4\text{, OH}^-\text{, heat}}$ 2CH$_3$(CH$_2$)$_3$COOH

(e) CH$_3$CH$_2$CH$_2$CH$_2$CHO $\xrightarrow[\text{(2) H}_3\text{O}^+]{\text{(1) Ag(NH}_3\text{)}_2{}^+\text{OH}^-}$ CH$_3$CH$_2$CH$_2$CH$_2$COOH

19.19

(a) CH$_3$COOH + HCl

(b) CH$_3$COOH + AgCl

(c) CH$_3$COOCH$_2$(CH$_2$)$_2$CH$_3$

(d) CH$_3$CONH$_2$

(e) $\bigcirc\begin{smallmatrix}\text{CH}_3\\\text{CCH}_3\\\|\\\text{O}\end{smallmatrix}$ + CH$_3$—$\bigcirc$—$\overset{\overset{\text{O}}{\|}}{\text{C}}CH_3$

(f) CH$_3$CHO

(g) CH$_3$COCH$_3$

(h) CH$_3$COCH$_2$CH$_3$

(i) CH$_3$CONHCH$_3$

(j) CH$_3$CONHC$_6$H$_5$

(k) CH$_3$CON(CH$_3$)$_2$

(l) CH$_3$COSCH$_2$CH$_3$

(m) (CH$_3$CO)$_2$O

(n) (CH$_3$CO)$_2$O

(o) CH$_3$COOC$_6$H$_5$

19.20

(a) CH_3CONH_2 + CH_3COONH_4

(b) $2CH_3COOH$

(c) $CH_3COOCH_2CH_2CH_3$ + CH_3COOH

(d) $C_6H_5COCH_3$ + CH_3COOH

(e) $CH_3CONHCH_2CH_3$ + CH_3COO^- $CH_3CH_2NH_3^+$

(f) $CH_3CON(CH_2CH_3)_2$ + CH_3COO^- $(CH_3CH_2)_2NH_2^+$

19.21

(a) 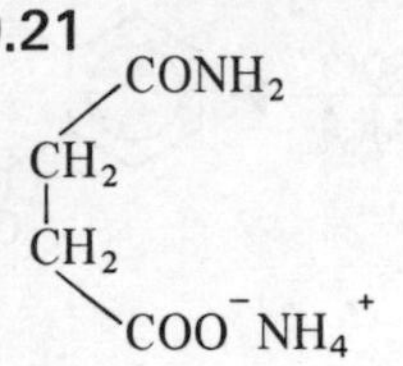
$$\begin{array}{l} CH_2-CONH_2 \\ | \\ CH_2 \\ COO^-NH_4^+ \end{array}$$

(b) 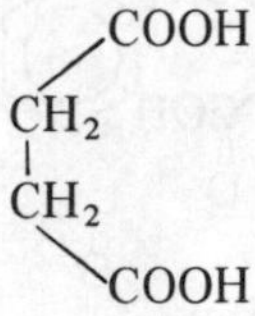
$$\begin{array}{l} CH_2-COOH \\ | \\ CH_2 \\ COOH \end{array}$$

(c) 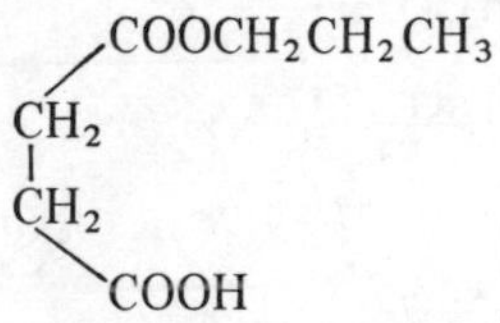
$$\begin{array}{l} CH_2-COOCH_2CH_2CH_3 \\ | \\ CH_2 \\ COOH \end{array}$$

(d)

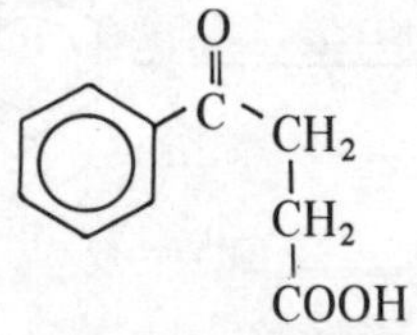

(e) 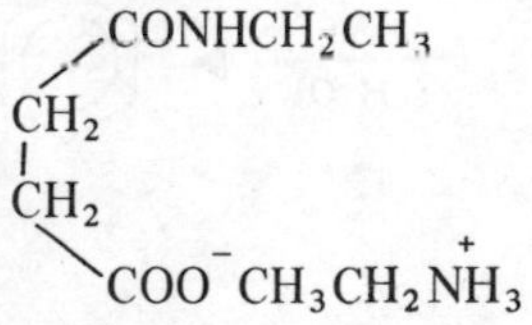
$$\begin{array}{l} CH_2-CONHCH_2CH_3 \\ | \\ CH_2 \\ COO^-CH_3CH_2NH_3^+ \end{array}$$

(f) 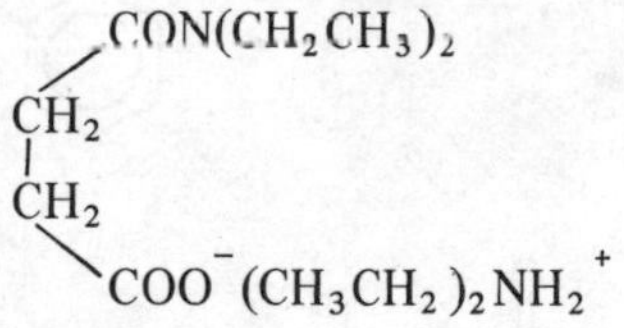
$$\begin{array}{l} CH_2-CON(CH_2CH_3)_2 \\ | \\ CH_2 \\ COO^-(CH_3CH_2)_2NH_2^+ \end{array}$$

19.22

(a)

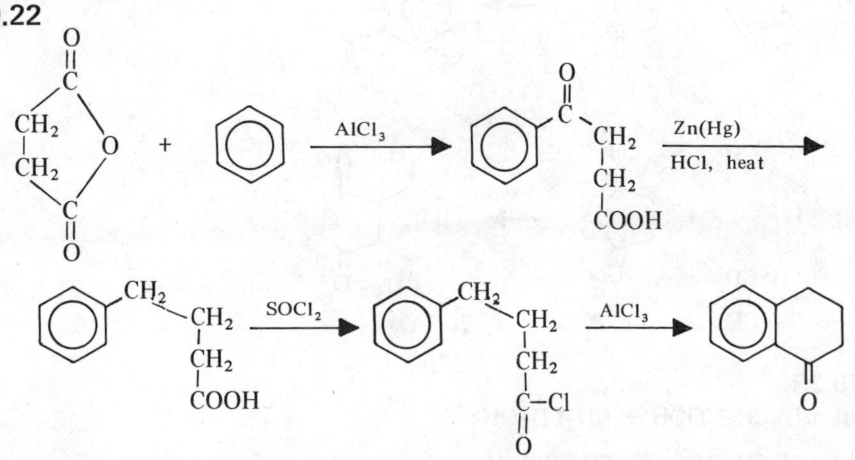

(b)

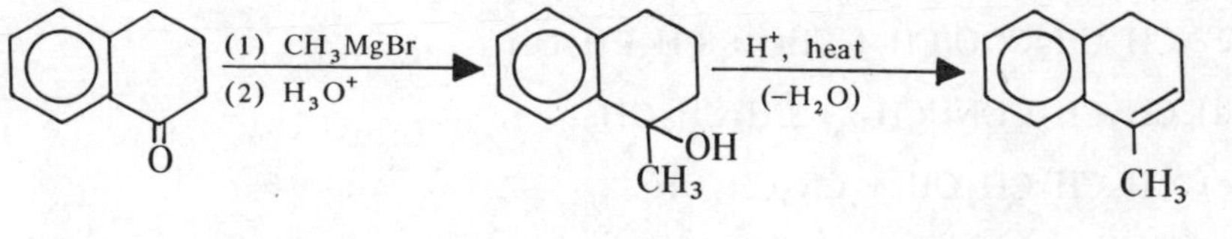

(c) 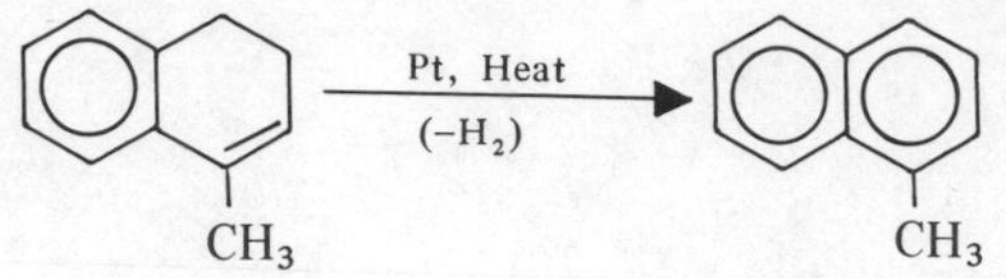 $\xrightarrow[\text{(}-H_2\text{)}]{\text{Pt, Heat}}$

(d) 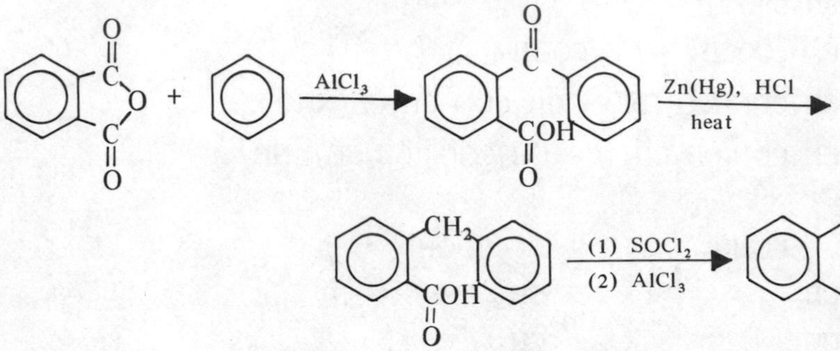 + $\xrightarrow{\text{AlCl}_3}$ $\xrightarrow[\text{heat}]{\text{Zn(Hg), HCl}}$

$\xrightarrow[\text{(2) AlCl}_3]{\text{(1) SOCl}_2}$

(e) 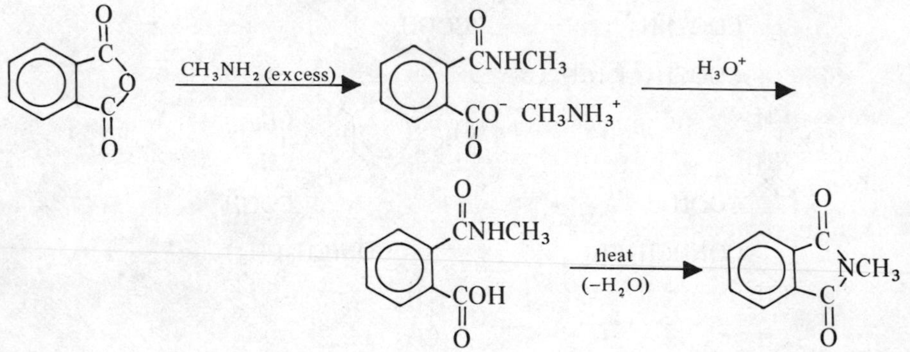 $\xrightarrow{\text{CH}_3\text{NH}_2\,(\text{excess})}$ $\xrightarrow{\text{H}_3\text{O}^+}$

$\xrightarrow[\text{(}-\text{H}_2\text{O}\text{)}]{\text{heat}}$

(f) 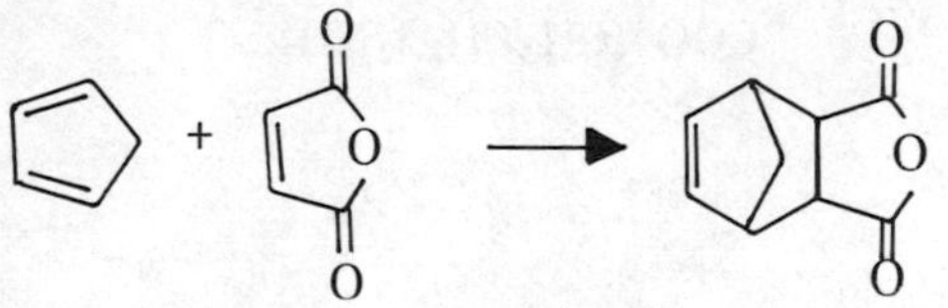 + $\longrightarrow$

(g) 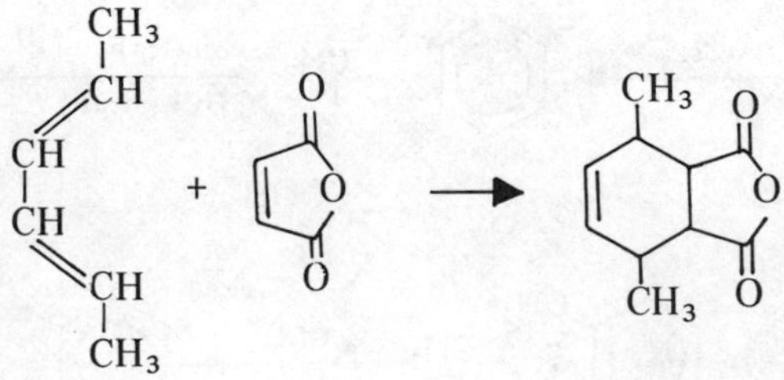 + $\longrightarrow$

19.23

(a) $CH_3CH_2COOH + CH_3CH_2OH$

(b) $CH_3CH_2COO^- + CH_3CH_2OH$

(c) $CH_3CH_2COO(CH_2)_7CH_3 + CH_3CH_2OH$

(d) $CH_3CH_2CONHCH_3 + CH_3CH_2OH$

(e) $CH_3CH_2CH_2OH + CH_3CH_2OH$

(f) $CH_3CH_2\overset{\underset{|}{C_6H_5}}{\underset{|}{\underset{OH}{C}}}-C_6H_5 + CH_3CH_2OH$

19.24

(a) $CH_3CH_2COOH + NH_4{}^+$

(b) $CH_3CH_2COO^- + NH_3$

(c) CH_3CH_2CN

19.25

(a) Benzoic acid dissolves in aqueous $NaHCO_3$. Methyl benzoate does not.

(b) Benzoyl chloride gives a precipitate (AgCl) when treated with alcoholic $AgNO_3$. Benzoic acid does not.

(c) Benzoic acid dissolves in aqueous $NaHCO_3$. Benzamide does not.

(d) Benzoic acid dissolves in aqueous $NaHCO_3$. p-Cresol does not.

(e) Refluxing benzamide with aqueous NaOH liberates NH_3 which can be detected in the vapors with moist red litmus paper. Ethyl benzoate does not liberate NH_3.

(f) Cinnamic acid, because it has a double bond, decolorizes Br_2 in CCl_4. Benzoic acid does not.

(g) Benzoyl chloride gives a precipitate (AgCl) when treated with alcoholic $AgNO_3$. Ethyl benzoate does not.

(h) α-Chlorobutanoic acid gives a precipitate (AgCl) when treated with alcoholic silver nitrate. Butanoic acid does not.

19.26

(a)
```
CH₂—CH₂
|         \
|          C=O
CH₂—O
```

(b) $CH_3CH=CHCOOH$

(c)
```
        O
        ||
        C—O
CH₃CH₂CH       CHCH₂CH₃
        O—C
           ||
           O
```

(d)
```
        CH₂—C
       /       \\O
CH₂             O
       \       /
        CH₂—C
              \\O
```

(e)
```
        CH₂—C
       /       \\O
CH₂             NH
       \       /
        CH₂—CH
              \
               CH₃
```

(f)
```
        O
        ||
        C   NH
            CH₂
```

19.27

(a)

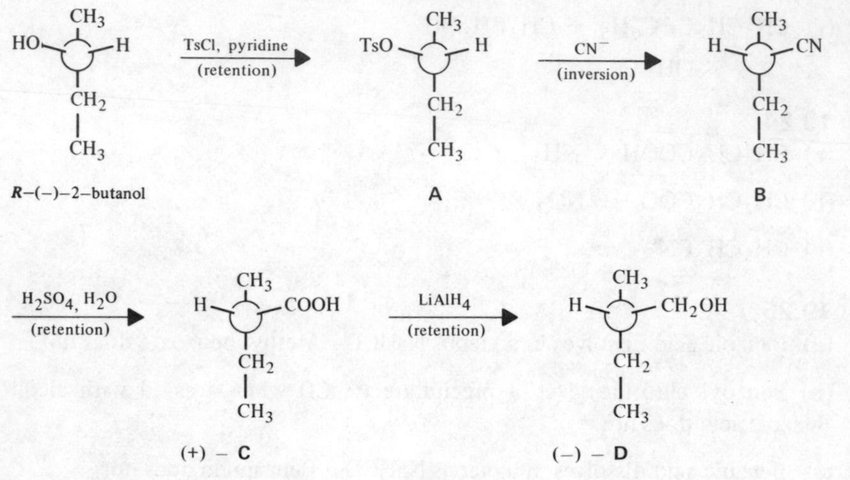
TsCl, pyridine
(retention)
CN⁻
(inversion)
R–(–)–2–butanol
A
B
H_2SO_4, H_2O
(retention)
$LiAlH_4$
(retention)
(+) – C
(–) – D

(b)

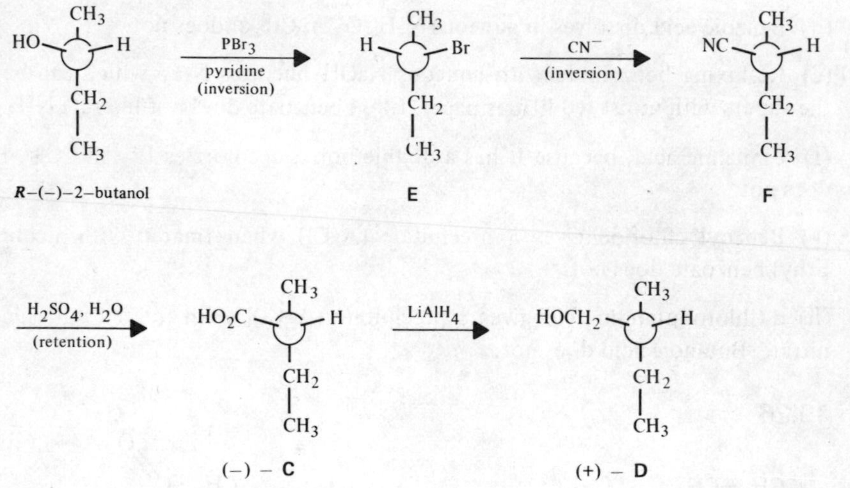
PBr_3
pyridine
(inversion)
CN⁻
(inversion)
R–(–)–2–butanol
E
F
H_2SO_4, H_2O
(retention)
$LiAlH_4$
(–) – C
(+) – D

(c)

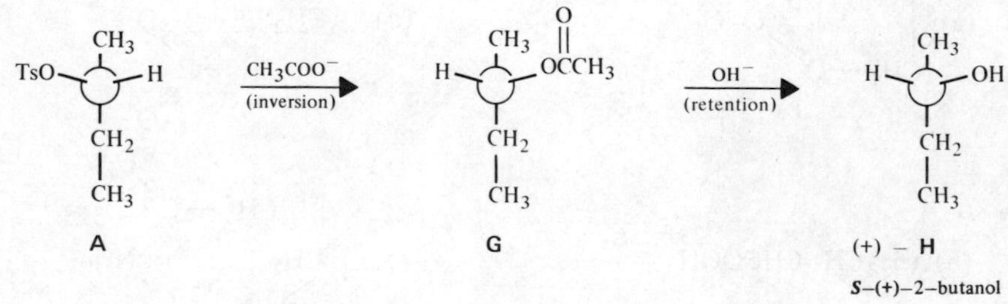
CH_3COO^-
(inversion)
OH^-
(retention)
A
G
(+) – H
S–(+)–2–butanol

(d)

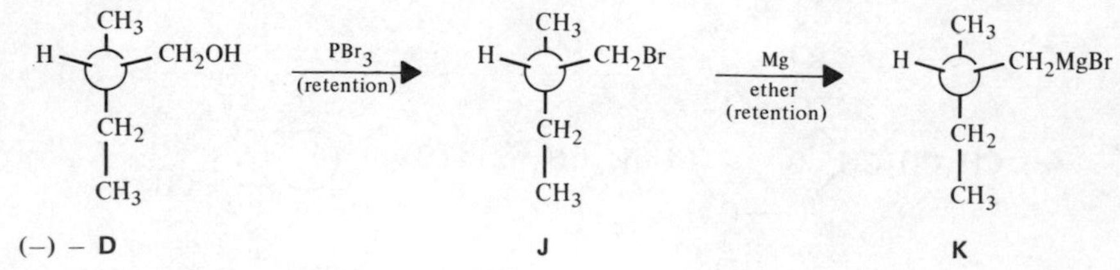
PBr_3
(retention)
Mg
ether
(retention)
(–) – D
J
K

$$\xrightarrow[\substack{(2)\ H^+ \\ (\text{retention})}]{(1)\ CO_2}$$

(structure **L**: central carbon bonded to CH_3, H, CH_2COOH, and CH_2CH_3)

L

(e) R–(+)–glyceraldehyde (structure with CHO, H, OH, CH_2OH) $\xrightarrow{HCN}$ **M** (structure with CN, H, OH, CH_2OH) + **N** (structure with CN, HO, H, CH_2OH)

M   **N**

(f) **M** $\xrightarrow[\substack{H_2O \\ heat}]{H_2SO_4}$ **P** (structure COOH, H, OH, CH_2OH) $\xrightarrow[HNO_3]{(O)}$ *meso*–tartaric acid (structure COOH, OH, COOH)

P   *meso*–tartaric acid

(g) **N** $\xrightarrow[\substack{H_2O \\ heat}]{H_2SO_4}$ (structure COOH, HO, H, CH_2OH) $\xrightarrow[HNO_3]{(O)}$ (–)–tartaric acid (structure COOH, HO, H, OH, COOH)

(–)–tartaric acid

19.28

(a) CH_3CHCHO (with CH_3) $+$ $H\overset{O}{C}H$ $\xrightarrow[H_2O]{K_2CO_3}$ CH_3CCHO (with CH_3 and CH_2OH) $\xrightarrow[(2)\ KCN]{(1)\ NaHSO_3}$

A

$CH_3C{-}CHCN$ (with CH_3, OH, CH_2OH) $\xrightarrow{H_3O^+}$ $\left[CH_3C{-}CHCOOH \text{ (with } CH_3,\ OH,\ CH_2OH) \right]$ $\xrightarrow{-H_2O}$

(±)-**B**   (±)-**C**

$CH_3{-}C{-}CH$ (with CH_3, OH, CH_2, $C{=}O$, O ring) $\xrightarrow{H_2NCH_2CH_2\overset{O}{C}OH}$ (±)-Pantothenic acid

(±)-**D**

$\xrightarrow{H_2NCH_2CH_2\overset{O}{C}NHCH_2CH_2SH}$ (±)-Pantetheine

(b)

$$(CH_3)_2C \underset{H}{\overset{CH_2OH}{\underset{|}{\overset{|}{\diamondsuit}}}} OH \; \overset{O}{\overset{\|}{C}} - NHCH_2CH_2 \overset{O}{\overset{\|}{C}} - NHCH_2CH_2SH$$

$$\xrightarrow[\text{heat}]{OH^-, \; H_2O} (CH_3)_2C \underset{H}{\overset{CH_2OH}{\underset{|}{\overset{|}{\diamondsuit}}}} OH \; COO^- \; + \; H_2NCH_2CH_2COO^- \; + \; H_2NCH_2CH_2S^-$$

19.29

$$CH_3CH_2O-\bigcirc-NH-\overset{O}{\overset{\|}{C}}-CH_3 \xrightarrow[\substack{H_2O \\ reflux}]{OH^-} CH_3CH_2O-\bigcirc-NH_2$$

Phenacetin   Phenetidine
                                                          +
                                                   CH_3COO^-

An interpretation of the spectral data for phenacetin is given in Fig. 19.1.

19.30

(a) $\underset{a \quad \quad c}{CH_3CH_2}-O-\overset{O}{\overset{\|}{C}}-\underset{b \quad \quad b}{CH_2CH_2}-\overset{O}{\overset{\|}{C}}-O-\underset{c \quad \quad a}{CH_2CH_3}$

Interpretation:

   a Triplet $\delta 1.2$ (6H) $2-\overset{O}{\overset{\|}{C}}-$, 1740 cm^{-1}

   b Singlet $\delta 2.5$ (4H)

   c Quartet $\delta 4.1$ (4H)

(b) $\bigcirc-\overset{O}{\overset{\|}{C}}-O-\underset{c}{CH_2}-\underset{b}{\overset{\overset{\displaystyle CH_3 \;\; a}{|}}{CH}}-\underset{a}{CH_3}$

   d

Interpretation:

   a Doublet $\delta 1.0$ (6H) $-\overset{O}{\overset{\|}{C}}-$, 1720 cm^{-1}

   b Multiplet $\delta 2.1$ (1H)

   c Doublet $\delta 4.1$ (2H)

   d Multiplet $\delta 7.8$ (5H)

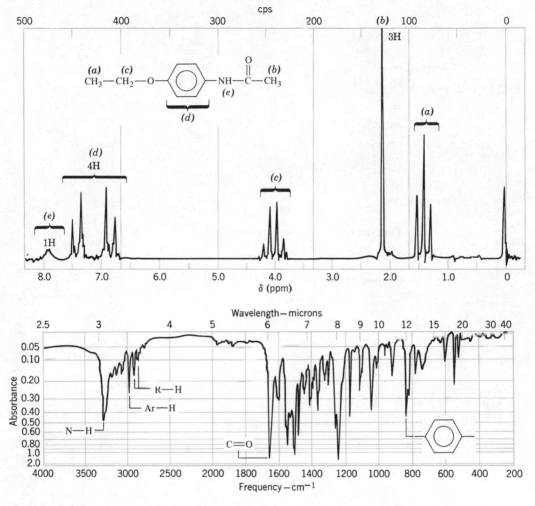

FIG. 19.1. The pmr and infrared spectra of phenacetin. (pmr spectrum courtesy of Varian Associates. IR spectrum courtesy of Sadtler Inc.)

(c)
$$
\text{C}_6\text{H}_5\text{—CH}_2\text{—}\overset{\displaystyle O}{\overset{\displaystyle \|}{\text{C}}}\text{—O—CH}_2\text{CH}_3
$$
$$
\qquad\qquad\ \ b \qquad\qquad\ \ c \quad a
$$
$$
\underbrace{\qquad\qquad}_{d}
$$

Interpretation:

    a Triplet $\delta\,1.2$ (3H) $-\overset{\displaystyle O}{\overset{\displaystyle \|}{\text{C}}}-$, 1740 cm^{-1}

    b Singlet $\delta\,3.5$ (2H)

    c Quartet $\delta\,4.1$ (2H)

    d Multiplet $\delta\,7.3$ (5H)

$$\begin{array}{c} \text{Cl} \\ | \\ \text{(d) Cl}-\underset{a}{\text{CH}}-\underset{b}{\text{COOH}} \end{array}$$

Interpretation:

a Singlet δ6.0 — $-OH$, 2500-2700 cm^{-1}

b Singlet δ11.70 — $-\overset{O}{\overset{\|}{C}}-$, 1705 cm^{-1}

$$\text{(e) Cl}-\underset{b}{CH_2}-\overset{O}{\overset{\|}{C}}-O\underset{c}{CH_2}\underset{a}{CH_3}$$

Interpretation:

a Triplet δ1.3 — $-\overset{O}{\overset{\|}{C}}-$, 1745 cm^{-1}

b Singlet δ4.0

c Quartet δ4.2

20

SYNTHESIS AND REACTIONS OF β-DICARBONYL COMPOUNDS

20.1

(a) Step 1

$$CH_3CH-\overset{O}{\overset{\|}{C}}OC_2H_5 \quad {}^-OC_2H_5 \longleftrightarrow CH_3\overset{..}{C}H-\overset{O}{\overset{\|}{C}}OC_2H_5 + C_2H_5OH$$
$$\underset{H}{|}$$

$$CH_3CH=\overset{O^-}{\overset{|}{C}}OC_2H_5$$

Step 2

$$CH_3CH_2\overset{O}{C}\overset{}{\underset{OC_2H_5}{\diagdown}} + {}^-\!:CHCOC_2H_5 \longleftrightarrow CH_3CH_2\overset{O^-}{\underset{C_2H_5O}{\overset{|}{C}}}\!\!-\!\!\underset{CH_3}{\overset{|}{CH}}\!\!-\!\!\overset{O}{\overset{\|}{C}}OC_2H_5$$
$$\underset{CH_3}{}$$

$$C_2H_5O^- + CH_3CH_2\overset{O}{\overset{\|}{C}}-\underset{CH_3}{\overset{|}{CH}}-\overset{O}{\overset{\|}{C}}OC_2H_5$$

Step 3

$$CH_3CH_2\overset{O}{\overset{\|}{C}}-\underset{CH_3}{\overset{H}{\underset{|}{C}}}-\overset{O}{\overset{\|}{C}}OC_2H_5 + {}^-OC_2H_5 \longleftrightarrow CH_3CH_2\overset{O}{\overset{\|}{C}}=\underset{CH_3}{\overset{\overline{}}{C}}=\overset{O}{\overset{\|}{C}}OC_2H_5$$
$$+ C_2H_5OH$$

(b)

$$CH_3CH_2\overset{O}{\overset{\|}{C}}\underset{CH_3}{\overset{O}{\overset{\|}{C}}}HCOC_2H_5 + CH_3CH_2\overset{OH}{\overset{|}{C}}=\underset{CH_3}{\overset{O}{\overset{\|}{C}}}COC_2H_5$$

20.2

(a)

$$C_2H_5O\overset{O}{\overset{\|}{C}}CH_2CH_2CH_2CH_2\overset{O}{\overset{\|}{C}}OC_2H_5 \underset{+H^+}{\overset{-H^+}{\rightleftharpoons}} \overset{C_2H_5O}{\underset{\underset{CH_2-\!-\!CH_2}{\overset{|}{CH_2}}}{\overset{|}{C}}}\overset{O}{\overset{}{\diagdown}} {}^-\!:CHCO_2C_2H_5$$

[Structure: cyclopentane ring with C_2H_5O and O^- substituent, H, $\overset{O}{\overset{\|}{C}}OC_2H_5$]

$\rightleftharpoons$ [cyclopentanone ring with H, $\overset{O}{\overset{\|}{C}}OC_2H_5$] $+ C_2H_5O^-$

$\rightleftharpoons$ [cyclopentanone ring with $\overset{O}{\cdots}\overset{O}{\cdots}\overset{\|}{C}OC_2H_5$] $\xrightarrow{H^+}$ [cyclopentanone ring with $\overset{O}{\overset{\|}{C}}OC_2H_5$] $+$ enol form

$+\ C_2H_5OH$

(b) [cyclohexanone ring with $\overset{O}{\overset{\|}{}}\ \overset{O}{\overset{\|}{C}}OC_2H_5$]

(c) To undergo a Dieckmann condensation, diethyl glutarate would have to form a highly strained four-membered ring.

20.3

$CH_3\overset{O}{\overset{\|}{C}}OC_2H_5 + C_2H_5O^- \rightleftharpoons {}^{-}{:}CH_2\overset{O}{\overset{\|}{C}}OC_2H_5 + C_2H_5OH$

$C_6H_5\overset{O}{\overset{\|}{C}}OC_2H_5 + {}^{-}{:}CH_2\overset{O}{\overset{\|}{C}}OC_2H_5 \rightleftharpoons C_6H_5\overset{O^-}{\underset{OC_2H_5}{\overset{|}{C}}}-CH_2\overset{O}{\overset{\|}{C}}OC_2H_5$

$\rightleftharpoons C_6H_5\overset{O}{\overset{\|}{C}}CH_2\overset{O}{\overset{\|}{C}}OC_2H_5 + C_2H_5O^- \rightleftharpoons C_6H_5\overset{O}{\overset{\|}{C}}{=}CH{=}\overset{O}{\overset{\|}{C}}OC_2H_5^{-}$

$+\ C_2H_5OH$

$\xrightarrow{H^+} C_6H_5\overset{O}{\overset{\|}{C}}CH_2\overset{O}{\overset{\|}{C}}OC_2H_5$

$C_6H_5CH_2\overset{O}{\overset{\|}{C}}OC_2H_5 + C_2H_5O^- \rightleftharpoons C_6H_5\overset{..}{C}H\overset{O}{\overset{\|}{C}}OC_2H_5 + C_2H_5OH$

$C_6H_5\overset{..}{C}H\overset{O}{\overset{\|}{C}}OC_2H_5 + C_2H_5O\overset{O}{\overset{\|}{C}}OC_2H_5 \rightleftharpoons C_6H_5\overset{\overset{\displaystyle C_2H_5O-\overset{O^-}{\overset{|}{C}}-OC_2H_5}{|}}{\underset{\underset{O}{\overset{\|}{C}}OC_2H_5}{C}H}$

$\rightleftharpoons C_6H_5\overset{\overset{\displaystyle \overset{O}{\overset{\|}{C}}OC_2H_5}{|}}{\underset{\underset{O}{\overset{\|}{C}}OC_2H_5}{C}H} + C_2H_5O^- \rightleftharpoons C_6H_5\overset{\overset{\displaystyle \overset{O}{\overset{\|}{C}}OC_2H_5}{|}}{\underset{\underset{O}{\overset{\|}{C}}OC_2H_5}{\overset{..}{C}}{:}} + C_2H_5OH$

resonance
stabilized

$$\xrightarrow{H^+} C_6H_5CH \begin{array}{c} COC_2H_5 \\ | \\ COC_2H_5 \end{array}$$

20.4

(a) $CH_3CH_2\overset{O}{\overset{\|}{C}}OC_2H_5 + C_2H_5O\overset{O}{\overset{\|}{C}}-\overset{O}{\overset{\|}{C}}OC_2H_5 \xrightarrow[\text{(2) H}^+]{\text{(1) NaOC}_2\text{H}_5} CH_3\overset{}{C}H\overset{O}{\overset{\|}{C}}OC_2H_5$

$$\begin{array}{c} | \\ C-COC_2H_5 \\ \| \ \| \\ O \ O \end{array}$$

(b) $CH_3\overset{O}{\overset{\|}{C}}OC_2H_5 + H\overset{O}{\overset{\|}{C}}OC_2H_5 \xrightarrow[\text{(2) H}^+]{\text{(1) NaOC}_2\text{H}_5} H\overset{O}{\overset{\|}{C}}CH_2\overset{O}{\overset{\|}{C}}OC_2H_5$

20.5

(a) $+ H\overset{O}{\overset{\|}{C}}OC_2H_5 \xrightarrow[\text{(2) H}^+]{\text{(1) NaOC}_2\text{H}_5}$

(b) $CH_3CH_2\overset{O}{\overset{\|}{C}}CH_2CH_2CH_2\overset{O}{\overset{\|}{C}}OC_2H_5 \xrightarrow[\text{(2) H}^+]{\text{(1) NaOC}_2\text{H}_5}$

(c) $C_2H_5O_2CCH_2\overset{CH_3}{\underset{CH_3}{\overset{|}{\underset{|}{C}}}}CH_2CO_2C_2H_5 + C_2H_5O\overset{O}{\overset{\|}{C}}-\overset{O}{\overset{\|}{C}}OC_2H_5 \xrightarrow[\text{(2) H}^+]{\text{(1) NaOC}_2\text{H}_5}$

$\xrightarrow[\text{(2) H}^+]{\text{(1) NaOC}_2\text{H}_5}$

20.6

(a) $CH_3\overset{O}{\overset{\|}{C}}CH_2CH_2CH_2CH_2\overset{O}{\overset{\|}{C}}OC_2H_5 + {}^-OC_2H_5 \underset{(-C_2H_5OH)}{\rightleftarrows}$

$\rightleftarrows$ $+ \ C_2H_5O^-$ (cont. on p. 212)

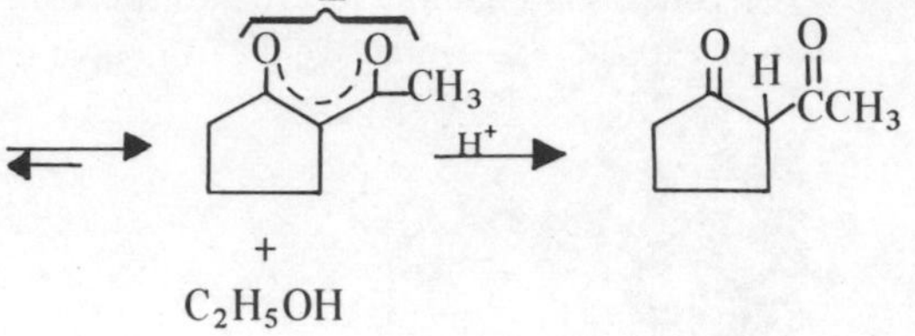

$+$

C_2H_5OH

(b) $CH_3\overset{O}{\overset{\|}{C}}CH_2CH_2CH_2\overset{O}{\overset{\|}{C}}OC_2H_5 \;+\; {}^-OC_2H_5 \underset{(-C_2H_5OH)}{\rightleftharpoons}$

$+\; {}^-OC_2H_5$

$+\; C_2H_5OH$

$\downarrow H^+$

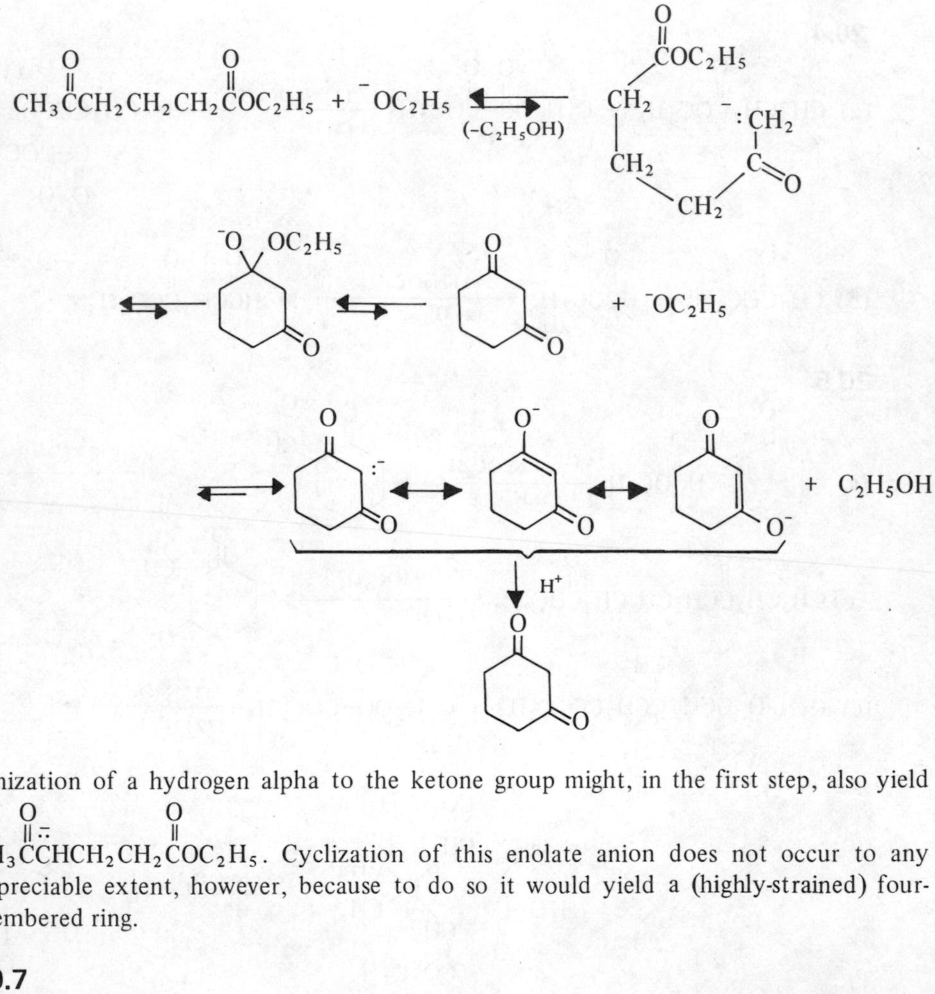

Ionization of a hydrogen alpha to the ketone group might, in the first step, also yield $CH_3\overset{O}{\overset{\|}{C}}\overset{\cdot\,-}{C}HCH_2CH_2\overset{O}{\overset{\|}{C}}OC_2H_5$. Cyclization of this enolate anion does not occur to any appreciable extent, however, because to do so it would yield a (highly-strained) four-membered ring.

20.7
(a) The iminium salt transfers hydrogen chloride to the enamine in the following way.

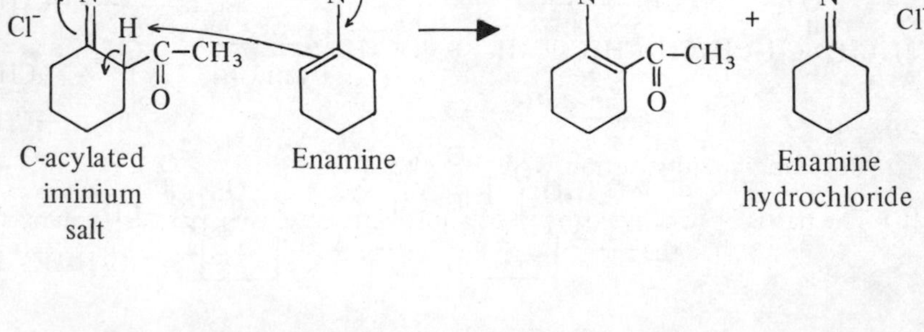

Cl^- H $\overset{\|}{C}-CH_3$

$+$ Cl^-

$\overset{\|}{C}-CH_3$

C-acylated
iminium
salt

Enamine

Enamine
hydrochloride

(b) Since the enamine hydrochloride formed in this acid-base reaction is not susceptible to acylation, the overall yield of the enamine synthesis will be decreased.

20.8

These syntheses are easy to see if we work backward.

(a)

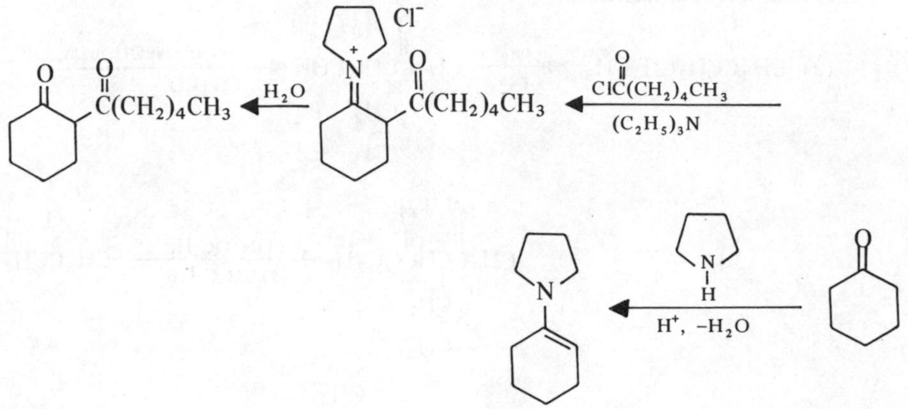

(b)

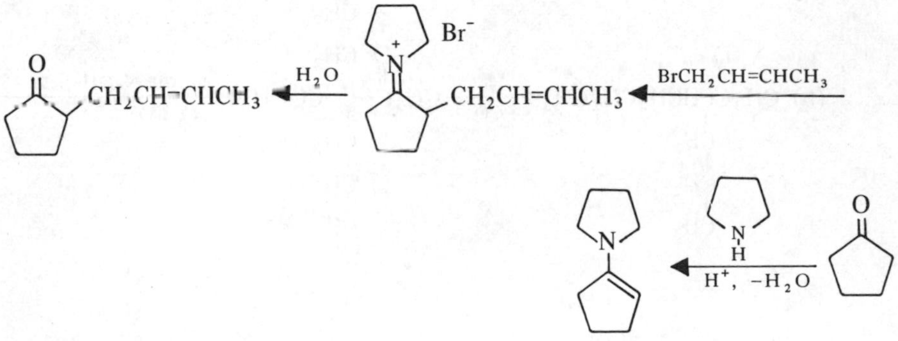

(c)

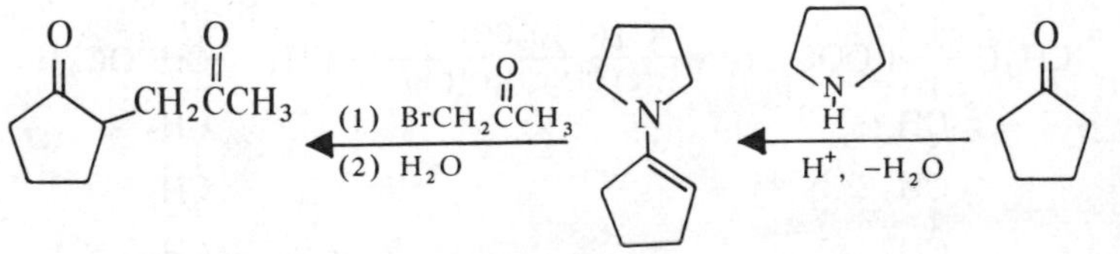

(d)

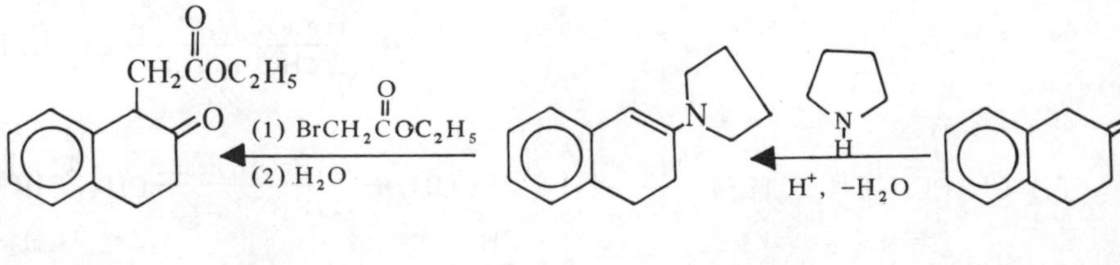

20.9

(a) Nucleophilic substitution, S_N2.

(b) The partially negative oxygen of sodioacetoacetic ester acts as the nucleophile.

$$\overset{\underset{\text{O}}{\|}}{CH_3C}-\overset{\cdot\cdot}{CH}-\overset{\underset{\text{O}}{\|}}{C}-OC_2H_5 \longleftrightarrow CH_3\overset{\underset{\text{O}^-}{|}}{C}=CH-\overset{\underset{\text{O}}{\|}}{C}-OC_2H_5$$

20.10

Again, working backward,

(a) $CH_3\overset{\underset{\text{O}}{\|}}{C}CH_2CH_2CH_3 \xleftarrow[-CO_2]{\text{heat}} CH_3\overset{\underset{\text{O}}{\|}}{C}\underset{\underset{\underset{\text{CH}_3}{|}}{\overset{|}{CH_2}}}{\overset{|}{CH}}-\overset{\underset{\text{O}}{\|}}{C}OH \xleftarrow[(2)\,H_3O^+]{(1)\,\text{dil. NaOH, heat}}$

$CH_3\overset{\underset{\text{O}}{\|}}{C}\underset{\underset{\underset{\text{CH}_3}{|}}{\overset{|}{CH_2}}}{\overset{|}{CH}}\overset{\underset{\text{O}}{\|}}{C}OC_2H_5 \xleftarrow[(2)\,CH_3CH_2Br]{(1)\,NaOC_2H_5} CH_3\overset{\underset{\text{O}}{\|}}{C}CH_2\overset{\underset{\text{O}}{\|}}{C}OC_2H_5$

(b) $CH_3\overset{\underset{\text{O}}{\|}}{C}\underset{\underset{\underset{\text{CH}_3}{|}}{\overset{|}{CH_2}}}{\overset{|}{CH}}CH_2CH_2CH_3 \xleftarrow[-CO_2]{\text{heat}} CH_3\overset{\underset{\text{O}}{\|}}{C}-\underset{\underset{\underset{\text{CH}_3}{|}}{\overset{|}{CH_2}}}{\overset{\overset{\overset{\displaystyle CH_3}{|}}{\overset{\displaystyle CH_2}{|}}}{\overset{\displaystyle CH_2}{|}}}{C}-COOH \xleftarrow[(2)\,H_3O^+]{(1)\,\text{dil. NaOH, heat}}$

$CH_3\overset{\underset{\text{O}}{\|}}{C}-\underset{\underset{\underset{\text{CH}_3}{|}}{\overset{|}{CH_2}}}{\overset{\overset{\overset{\displaystyle CH_3}{|}}{\overset{\displaystyle CH_2}{|}}}{\overset{\displaystyle CH_2}{|}}}{C}-COOC_2H_5 \xleftarrow[(2)\,CH_3CH_2CH_2Br]{(1)\,(CH_3)_3COK} CH_3\overset{\underset{\text{O}}{\|}}{C}-\underset{\underset{\underset{\text{CH}_3}{|}}{\overset{|}{CH_2}}}{\overset{|}{CH}}\overset{\underset{\text{O}}{\|}}{C}OC_2H_5$

$\xleftarrow[(2)\,CH_3CH_2CH_2Br]{(1)\,NaOC_2H_5} CH_3\overset{\underset{\text{O}}{\|}}{C}CH_2\overset{\underset{\text{O}}{\|}}{C}OC_2H_5$

(c) $CH_3\overset{\underset{\text{O}}{\|}}{C}CH_2CH_2C_6H_5 \xleftarrow[-CO_2]{\text{heat}} CH_3\overset{\underset{\text{O}}{\|}}{C}\underset{\underset{\underset{\text{C}_6\text{H}_5}{|}}{\overset{|}{CH_2}}}{\overset{|}{CH}}\overset{\underset{\text{O}}{\|}}{C}OH \xleftarrow[(2)\,H_3O^+]{(1)\,NaOH, heat} CH_3\overset{\underset{\text{O}}{\|}}{C}\underset{\underset{\underset{\text{C}_6\text{H}_5}{|}}{\overset{|}{CH_2}}}{\overset{|}{CH}}\overset{\underset{\text{O}}{\|}}{C}OC_2H_5$

$\xleftarrow[(2)\,C_6H_5CH_2Br]{(1)\,NaOC_2H_5} CH_3\overset{\underset{\text{O}}{\|}}{C}CH_2\overset{\underset{\text{O}}{\|}}{C}OC_2H_5$

20.11

(a) The first alkylation produces ethyl methylacetoacetate; it then reacts with sodium ethoxide to produce an anion that can undergo a second alkylation.

$$CH_3\overset{O}{\overset{\|}{C}}-\overset{H}{\underset{CH_3}{\overset{|}{C}}}-\overset{O}{\overset{\|}{C}}OC_2H_5 \xrightarrow{NaOC_2H_5} CH_3\overset{O}{\overset{\|}{C}}-\overset{\ddots}{\underset{CH_3}{C}}-\overset{O}{\overset{\|}{C}}OC_2H_5 \xrightarrow{CH_3I} CH_3\overset{O}{\overset{\|}{C}}-\overset{CH_3}{\underset{CH_3}{\overset{|}{C}}}-COOC_2H_5$$

Ethyl methyl-   Ethyl dimethyl-
acetoacetate   acetoacetate

(b) It favors monoalkylation.

(c) Methyl and ethyl halides present less steric hindrance to the attacking anion.

(d) Carry out the first alkylation with the larger alkyl halide, that is, with propyl bromide. This will minimize dialkylation in the first step.

20.12

(a) Reactivity is the same as with any S_N2 reaction. With primary halides substitution is highly favored, with secondary halides elimination competes with substitution, and with tertiary halides elimination is the exclusive course of reaction.

(b) Acetoacetic ester and 2-methylpropene.

(c) Bromobenzene is unreactive toward nucleophilic substitution (Cf. Section 17.11 of the text).

20.13

$$CH_3CH_2CH_2\overset{O}{\overset{\|}{C}}OC_2H_5 \xrightarrow[(2)\,H^+]{(1)\,NaOC_2H_5} CH_3CH_2CH_2\overset{O}{\overset{\|}{C}}\overset{O}{\underset{\underset{CH_3}{\overset{|}{CH_2}}}{\overset{\|}{CH}}C}OC_2H_5$$

$$\xrightarrow[(2)\,H_3O^+]{(1)\,NaOH,\,H_2O,\,heat} CH_3CH_2CH_2\overset{O}{\overset{\|}{C}}\overset{O}{\underset{\underset{CH_3}{\overset{|}{CH_2}}}{\overset{\|}{CH}}C}OH \xrightarrow[-CO_2]{heat} CH_3CH_2CH_2\overset{O}{\overset{\|}{C}}CH_2CH_2CH_3$$

20.14

The carboxyl group that is lost most readily is the one that is *beta* to the keto group (cf. page 797 of the text).

20.15

$$CH_3\overset{O}{\overset{\|}{C}}CH_2CH_2\overset{O}{\overset{\|}{C}}C_6H_5 \xleftarrow[-CO_2]{heat} CH_3\overset{O}{\overset{\|}{C}}\overset{O}{\underset{\underset{\underset{\overset{|}{C_6H_5}}{\overset{|}{C=O}}}{\overset{|}{CH_2}}}{\overset{\|}{CH}}C}OH \xleftarrow[(2)\,H_3O^+]{(1)\,OH^-,\,H_2O,\,heat}$$

$$\underset{\begin{array}{c}\text{CH}_2\\|\\\text{C}=\text{O}\\|\\\text{C}_6\text{H}_5\end{array}}{\text{CH}_3\overset{\text{O}}{\overset{||}{\text{C}}}\text{CH}\overset{\text{O}}{\overset{||}{\text{C}}}\text{OC}_2\text{H}_5} \quad \xleftarrow[\text{(2)}\text{C}_6\text{H}_5\text{COCH}_2\text{Br}]{\text{(1) NaOC}_2\text{H}_5} \quad \text{CH}_3\overset{\text{O}}{\overset{||}{\text{C}}}\text{CH}_2\overset{\text{O}}{\overset{||}{\text{C}}}\text{OC}_2\text{H}_5$$

20.16

$$\text{CH}_3\overset{\text{O}}{\overset{||}{\text{C}}}\text{CH}_2\overset{\text{O}}{\overset{||}{\text{C}}}\text{C}_6\text{H}_5 \xleftarrow[-\text{CO}_2]{\text{heat}} \underset{\begin{array}{c}\text{C}=\text{O}\\|\\\text{C}_6\text{H}_5\end{array}}{\text{CH}_3\overset{\text{O}}{\overset{||}{\text{C}}}\text{CH}\overset{\text{O}}{\overset{||}{\text{C}}}\text{OH}} \xleftarrow[\text{(2)}\text{H}_3\text{O}^+]{\text{(1) OH}^-,\,\text{H}_2\text{O, heat}}$$

$$\underset{\begin{array}{c}\text{C}=\text{O}\\|\\\text{C}_6\text{H}_5\end{array}}{\text{CH}_3\overset{\text{O}}{\overset{||}{\text{C}}}\text{CH}\overset{\text{O}}{\overset{||}{\text{C}}}\text{OC}_2\text{H}_5} \xleftarrow[\text{(2)}\text{C}_6\text{H}_5\text{COCl}]{\text{(1) NaOC}_2\text{H}_5} \text{CH}_3\overset{\text{O}}{\overset{||}{\text{C}}}\text{CH}_2\overset{\text{O}}{\overset{||}{\text{C}}}\text{OC}_2\text{H}_5$$

20.17

Here we acylate the dianion,

$$\text{CH}_3-\overset{\text{O}}{\overset{||}{\text{C}}}-\text{CH}_2-\overset{\text{O}}{\overset{||}{\text{C}}}\text{OC}_2\text{H}_5 \xrightarrow[\text{liq NH}_3]{\text{2 KNH}_2} \overset{..}{:}\text{CH}_2-\overset{\text{O}}{\overset{||}{\text{C}}}-\overset{..}{\text{CH}}-\overset{\text{O}}{\overset{||}{\text{C}}}\text{OC}_2\text{H}_5$$

$$\xrightarrow[\text{(2) NH}_4\text{Cl}]{\text{(1) C}_6\text{H}_5\overset{\text{O}}{\overset{||}{\text{C}}}\text{Cl}} \text{C}_6\text{H}_5\overset{\text{O}}{\overset{||}{\text{C}}}\text{CH}_2\overset{\text{O}}{\overset{||}{\text{C}}}\text{CH}_2\overset{\text{O}}{\overset{||}{\text{C}}}\text{OC}_2\text{H}_5$$

20.18

Working backward,

(a) $\text{CH}_3\text{CH}_2\text{CH}_2\text{CH}_2\text{COOH} \xleftarrow[-\text{CO}_2]{\text{heat}} \text{CH}_3\text{CH}_2\text{CH}_2\overset{\displaystyle\diagup\text{COOH}}{\underset{\diagdown\text{COOH}}{\text{CH}}} \xleftarrow[\text{(2)}\text{H}_3\text{O}^+]{\text{(1) OH}^-,\,\text{H}_2\text{O, heat}}$

$$\text{CH}_3\text{CH}_2\text{CH}_2\overset{\displaystyle\diagup\text{COOC}_2\text{H}_5}{\underset{\diagdown\text{COOC}_2\text{H}_5}{\text{CH}}} \xleftarrow[\text{CH}_3\text{CH}_2\text{CH}_2\text{Br}]{\text{NaOC}_2\text{H}_5} \underset{\begin{array}{c}|\\\text{COOC}_2\text{H}_5\end{array}}{\overset{\text{COOC}_2\text{H}_5}{\overset{|}{\text{CH}_2}}}$$

(b) $\text{CH}_3\text{CH}_2\text{CH}_2\underset{\begin{array}{c}|\\\text{CH}_3\end{array}}{\text{CHCOOH}} \xleftarrow[-\text{CO}_2]{\text{heat}} \underset{\diagup\diagdown}{\overset{\text{CH}_3\text{CH}_2\text{CH}_2\quad\diagdown\quad\diagup\text{COOH}}{\underset{\text{CH}_3\qquad\diagdown\text{COOH}}{\text{C}}}} \xleftarrow[\text{(2)}\text{H}_3\text{O}^+]{\text{(1) OH}^-,\,\text{H}_2\text{O, heat}}$

$$\underset{\diagup\quad\diagdown}{\overset{\text{CH}_3\text{CH}_2\text{CH}_2\quad\diagdown\quad\diagup\text{COOC}_2\text{H}_5}{\underset{\text{CH}_3\qquad\diagdown\text{COOC}_2\text{H}_5}{\text{C}}}} \xleftarrow[\text{NaOC}_2\text{H}_5]{\text{CH}_3\text{I}} \text{CH}_3\text{CH}_2\text{CH}_2\overset{\displaystyle\diagup\text{COOC}_2\text{H}_5}{\underset{\diagdown\text{COOC}_2\text{H}_5}{\text{CH}}}$$

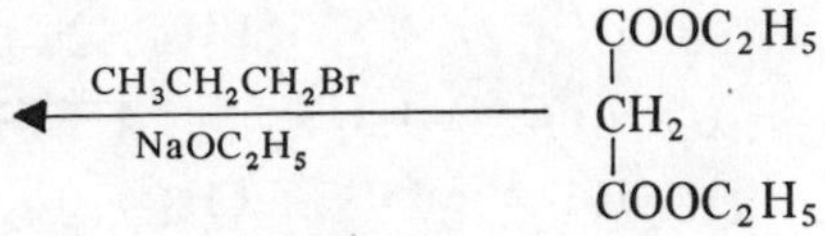

$$\begin{array}{c} COOC_2H_5 \\ | \\ CH_2 \\ | \\ COOC_2H_5 \end{array} \xrightarrow[NaOC_2H_5]{CH_3CH_2CH_2Br}$$

(c) $CH_3CHCH_2CH_2COOH \xleftarrow[-CO_2]{heat}$ $CH_3CHCH_2CH \xleftarrow[(2)\ H_3O^+]{(1)\ OH^-,\ H_2O,\ heat}$
$\quad\ \ |$ $\qquad\qquad\qquad\qquad |\qquad\ \searrow COOH$
$\quad\ \ CH_3$ $\qquad\qquad\qquad\qquad CH_3\qquad COOH$

$CH_3CHCH_2CH \xleftarrow[CH_3CHCH_2Br]{NaOC_2H_5}$ $\begin{array}{c} COOC_2H_5 \\ | \\ CH_2 \\ | \\ COOC_2H_5 \end{array}$
$\quad |\qquad\ \searrow COOC_2H_5$
$\quad CH_3$
$\qquad\ \nearrow COOC_2H_5$
with CH_3CHCH_2 and CH_3

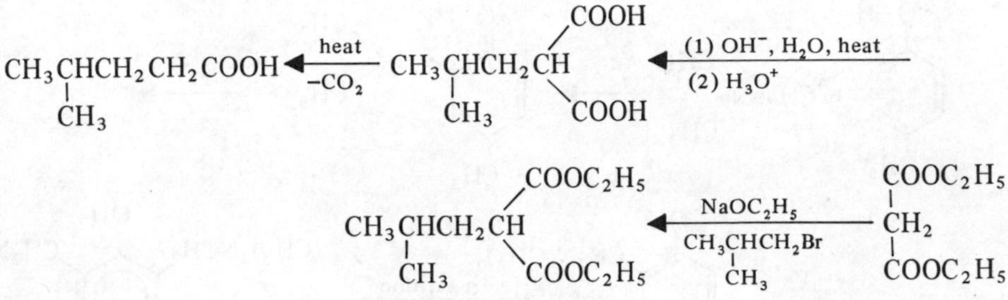

20.19

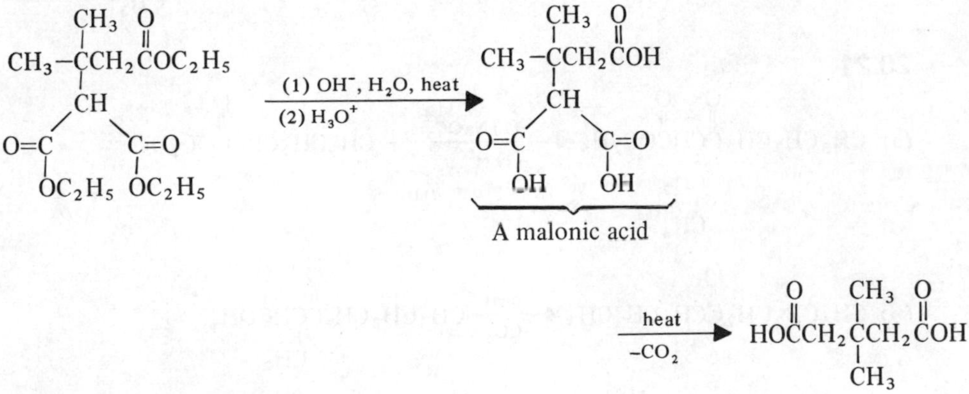

$$CH_3-\underset{\displaystyle \underset{\displaystyle O=C \quad\quad C=O}{\underset{\displaystyle \overset{|}{OC_2H_5}\ \overset{|}{OC_2H_5}}{CH}}}{\overset{CH_3\ \ O}{\underset{|}{\overset{||}{C}}CH_2\overset{||}{C}OC_2H_5}} \xrightarrow[(2)\ H_3O^+]{(1)\ OH^-,\ H_2O,\ heat} CH_3-\underset{\displaystyle \underset{\displaystyle O=C \quad\quad C=O}{\underset{\displaystyle \overset{|}{OH}\ \ \overset{|}{OH}}{CH}}}{\overset{CH_3\ \ O}{\underset{|}{\overset{||}{C}}CH_2\overset{||}{C}OH}}$$

A malonic acid

$$\xrightarrow[-CO_2]{heat} \overset{O\qquad CH_3\ \ O}{HO\overset{||}{C}CH_2\underset{\displaystyle \overset{|}{CH_3}}{C}CH_2\overset{||}{C}OH}$$

20.20

(a) $\underset{H}{\overset{H}{>}}C=O + HN(CH_3)_2 \underset{}{\overset{H^+}{\rightleftharpoons}} CH_2=\overset{+}{N}\underset{CH_3}{\overset{CH_3}{<}} + H_2O$

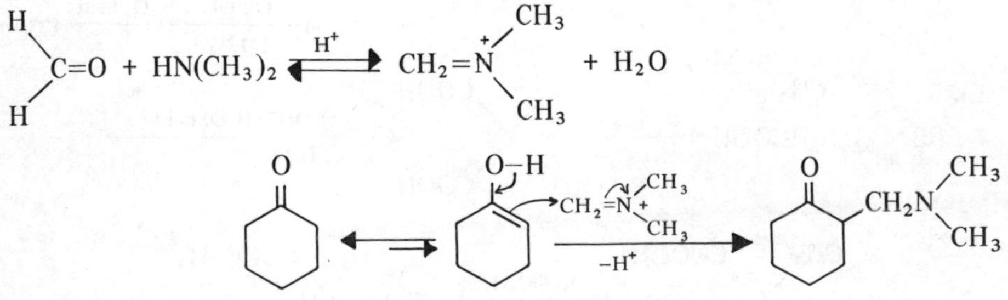

(b) $\underset{H}{\overset{H}{>}}C=O + H-N\bigcirc \overset{H^+}{\rightleftharpoons} CH_2=\overset{+}{N}\bigcirc + H_2O$

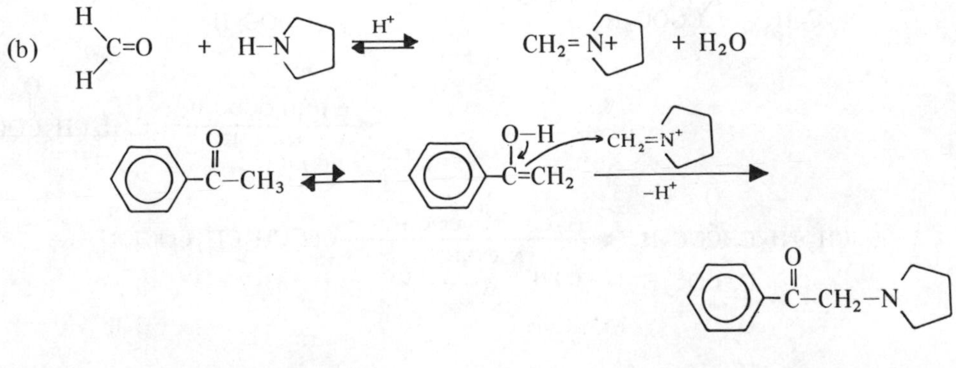

(c)
$$\begin{array}{c} H \\ H \end{array} C=O + HN \begin{array}{c} CH_3 \\ CH_3 \end{array} \underset{}{\overset{H^+}{\rightleftharpoons}} CH_2 = \overset{+}{N} \begin{array}{c} CH_3 \\ CH_3 \end{array} + H_2O$$

$$\overset{OH}{\underset{CH_3}{\bigodot}} CH_2N(CH_3)_2 \xrightarrow[\text{of similar} \atop \text{steps}]{\text{repetition}} (CH_3)_2NCH_2 \overset{OH}{\underset{CH_3}{\bigodot}} CH_2N(CH_3)_2$$

20.21

(a)
$$CH_3CH_2CH_2\overset{O}{\overset{\|}{C}}\overset{O}{\overset{\|}{C}}HCOC_2H_5 \xleftarrow[\text{(2) H}^+]{\text{(1) NaOC}_2\text{H}_5} CH_3CH_2CH_2\overset{O}{\overset{\|}{C}}OC_2H_5$$
with $\underset{CH_3}{\overset{|}{CH_2}}$ branch

(b)
$$CH_3CH_2CH_2\overset{O}{\overset{\|}{C}}CH_2CH_2CH_3 \xleftarrow[-CO_2]{\text{heat}} CH_3CH_2CH_2\overset{O}{\overset{\|}{C}}\overset{O}{\overset{\|}{C}}HCOH$$
with $\underset{CH_3}{\overset{|}{CH_2}}$ branch

$$\xleftarrow[\text{(2) H}_3\text{O}^+]{\text{(1) OH}^-, \text{H}_2\text{O, heat}} \text{Product of (a)}$$

(c)
$$\underset{C_6H_5}{\overset{CH_3}{\underset{|}{C}HCOOH}} \xleftarrow[-CO_2]{\text{heat}} \begin{array}{c} CH_3 \quad COOH \\ C \\ C_6H_5 \quad COOH \end{array} \xleftarrow[\text{(2) H}_3\text{O}^+]{\text{(1) OH}^-, \text{H}_2\text{O, heat}}$$

$$\begin{array}{c} CH_3 \quad COOC_2H_5 \\ C \\ C_6H_5 \quad COOC_2H_5 \end{array} \xleftarrow[\text{CH}_3\text{I}]{\text{NaOC}_2\text{H}_5} \begin{array}{c} COOC_2H_5 \\ C_6H_5-CH \\ COOC_2H_5 \end{array}$$

$$\xleftarrow[\substack{\text{NaOC}_2\text{H}_5 \\ \text{(2) H}^+}]{\text{(1) C}_2\text{H}_5\text{O}\overset{O}{\overset{\|}{C}}\text{OC}_2\text{H}_5} C_6H_5CH_2\overset{O}{\overset{\|}{C}}OC_2H_5$$

(d)
$$CH_3CH_2\overset{O}{\underset{\underset{O\ O}{\overset{|}{C}-COC_2H_5}}{\overset{|}{C}HCOC_2H_5}} \xleftarrow[\substack{\text{NaOC}_2\text{H}_5 \\ \text{(2) H}^+}]{\text{(1) C}_2\text{H}_5\text{O}\overset{O\ O}{\overset{\|\ \|}{C-C}}\text{OC}_2\text{H}_5} CH_3CH_2CH_2\overset{O}{\overset{\|}{C}}OC_2H_5$$

(e) $\underset{\substack{\| \quad \| \\ O \quad O}}{CH_3CH_2CH_2C-COC_2H_5} \xleftarrow[C_2H_5OH]{H^+} \underset{\substack{\| \quad \| \\ O \quad O}}{CH_3CH_2CH_2C-COH}$

$\xleftarrow[-CO_2]{heat} \underset{\substack{| \\ C-COH \\ \| \quad \| \\ O \quad O}}{CH_3CH_2CHCOOH} \xleftarrow[\substack{(1)\ OH^-,\ H_2O,\ heat \\ (2)\ H_3O^+}]{} \text{Product of (d)}$

(f) $\underset{\substack{| \\ CH \\ \| \\ O}}{C_6H_5CHCOC_2H_5} \xleftarrow[\substack{(1)\ HCOC_2H_5 \\ NaOC_2H_5 \\ (2)\ H^+}]{} \underset{\| \\ O}{C_6H_5CH_2COC_2H_5}$

(g)

(h)

Product of (g)

(i)

20.22

(a) $\underset{\substack{| \\ CH_3}}{\underset{\| \\ O}{CH_3C}-\underset{\substack{| \\ CH_3}}{C}-CH_3} \xleftarrow[]{Zn,\ H^+} \underset{\substack{| \\ CH_3}}{\underset{\| \\ O}{CH_3C}-\underset{\substack{| \\ CH_3}}{C}-CH_2Br} \xleftarrow[]{PBr_3} \underset{\substack{| \\ CH_3}}{\underset{\| \\ O}{CH_3C}-\underset{\substack{| \\ CH_3}}{C}-CH_2OH}$

$\xleftarrow[\substack{(1)\ LiAlH_4 \\ (2)\ H_3O^+}]{} \underset{\substack{| \\ CH_3}}{CH_3C-CCOOC_2H_5} \xleftarrow[H^+]{} \underset{\substack{| \\ CH_3}}{\underset{\| \\ O}{CH_3C}-\underset{\substack{| \\ CH_3}}{C}-COOC_2H_5} \xleftarrow[NaOC_2H_5]{CH_3I}$

$\underset{\substack{| \\ CH_3}}{\underset{\| \\ O}{CH_3C}-CH-COOC_2H_5} \xleftarrow[NaOC_2H_5]{CH_3I} \underset{\substack{\| \quad \| \\ O \quad O}}{CH_3CCH_2COC_2H_5}$

(b) 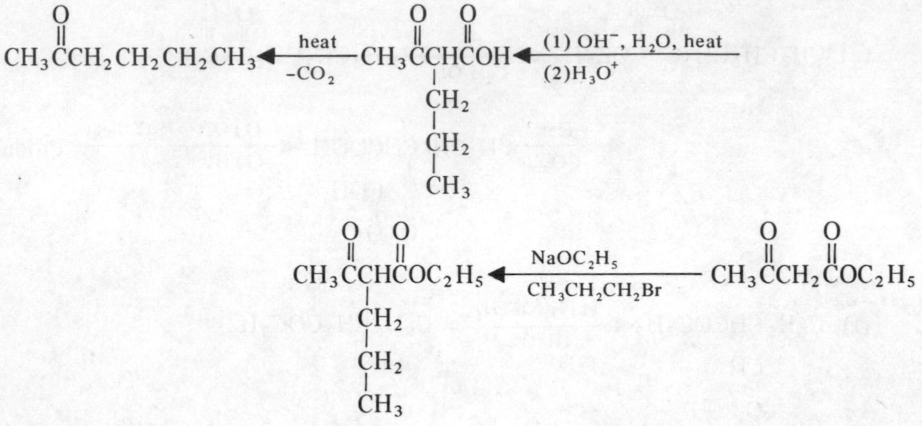
$$CH_3\overset{O}{\overset{\|}{C}}CH_2CH_2CH_2CH_3 \xleftarrow[-CO_2]{heat} CH_3\overset{O}{\overset{\|}{C}}\overset{O}{\overset{\|}{C}}HCOH \xleftarrow[(2)H_3O^+]{(1)\ OH^-,\ H_2O,\ heat}$$

with side chain on the central carbon:
$$\begin{array}{c} CH_2 \\ | \\ CH_2 \\ | \\ CH_3 \end{array}$$

$$CH_3\overset{O}{\overset{\|}{C}}\overset{O}{C}HCOC_2H_5 \xleftarrow[CH_3CH_2CH_2Br]{NaOC_2H_5} CH_3\overset{O}{\overset{\|}{C}}CH_2\overset{O}{\overset{\|}{C}}OC_2H_5$$

side chain:
$$\begin{array}{c} CH_2 \\ | \\ CH_2 \\ | \\ CH_3 \end{array}$$

(c)
$$CH_3\overset{O}{\overset{\|}{C}}CH_2CH_2\overset{O}{\overset{\|}{C}}CH_3 \xleftarrow[-CO_2]{heat} CH_3\overset{O}{\overset{\|}{C}}\overset{O}{\overset{\|}{C}}HCOH \xleftarrow[(2)\ H_3O^+]{(1)\ OH^-,\ H_2O,\ heat}$$

side chain:
$$\begin{array}{c} CH_2 \\ | \\ C=O \\ | \\ CH_3 \end{array}$$

$$CH_3\overset{O}{\overset{\|}{C}}\overset{O}{C}HCOC_2H_5 \xleftarrow[CH_3COCH_2Br]{NaOC_2H_5} CH_3\overset{O}{\overset{\|}{C}}CH_2\overset{O}{\overset{\|}{C}}OC_2H_5$$

side chain:
$$\begin{array}{c} CH_2 \\ | \\ C=O \\ | \\ CH_3 \end{array}$$

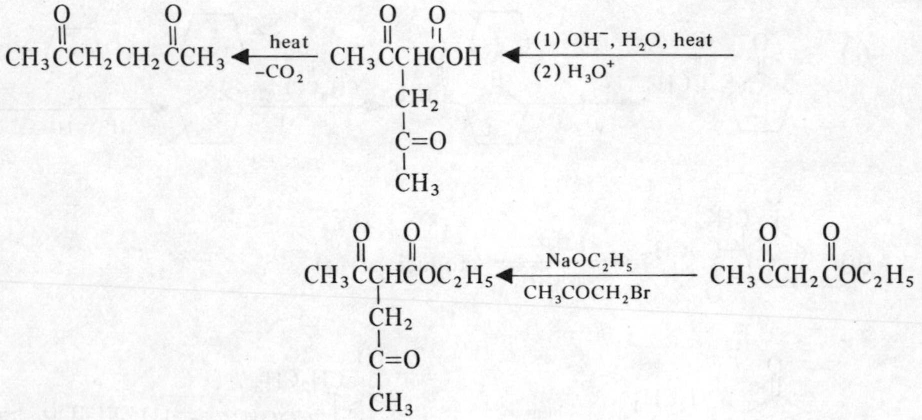

(d) 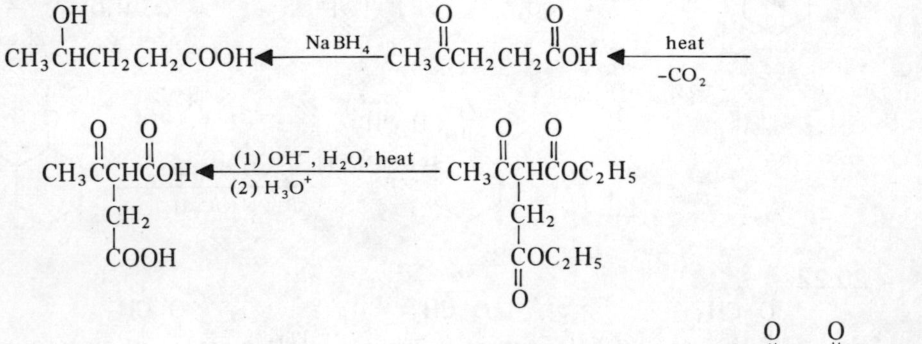
$$CH_3\overset{OH}{\overset{|}{C}}HCH_2CH_2COOH \xleftarrow{NaBH_4} CH_3\overset{O}{\overset{\|}{C}}CH_2CH_2\overset{O}{\overset{\|}{C}}OH \xleftarrow[-CO_2]{heat}$$

$$CH_3\overset{O}{\overset{\|}{C}}\overset{O}{\overset{\|}{C}}HCOH \xleftarrow[(2)\ H_3O^+]{(1)\ OH^-,\ H_2O,\ heat} CH_3\overset{O}{\overset{\|}{C}}\overset{O}{C}HCOC_2H_5$$

left side chain:
$$\begin{array}{c} CH_2 \\ | \\ COOH \end{array}$$

right side chain:
$$\begin{array}{c} CH_2 \\ | \\ COC_2H_5 \\ \| \\ O \end{array}$$

$$\xleftarrow[BrCH_2COOC_2H_5]{NaOC_2H_5} CH_3\overset{O}{\overset{\|}{C}}CH_2\overset{O}{\overset{\|}{C}}OC_2H_5$$

(e) 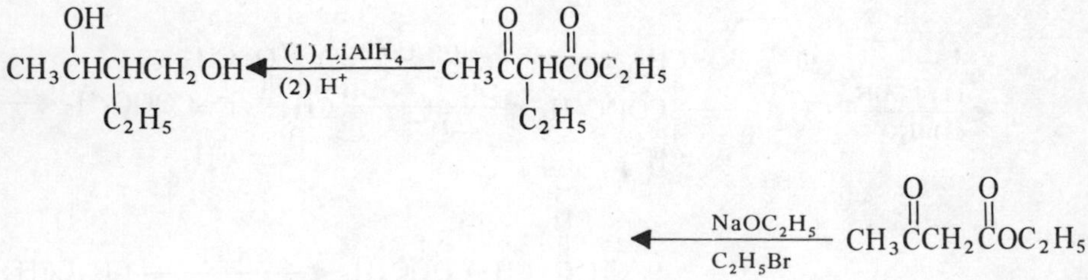
$$CH_3\overset{OH}{\overset{|}{C}}HCHCH_2OH \xleftarrow[(2)\ H^+]{(1)\ LiAlH_4} CH_3\overset{O}{\overset{\|}{C}}\overset{O}{\overset{\|}{C}}HCOC_2H_5$$

left side chain: C_2H_5 ; right side chain: C_2H_5

$$\xleftarrow[C_2H_5Br]{NaOC_2H_5} CH_3\overset{O}{\overset{\|}{C}}CH_2\overset{O}{\overset{\|}{C}}OC_2H_5$$

(f) 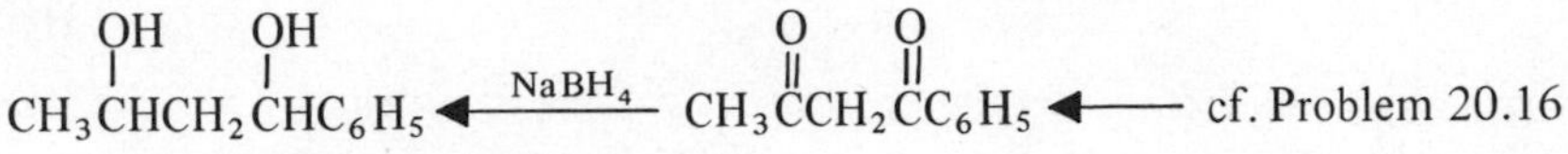 $\underset{\substack{| \\ OH}}{CH_3CH} \underset{\substack{}}{CH_2} \underset{\substack{| \\ OH}}{CHC_6H_5} \xleftarrow{NaBH_4} \underset{\substack{\parallel \\ O}}{CH_3C} CH_2 \underset{\substack{\parallel \\ O}}{CC_6H_5} \longleftarrow$ cf. Problem 20.16

(g) 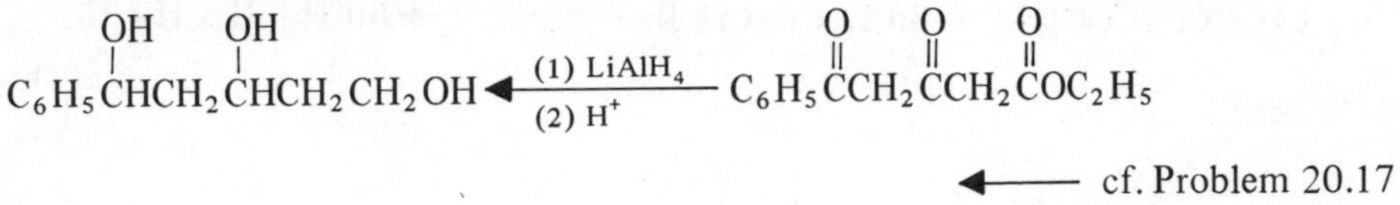 $\underset{\substack{| \\ OH}}{C_6H_5CH} CH_2 \underset{\substack{| \\ OH}}{CHCH_2} CH_2OH \xleftarrow[\text{(2) } H^+]{\text{(1) LiAlH}_4} \underset{\substack{\parallel \\ O}}{C_6H_5C} CH_2 \underset{\substack{\parallel \\ O}}{CCH_2} \underset{\substack{\parallel \\ O}}{COC_2H_5}$

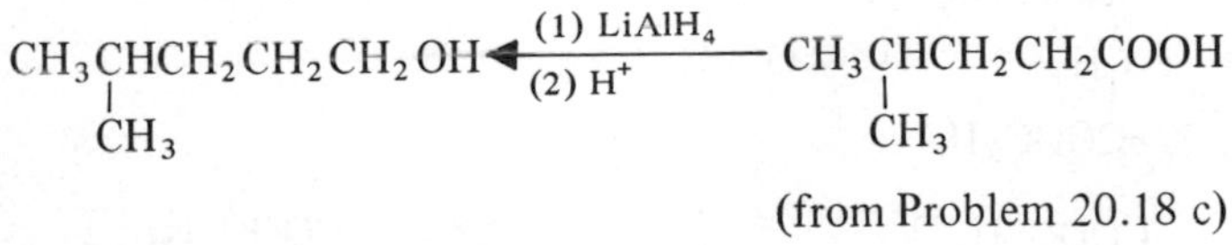

$\longleftarrow$ cf. Problem 20.17

20.23

(a) 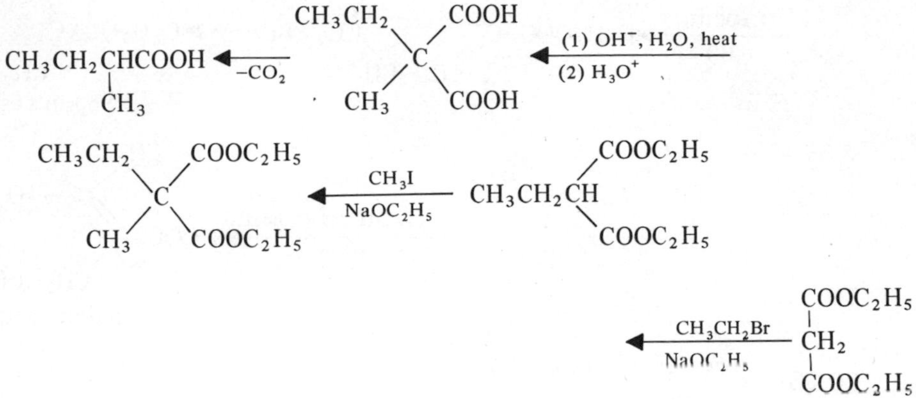 $\underset{\substack{| \\ CH_3}}{CH_3CH_2CH}COOH \xleftarrow{-CO_2}$ $\underset{\substack{CH_3CH_2 \diagup \diagdown COOH \\ C \\ CH_3 \diagup \diagdown COOH}}{} \xleftarrow[\text{(2) } H_3O^+]{\text{(1) } OH^-, H_2O, \text{ heat}}$

$\underset{\substack{CH_3CH_2 \diagup \diagdown COOC_2H_5 \\ C \\ CH_3 \diagup \diagdown COOC_2H_5}}{} \xleftarrow[\text{NaOC}_2H_5]{\text{CH}_3\text{I}} \underset{\substack{COOC_2H_5 \\ CH_3CH_2CH \\ COOC_2H_5}}{}$

$\xleftarrow[\text{NaOC}_2H_5]{\text{CH}_3\text{CH}_2\text{Br}} \underset{\substack{COOC_2H_5 \\ | \\ CH_2 \\ | \\ COOC_2H_5}}{}$

(b) $\underset{\substack{| \\ CH_3}}{CH_3CHCH_2CH_2CH_2OH} \xleftarrow[\text{(2) } H^+]{\text{(1) LiAlH}_4} \underset{\substack{| \\ CH_3}}{CH_3CHCH_2CH_2COOH}$

(from Problem 20.18 c)

(c) $\underset{\substack{| \\ CH_2OH}}{CH_3CH_2CHCH_2OH} \xleftarrow[\text{(2) } H^+]{\text{(1) LiAlH}_4} \underset{\substack{COOC_2H_5 \\ CH_3CH_2CH \\ COOC_2H_5}}{} \longleftarrow$ cf. p. 804

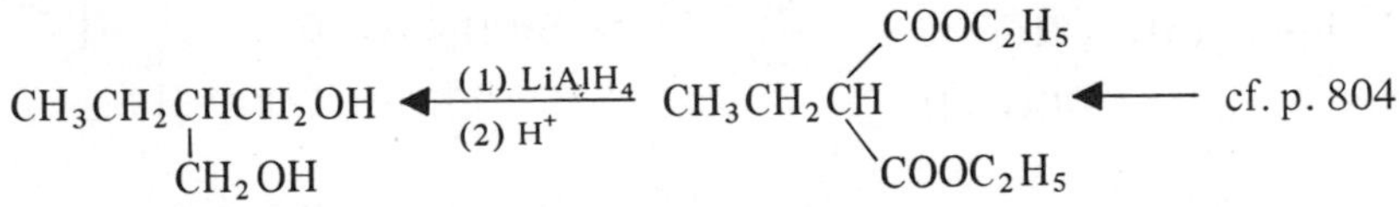

(d) 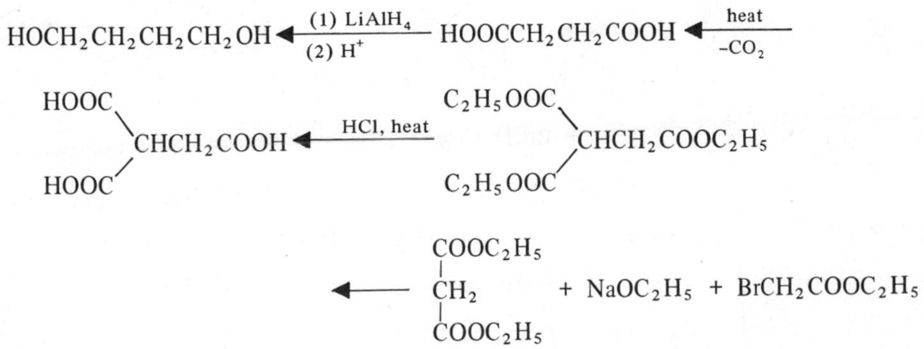 $HOCH_2CH_2CH_2CH_2OH \xleftarrow[\text{(2) } H^+]{\text{(1) LiAlH}_4} HOOCCH_2CH_2COOH \xleftarrow[\text{-CO}_2]{\text{heat}}$

$\underset{\substack{HOOC \diagdown \\ \diagup CHCH_2COOH \\ HOOC}}{} \xleftarrow{\text{HCl, heat}} \underset{\substack{C_2H_5OOC \diagdown \\ \diagup CHCH_2COOC_2H_5 \\ C_2H_5OOC}}{}$

$\longleftarrow \underset{\substack{COOC_2H_5 \\ | \\ CH_2 \\ | \\ COOC_2H_5}}{} + NaOC_2H_5 + BrCH_2COOC_2H_5$

20.24
The following reaction took place,

$$CH_3\overset{O}{\overset{\|}{C}}CH_2\overset{O}{\overset{\|}{C}}OC_2H_5 \ + \ BrCH_2CH_2CH_2Br \ \xrightarrow{NaOC_2H_5} \ BrCH_2CH_2CH_2\overset{\overset{\textstyle CH_3}{|}}{\underset{\underset{\textstyle O}{|}}{\overset{\overset{\textstyle C=O}{}}{C}}H}COC_2H_5$$

$$\xrightarrow[(-H^+)]{NaOC_2H_5} \ C_2H_5OOC-C \cdots CH_2-Br \ \longrightarrow \ \text{Perkin's ester}$$

$$\xrightarrow[(2)\ H_3O^+]{(1)\ OH^-,\ H_2O,\ heat} \ HOOC \cdots \text{Perkin's acid}$$

20.25

(a) $BrCH_2CH_2Br \ + \ \underset{\underset{\textstyle COOC_2H_5}{|}}{\overset{\overset{\textstyle COOC_2H_5}{|}}{C}}H_2 \ + \ NaOC_2H_5 \longrightarrow$

$$\left[BrCH_2CH_2-\underset{\underset{\textstyle COOC_2H_5}{|}}{\overset{\overset{\textstyle COOC_2H_5}{|}}{C}}H\right] \xrightarrow[(-H^+)]{NaOC_2H_5} \left[BrCH_2CH_2-\underset{\underset{\textstyle COOC_2H_5}{|}}{\overset{\overset{\textstyle COOC_2H_5}{|}}{C}}{:}^-\right]$$

$$\longrightarrow \ \underset{CH_2}{\overset{CH_2}{>}}C\underset{CO_2C_2H_5}{\overset{CO_2C_2H_5}{<}} \ \xrightarrow[\substack{(2)\ H_3O^+ \\ (3)\ heat,\ -CO_2}]{(1)\ OH^-,\ H_2O,\ heat} \ \triangleright\!\!-COOH$$

(b) $2NaCH(CO_2C_2H_5)_2 \ + \ BrCH_2CH_2CH_2Br$

$$\longrightarrow \ \underset{C_2H_5O_2C}{\overset{C_2H_5O_2C}{>}}H-CCH_2CH_2CH_2C-H\underset{CO_2C_2H_5}{\overset{CO_2C_2H_5}{<}}$$

A

$$\xrightarrow[Br_2]{NaOC_2H_5} \ \left[\underset{C_2H_5O_2C}{\overset{C_2H_5O_2C}{>}}H-CCH_2CH_2CH_2C-Br\underset{CO_2C_2H_5}{\overset{CO_2C_2H_5}{<}}\right] \xrightarrow{NaOC_2H_5}$$

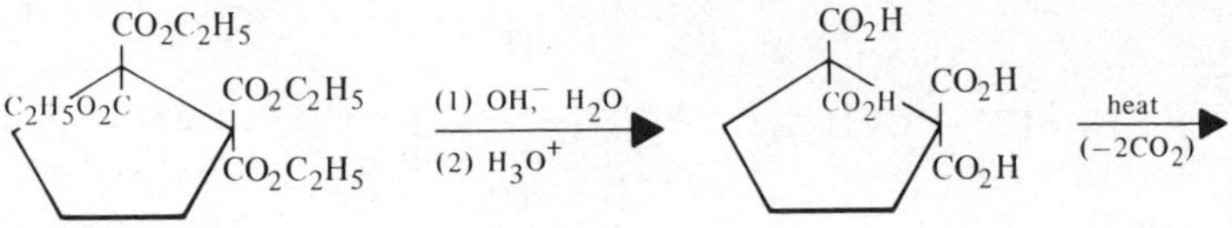

B

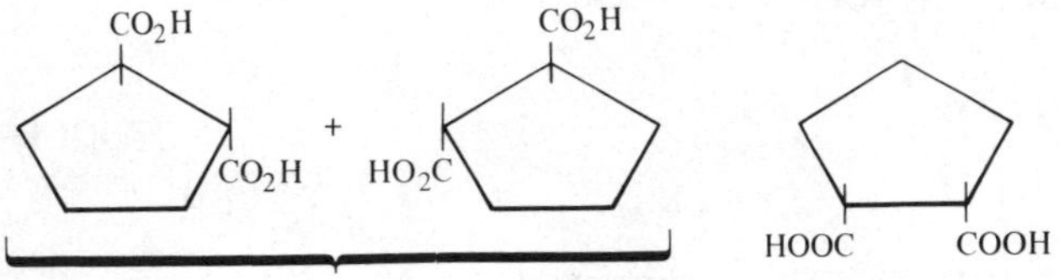

D

racemic modification

E

meso-compound

(c) $BrCH_2CH_2CH_2CH_2Br$ $\xrightarrow{NaCH(CO_2C_2H_5)_2}$ $BrCH_2CH_2CH_2CH_2CH\begin{smallmatrix} CO_2C_2H_5 \\ CO_2C_2H_5 \end{smallmatrix}$

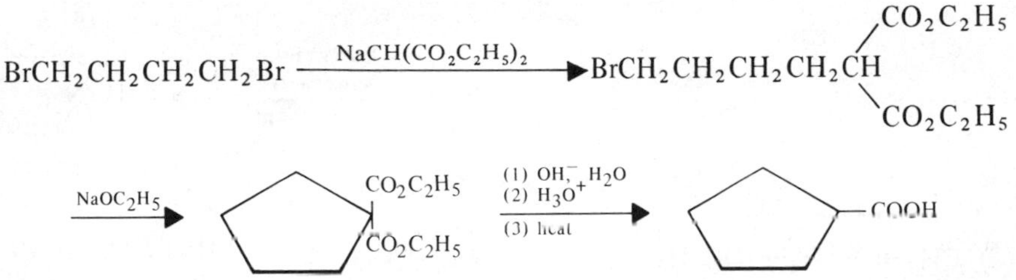

20.26
The α-hydrogens of a thiol ester are more acidic than those of an ordinary ester, and the RS^- group is a better leaving group than RO^-. (See pages 772 and 773 of the text for further explanation.)

20.27
(a) $CH_2(COOC_2H_5)_2$ + $^-OC_2H_5$ $\rightleftharpoons$ $^-:CH_2(COOC_2H_5)_2$ + C_2H_5OH

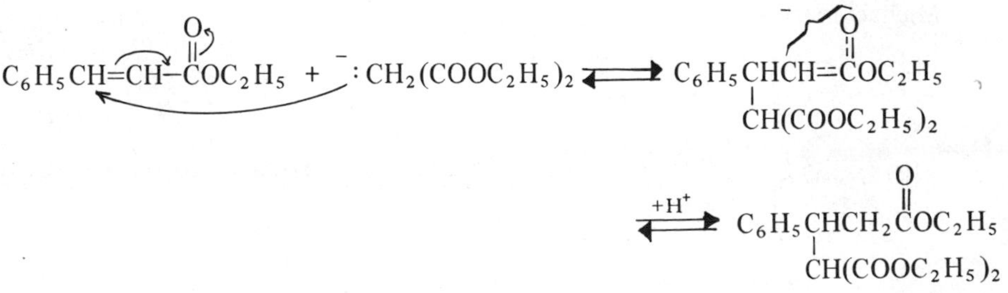

$C_6H_5CH=CH-\overset{\overset{O}{\|}}{C}OC_2H_5$ + $^-:CH_2(COOC_2H_5)_2$ $\rightleftharpoons$ $C_6H_5\overset{}{CH}CH=\overset{\overset{-O}{/}}{C}OC_2H_5$
$\underset{CH(COOC_2H_5)_2}{|}$

$\underset{+H^+}{\rightleftharpoons}$ $C_6H_5\underset{\underset{CH(COOC_2H_5)_2}{|}}{CH}CH_2\overset{\overset{O}{\|}}{C}OC_2H_5$

(b) $CH_3NH_2 + CH_2=CH-\overset{O}{\overset{\|}{C}}OCH_3 \rightleftharpoons CH_3-\overset{H}{\overset{+}{\underset{H}{N}}}-CH_2-CH=\overset{O^-}{\overset{\|}{C}}OCH_3$

$\rightleftharpoons CH_3\overset{}{\underset{H}{N}}-CH_2-CH_2-\overset{O}{\overset{\|}{C}}OCH_3 \overset{CH_2=CH-\overset{O}{\overset{\|}{C}}OCH_3}{\longleftarrow}$

$CH_3N(CH_2CH_2COOCH_3)_2 \overset{base}{\rightleftharpoons} CH_3-N \begin{array}{c} CH_2-CH- \text{—COOCH}_3 \\ \\ CH_2-CH_2 \overset{}{\underset{}{\searrow}} C \overset{OCH_3}{\underset{O}{}} \end{array}$

$\overset{\text{Dieckmann}}{\underset{\text{(several steps)}}{\overset{\text{condensation}}{\longrightarrow}}} CH_3-N \begin{array}{c} COOCH_3 \\ \\ =O \end{array}$

(c) $CH_3-\overset{CH_3}{\underset{CH(CO_2C_2H_5)_2}{\overset{|}{C}}}-CH_2-\overset{O}{\overset{\|}{C}}OC_2H_5 + C_2H_5O^- \rightleftharpoons CH_3-\overset{CH_3}{\underset{CH(CO_2C_2H_5)_2}{\overset{|}{C}}}\text{—}CH=\overset{O^-}{\overset{\|}{C}}OC_2H_5$

$+ C_2H_5OH$

$CH_3-\overset{CH_3}{\underset{CH(CO_2C_2H_5)_2}{\overset{|}{C}}}\text{—}CH=\overset{O^-}{\overset{\|}{C}}OC_2H_5 \rightleftharpoons CH_3-\overset{CH_3}{\overset{|}{C}}=CH-\overset{O}{\overset{\|}{C}}OC_2H_5 + \,^-\!:CH(CO_2C_2H_5)_2$

The Michael reaction is reversible and the reaction just given is an example of a reverse Michael reaction.

(d)

(e) This one is a real challenge.

20.28

Two reactions take place. The first is a normal Knoevenagel condensation,

Then the α,β-unsaturated diketone reacts with a second mole of the active methylene compound in a Michael addition.

21

AMINES

21.1

Because it has two phenyl groups attached to the amine nitrogen, diphenylamine is stabilized by resonance to a much greater extent than aniline. In addition to structures involving different Kekulé structures for the phenyl groups, structures such as the following delocalize the nitrogen electron pair into both benzene rings:

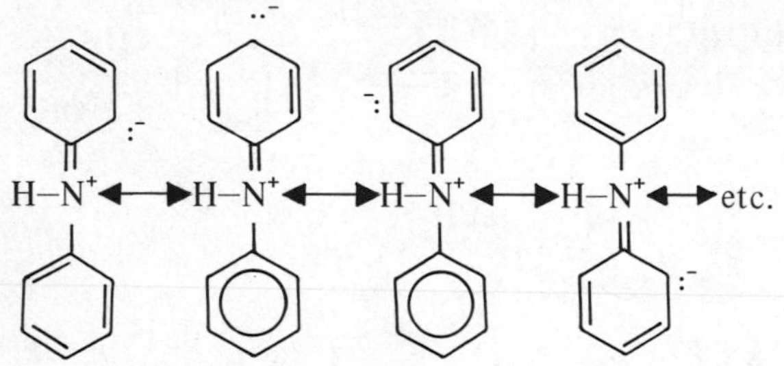

The diphenylammonium ion does not have this type of resonance stabilization. This greater resonance stabilization of diphenylamine means that the reaction,

$$(C_6H_5)_2\ddot{N}H + H_2O \longrightarrow (C_6H_5)_2\overset{+}{N}H_2 + \overset{-}{O}H$$

is even more endothermic than the reaction,

$$C_6H_5\ddot{N}H_2 + H_2O \longrightarrow C_6H_5\overset{+}{N}H_3 + \overset{-}{O}H$$

and that diphenylamine, consequently, is a weaker base.

21.2

The electron-releasing groups (CH_3- and CH_3O-) of p-toluidine and p-anisidine stabilize the cations that are produced:

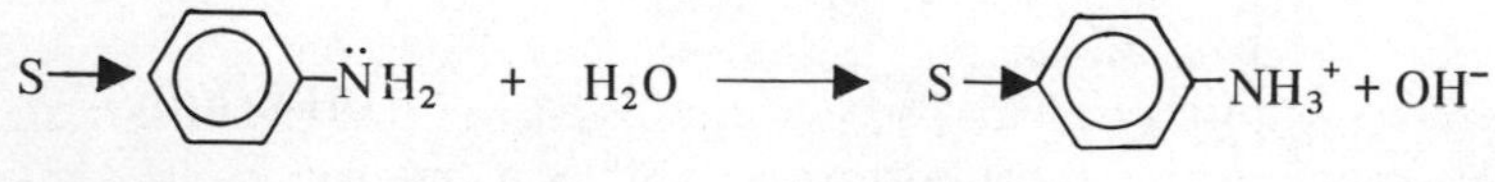

($S = CH_3-$ or CH_3O-)

S releases electrons and stabilizes the product of the acid-base reaction.

Greater stabilization of the product $(S-C_6H_4NH_3^+)$ relative to the reactant $(S-C_6H_4NH_2)$ means that these reactions will be less endothermic than the corresponding reaction of aniline. p-Toluidine and p-anisidine, consequently, are stronger bases.

With p-chloroaniline and p-nitroaniline the substituents (Cl− and NO$_2$−) are electron-withdrawing. These groups destabilize the cations that are the products of an acid-base reaction:

$$S \longleftarrow \bigcirc -\ddot{N}H_2 \ + \ H_2O \ \longrightarrow \ S \longleftarrow \bigcirc -NH_3^+ \ + \ OH^-$$

(S = Cl or NO$_2$)   S withdraws electrons and destabilizes
                                      the product of the acid-base reaction.

The reactions, therefore, are more endothermic than the corresponding reaction of aniline, and p-chloroaniline and p-nitroaniline are weaker bases.

21.3

Dissolve both compounds in ether and extract with aqueous HCl. This gives an ether layer that contains cyclohexane and an aqueous layer that contains cyclohexylammonium chloride. Cyclohexane may then be recovered from the ether layer by distillation. Cyclohexylamine may be recovered from the aqueous layer by adding aqueous NaOH (to convert cyclohexylammonium chloride to cyclohexylamine) and then by ether extraction and distillation

$$C_6H_{12} \ + \ C_6H_{11}NH_2$$
(in ether)

|$H_3O^+Cl^-/H_2O$

ether layer ↓   aqueous layer ↓

C_6H_{12} $C_6H_{11}NH_3^+ \ Cl^-$
(evaporate ether
and distill) | OH$^-$

                                                   ↓

                                                   $C_6H_{11}NH_2$
                                                   (extract into ether
                                                   and distill)

21.4

We begin by dissolving the mixture in a water-immiscible organic solvent such as CH_2Cl_2 or ether. Then, extractions with aqueous acids and bases allow us to separate the components. (We separate p-cresol from benzoic acid by taking advantage of benzoic acid's solubility in the more weakly basic aqueous NaHCO$_3$, whereas, p-cresol requires the more strongly basic, aqueous NaOH (cf., Problem 16.21, page 625 of the text)).

$$C_6H_5COOH, p\text{-}CH_3C_6H_4OH, C_6H_5NH_2, C_6H_6$$
$$(\text{in } CH_2Cl_2)$$

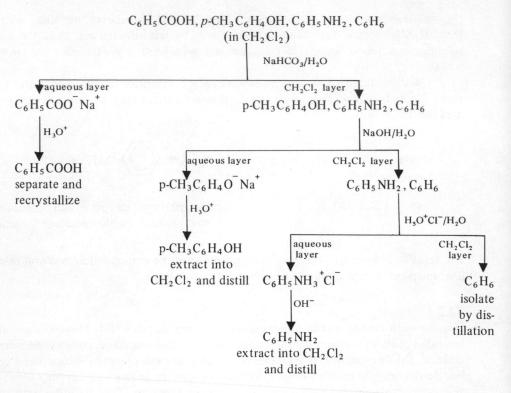

21.5

(a) Neglecting Kekulé' forms of the ring, we can write the following resonance structures for the phthalimide anion.

(b) Phthalimide is more acidic than benzamide because its anion is stabilized by resonance to a greater extent than the anion of benzamide. (Benzamide has only one carbonyl group attached to the nitrogen and thus fewer resonance contributors are possible.)

21.6

21.7

(a) $CH_3(CH_2)_3CHO + NH_3 \xrightarrow{H_2,\ Ni} CH_3(CH_2)_3CH_2NH_2$

(b) $C_6H_5\overset{\underset{\|}{O}}{C}CH_3 + NH_3 \xrightarrow{H_2,\ Ni} C_6H_5\underset{\underset{NH_2}{|}}{C}HCH_3$

(c) $CH_3(CH_2)_4CHO + C_6H_5NH_2 \xrightarrow[CH_3OH]{LiBH_3CN} CH_3(CH_2)_4CH_2NHC_6H_5$

21.8

The reaction of a secondary halide with ammonia would inevitably be accompanied by considerable elimination thus decreasing the yield.

$$\underset{RCH-X}{\overset{\overset{R'}{|}}{}} + NH_3 \underset{excess}{\longrightarrow}
\begin{cases}
\xrightarrow{\text{Substitution}} \underset{RCHNH_2}{\overset{\overset{R'}{|}}{}} \\
\xrightarrow{\text{Elimination}} \text{alkene}
\end{cases}$$

21.9

(a) $C_6H_5COOH \xrightarrow{SOCl_2} C_6H_5COCl \xrightarrow{CH_3CH_2NH_2}$

$C_6H_5CONHCH_2CH_3 \xrightarrow{LiAlH_4} C_6H_5CH_2NHCH_2CH_3$

(b) $CH_3CH_2CH_2CH_2CH_2Br \xrightarrow{NaCN} CH_3CH_2CH_2CH_2CH_2CN$

$\xrightarrow{LiAlH_4} CH_3CH_2CH_2CH_2CH_2CH_2NH_2$

(c) $CH_3CH_2COOH \xrightarrow{SOCl_2} CH_3CH_2COCl \xrightarrow{(CH_3CH_2CH_2)_2NH}$

$CH_3CH_2CON(CH_2CH_2CH_3)_2 \xrightarrow{LiAlH_4} (CH_3CH_2CH_2)_3N$

21.10

(a) $CH_3O-\langle\bigcirc\rangle \xrightarrow[H_2SO_4]{HNO_3} CH_3O-\langle\bigcirc\rangle-NO_2 \xrightarrow[HCl]{Fe}$

$CH_3O-\langle\bigcirc\rangle-NH_2$

(b) $CH_3O-\langle\bigcirc\rangle \xrightarrow[AlCl_3]{CH_3COCl} CH_3O-\langle\bigcirc\rangle-\overset{\overset{O}{\|}}{C}CH_3 \xrightarrow[H_2,\ Ni]{NH_3}$

$CH_3O-\langle\bigcirc\rangle-\underset{\underset{NH_2}{|}}{C}HCH_3$

(c) $\langle\bigcirc\rangle$-CH$_3$ $\xrightarrow{\text{Cl}_2, \ h\nu}$ $\langle\bigcirc\rangle$-CH$_2$Cl

$\xrightarrow{(\text{CH}_3)_3\text{N}}$ $\langle\bigcirc\rangle$-CH$_2\overset{+}{\text{N}}$(CH$_3$)$_3$ Cl$^-$

(d) NO$_2$-$\langle\bigcirc\rangle$-CH$_3$ $\xrightarrow[\text{(2) H}_3\text{O}^+]{\text{(1) KMnO}_4, \ \text{OH}^-}$ NO$_2$-$\langle\bigcirc\rangle$-COOH $\xrightarrow{\text{SOCl}_2}$

NO$_2$-$\langle\bigcirc\rangle$-$\overset{\overset{\text{O}}{\|}}{\text{C}}$-Cl $\xrightarrow{\text{NH}_3}$ NO$_2$-$\langle\bigcirc\rangle$-$\overset{\overset{\text{O}}{\|}}{\text{C}}NH_2$ $\xrightarrow{\text{Br}_2, \ \text{OH}^-}$ NO$_2$-$\langle\bigcirc\rangle$-NH$_2$

(e) CH$_3$-$\langle\bigcirc\rangle$ + Tl(OOCCF$_3$)$_3$ $\xrightarrow{\text{CF}_3\text{COOH}}$ CH$_3$-$\langle\bigcirc\rangle$-Tl(OOCCF$_3$)$_2$

$\xrightarrow{\text{KCN, H}_2\text{O}, \ h\nu}$ CH$_3$-$\langle\bigcirc\rangle$-CN $\xrightarrow{\text{LiAlH}_4}$ CH$_3$-$\langle\bigcirc\rangle$-CH$_2$NH$_2$

21.11

An amine acting as a base.

$$\text{CH}_3\text{CH}_2\overset{..}{\text{N}}\text{H}_2 \ +\text{H}_3\text{O}^+ \rightleftharpoons \text{CH}_3\text{CH}_2\overset{+}{\text{N}}\text{H}_3 \ + \text{H}_2\text{O}$$

An amine acting as a nucleophile in an alkylation reaction.

$$(\text{CH}_3\text{CH}_2)_3\text{N}: + \ \text{CH}_3 \text{I} \longrightarrow (\text{CH}_3\text{CH}_2)_3\overset{+}{\text{N}}\text{-CH}_3 \ \text{I}^-$$

An amine acting as a nucleophile in an acylation reaction.

$$(\text{CH}_3)_2\overset{..}{\text{N}}\text{H} + \text{CH}_3\overset{\overset{\text{O}}{\|}}{\text{C}} \overset{}{\underset{\text{Cl}}{}} \longrightarrow (\text{CH}_3)_2\text{N}\overset{\overset{\text{O}}{\|}}{\text{C}}\text{CH}_3 + (\text{CH}_3)_2\text{NH}_2\text{Cl}$$
$$\text{(excess)}$$

An enamine acting as a carbon nucleophile in an acylation reaction.

An enamine acting as a carbon nucleophile in an alkylation reaction.

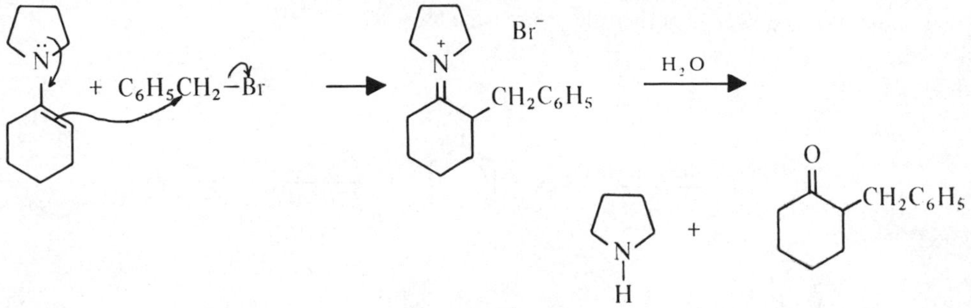

$$+ \ C_6H_5CH_2\text{—Br} \quad \longrightarrow \quad \xrightarrow{H_2O}$$

$CH_2C_6H_5$

Br^-

$+$

An amino group acting as an activating group and as an ortho-para director in electrophilic aromatic substituion.

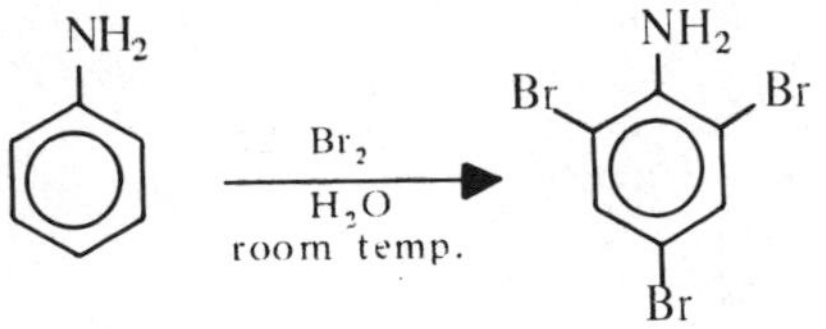

$$\xrightarrow[\substack{H_2O \\ room \ temp.}]{Br_2}$$

21.12

$$CH_3CH_2CH_2CH_2^+ \begin{cases} \xrightarrow{-H^+} CH_3CH_2CH=CH_2 \\ \xrightarrow[(-H^+)]{H_2O} CH_3CH_2CH_2CH_2OH \\ \xrightarrow{Cl^-} CH_3CH_2CH_2CH_2Cl \end{cases}$$

$\downarrow$ hydride migration

$$CH_3CH_2\overset{+}{C}HCH_3 \begin{cases} \xrightarrow{-H^+} CH_3CH=CHCH_3 \\ \xrightarrow[(-H^+)]{H_2O} CH_3CH_2CHCH_3 \ \overset{|}{OH} \\ \xrightarrow{Cl^-} CH_3CH_2CHCH_3 \ \overset{|}{Cl} \end{cases}$$

21.13

(a) $^-O\text{—}N{=}O + H_3O^+ \rightleftarrows HO\text{—}N{=}O$

$HO\text{—}N{=}O + H_3O^+ \rightleftarrows \overset{+}{HO}\text{—}N{=}O$
$\qquad\qquad\qquad\qquad\quad \overset{|}{H}$

$\overset{+}{HO}\text{—}N{=}O \rightleftarrows H_2O + \overset{+}{N}{=}O$
$\overset{|}{H}$

(b) $\qquad + \ + \ \overset{+}{N}O \quad \longrightarrow \quad \xrightarrow{-H^+}$

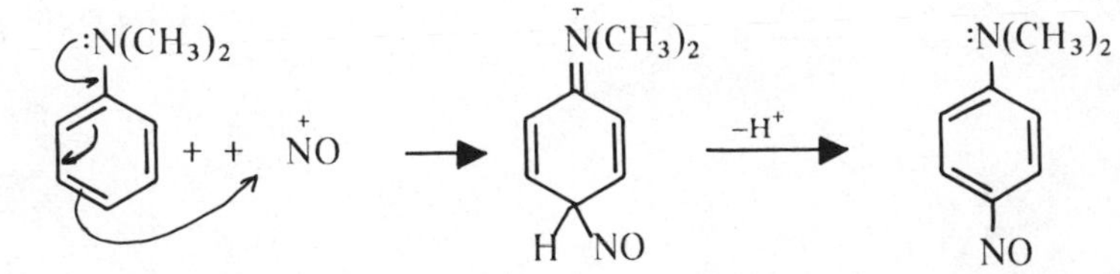

(c) The $\overset{+}{N}O$ ion is a weak electrophile. For it to react with an aromatic ring, the ring must have a powerful activating group such as —OH or —NR$_2$.

21.14

(a)

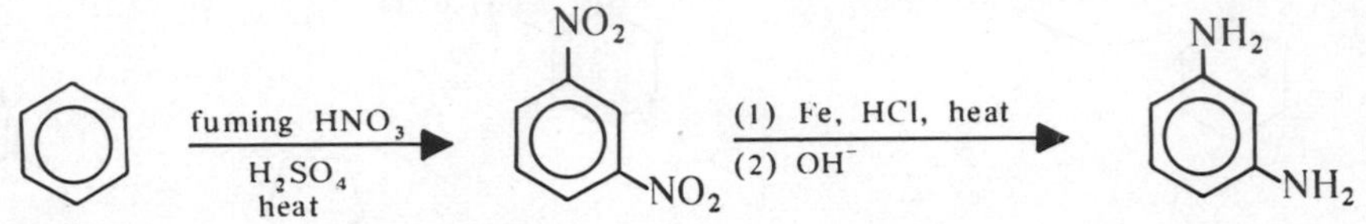

(b)

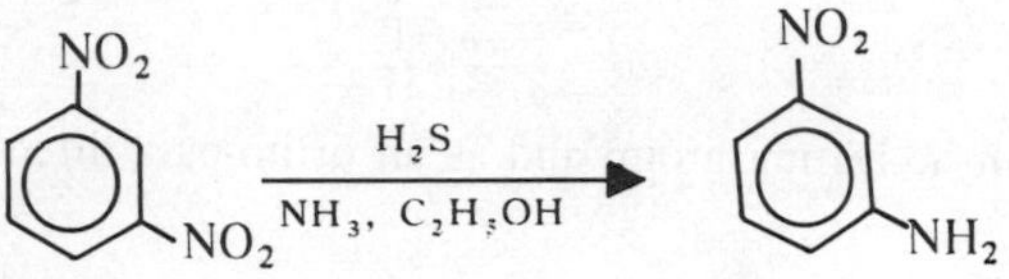

(from part a)

(c)

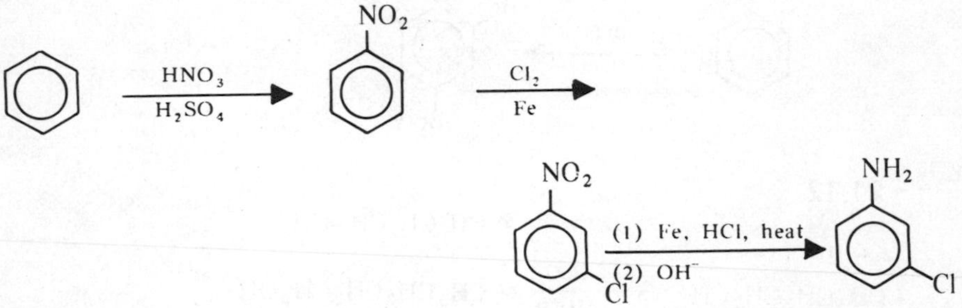

(d)

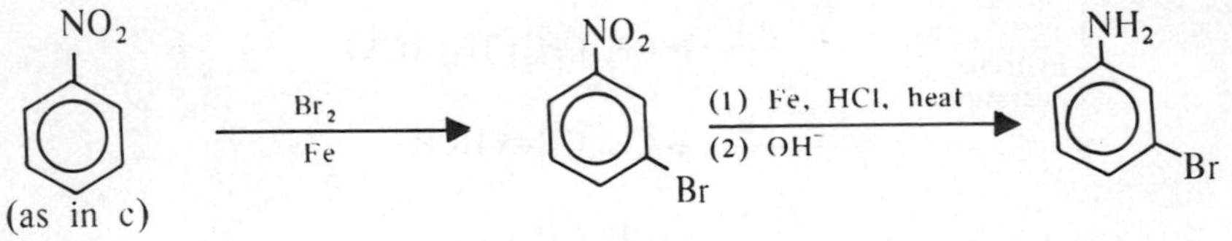

(as in c)

(e)

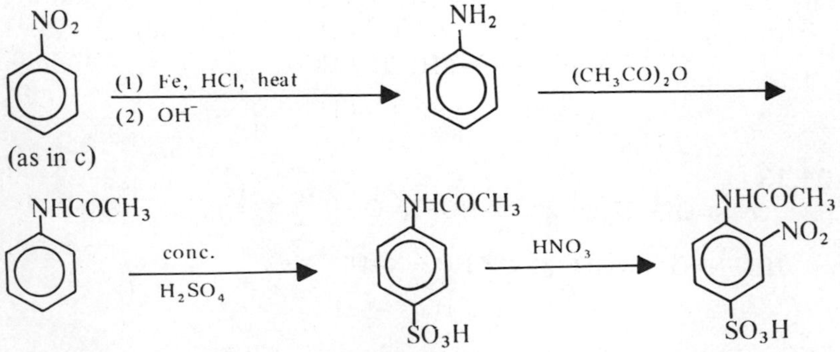

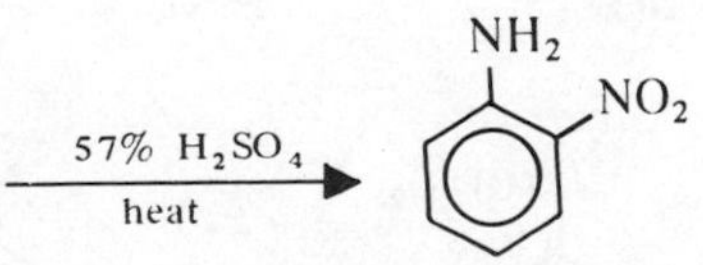

(as in c)

(56% yield from acetanilide cf., page 498 of text)

(f)

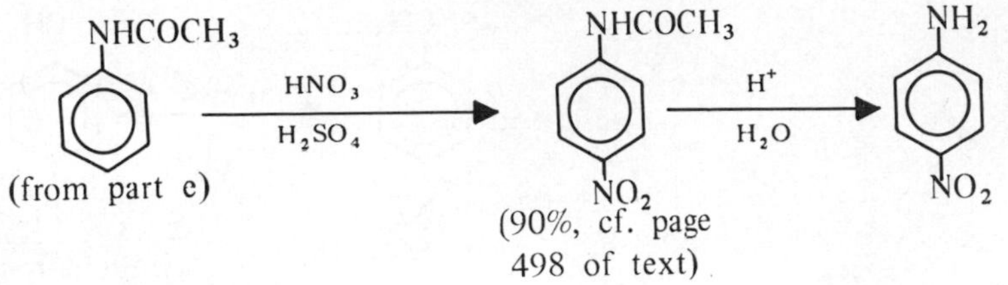
NHCOCH3
HNO3
H2SO4
(from part e)
NHCOCH3
NO2
(90%, cf. page
498 of text)
H+
H2O
NH2
NO2

21.15

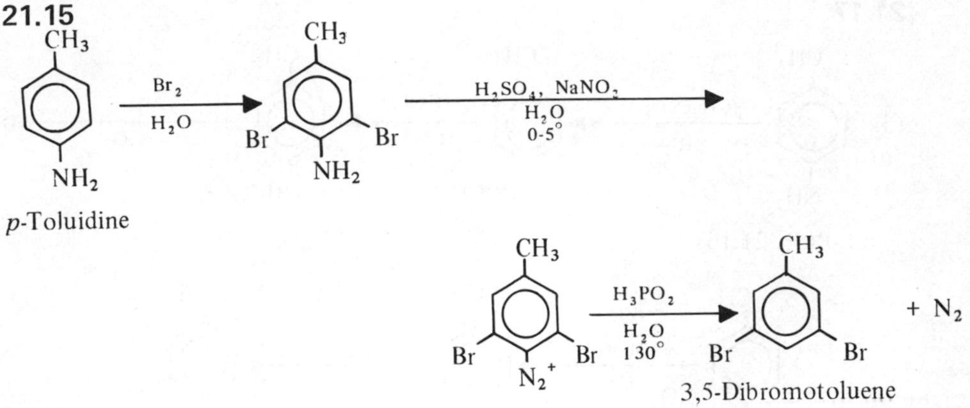
CH3
NH2
p-Toluidine
Br2
H2O
CH3
Br
Br
NH2
H2SO4, NaNO2
H2O
0-5°
CH3
Br
Br
N2+
H3PO2
H2O
130°
CH3
Br
Br
3,5-Dibromotoluene
+ N2

21.16

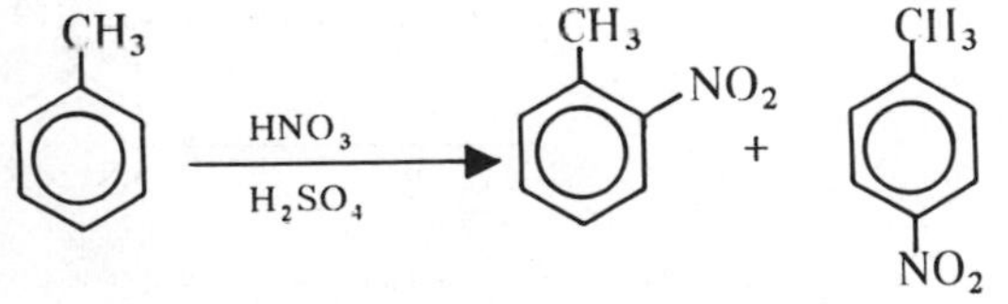
CH3
HNO3
H2SO4
CH3
NO2
+
CH3
NO2

(separate by distillation)

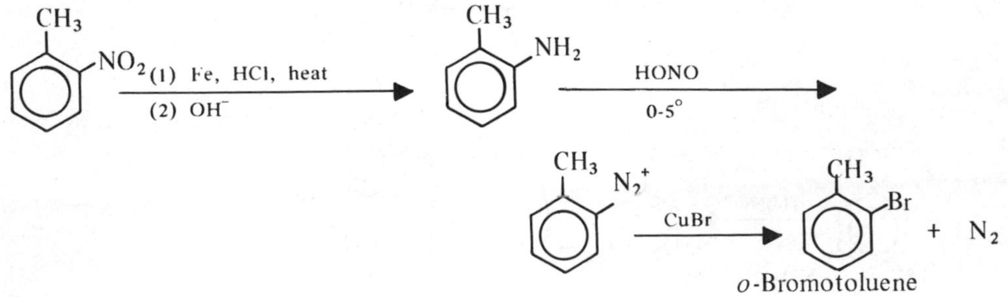
CH3
NO2
(1) Fe, HCl, heat
(2) OH−
CH3
NH2
HONO
0-5°
CH3
N2+
CuBr
CH3
Br
+ N2
o-Bromotoluene

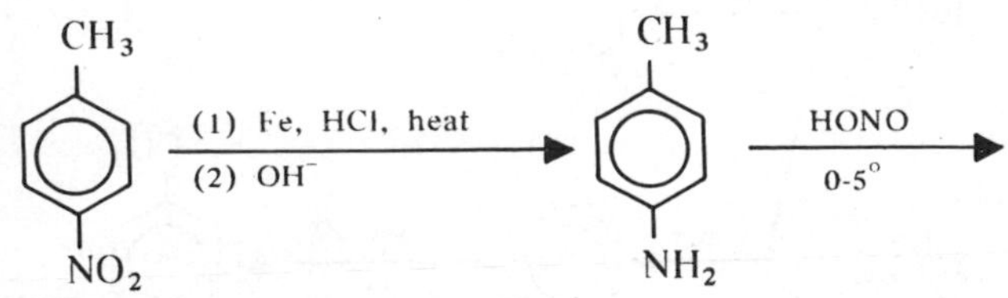
CH3
NO2
(1) Fe, HCl, heat
(2) OH−
CH3
NH2
HONO
0-5°

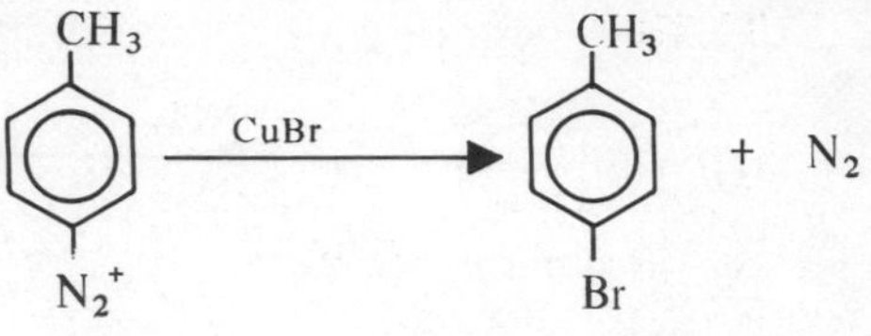

p-Bromotoluene

21.17

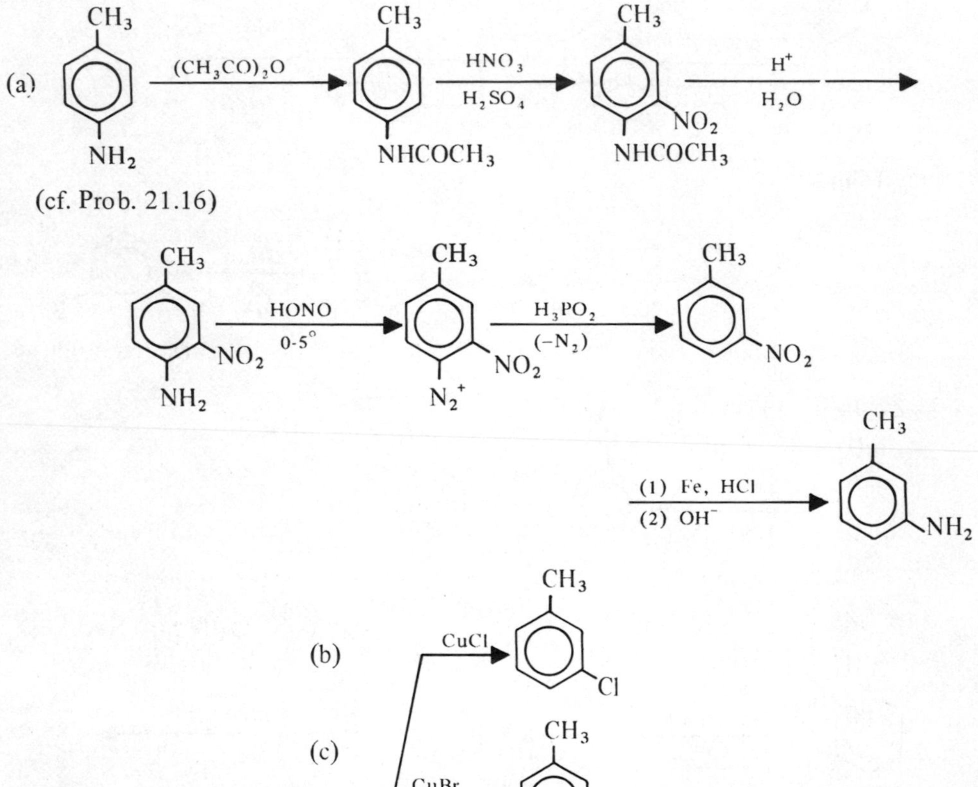

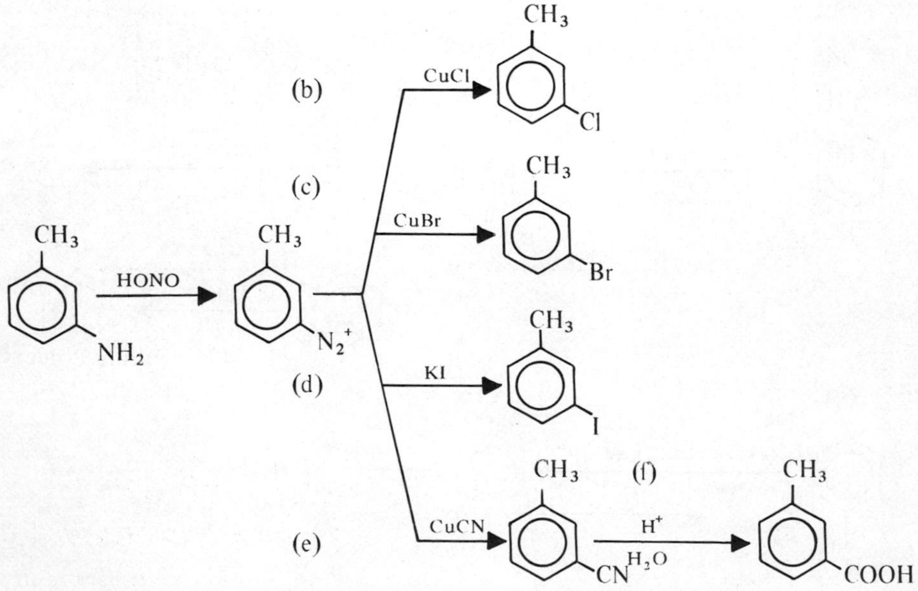

(a)

(cf. Prob. 21.16)

(b)

(c)

(d)

(e)

(f)

21.18

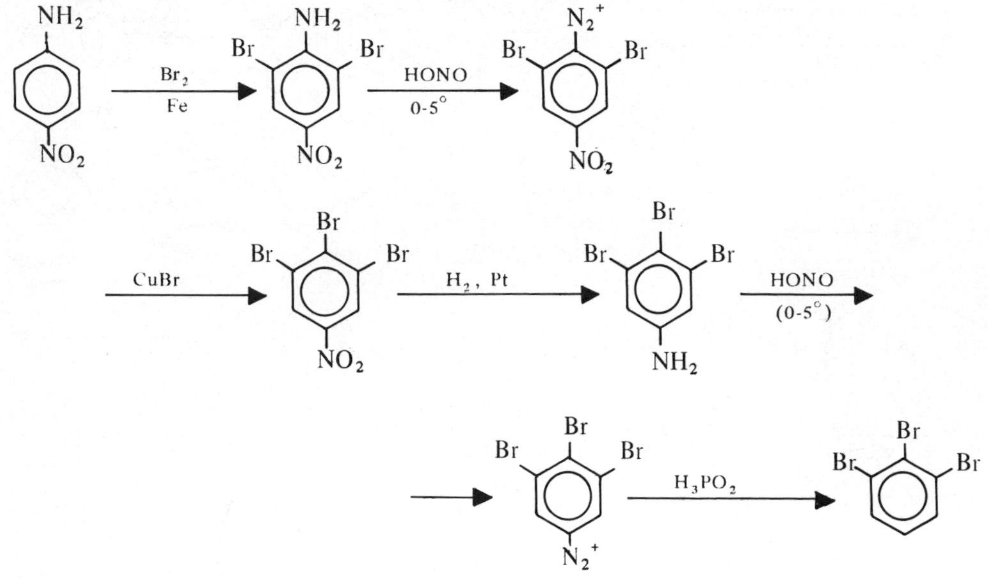

21.19

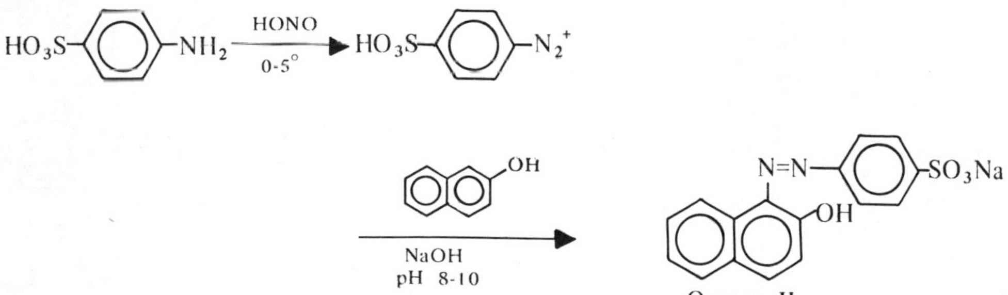

Orange II

21.20

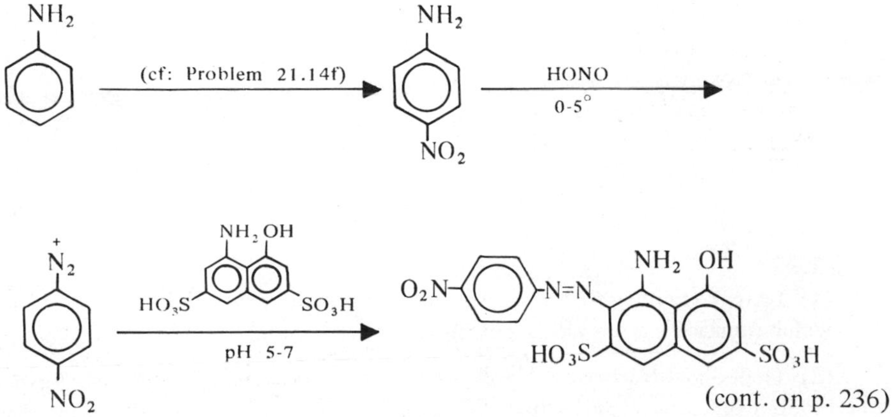

(cont. on p. 236)

N_2^+—⬡—⬡—N_2^+
from benzidine
+ HONO
——————→
pH 8-10
(NaOH)

O_2N—⬡—N=N— NH$_2$ OH ⬡⬡ —N=N—⬡—⬡—N_2^+
NaO$_3$S   SO$_3$Na

⬡—OH O_2N—⬡—N=N— NH$_2$ OH ⬡⬡ —N=N—⬡—⬡—N=N—⬡—OH
——————→
NaO$_3$S   SO$_3$Na
Diamine Green B

21.21

⬡—$\ddot{N}H_2$ + ⬡—SO_2Cl $\xrightarrow[H_2O]{KOH}$ ⬡—$\overset{K^+}{\underset{\ddot{}}{N}}$—$SO_2$—⬡
water soluble

$\xrightarrow{HCl}$ ⬡—$\overset{H}{\underset{\ddot{}}{N}}$—$SO_2$—⬡
precipitate

(b) ⬡—$\overset{CH_3}{\underset{\ddot{}}{N}}$—H + ⬡—$SO_2Cl$ $\xrightarrow[H_2O]{KOH}$ ⬡—$\overset{CH_3}{\underset{\ddot{}}{N}}$—$SO_2$—⬡
precipitate

$\xrightarrow{HCl}$ No reaction (precipitate remains)

(c) ⬡—$\overset{CH_3}{\underset{\ddot{}}{N}}$—$CH_3$ + ⬡—SO_2Cl $\xrightarrow[H_2O]{KOH}$ No reaction
(insoluble amine)

$\xrightarrow{HCl}$ ⬡—$\overset{CH_3}{\underset{H}{\overset{+}{N}}}$—$CH_3$ Cl$^-$
amine dissolves

21.22

(1) That A reacts with benzenesulfonyl chloride in aqueous KOH to give a clear solution which on acidification yields a precipitate shows that A is a primary amine.

(2) That diazotization of A followed by treatment with 2-naphthol gives an intensely colored precipitate shows that A is a primary aromatic amine, that is, A is a substituted aniline.

(3) Consideration of the molecular formula of A leads us to conclude that A is a toluidine

$$\begin{array}{l} C_7H_9N \\ - \underline{C_6H_6N} \\ CH_3 \end{array} = \text{—◯—NH}_2$$

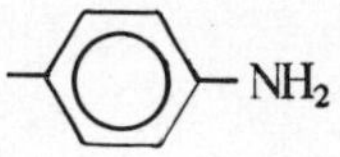

But is A *o*-toluidine, *m*-toluidine, or *p*-toluidine?

(4) This question is answered by the infrared data. A single absorption peak in the 680-840 cm^{-1} region at 815 cm^{-1} is indicative of a *para* substituted benzene. Thus A is *p*-toluidine.

A

21.23

(a)

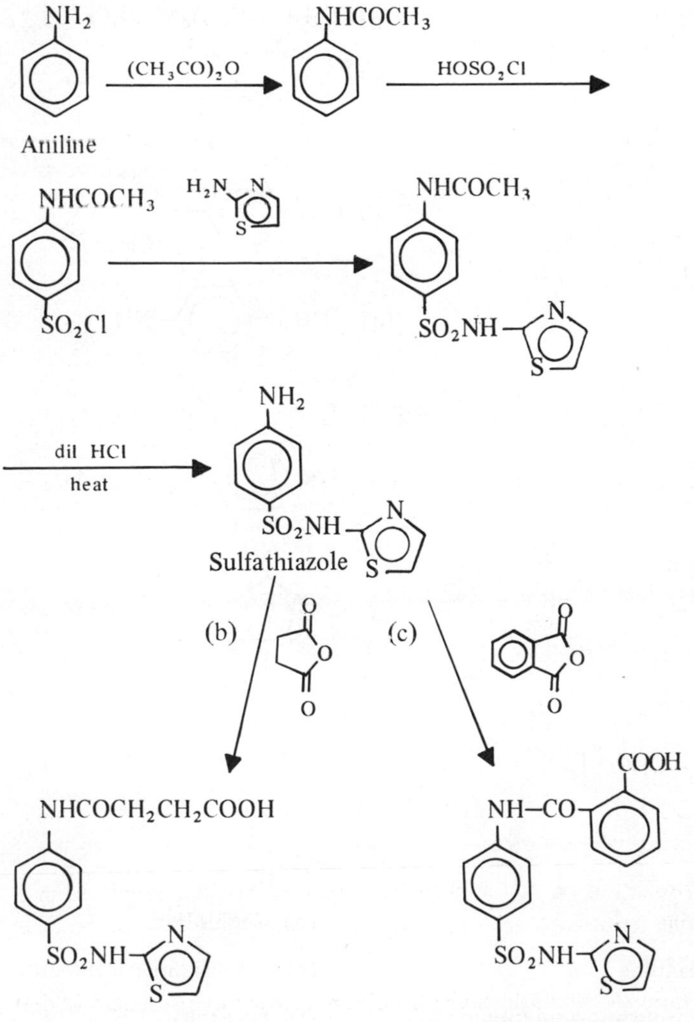

Aniline

Sulfathiazole

(b)

(c)

Succinoylsulfathiazole

Phthalylsulfathiazole

21.24

(a) $C_6H_5CH_2NCH_3$

(b) 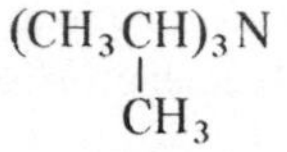 $(CH_3CH)_3N$
         |
        CH_3

(c)

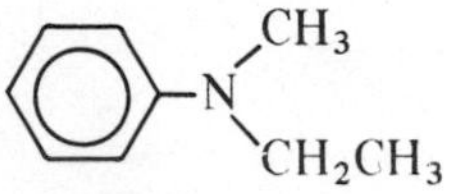

N-methyl-N-ethylaniline structure

(d)

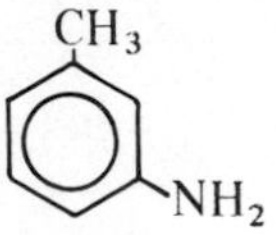

3-methylaniline structure

(e)

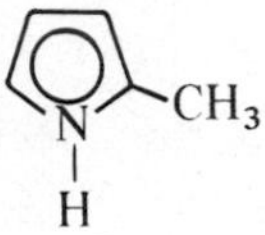

2-methylpyrrole structure

(f)

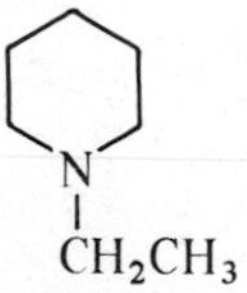

N-ethylpiperidine structure

(g)

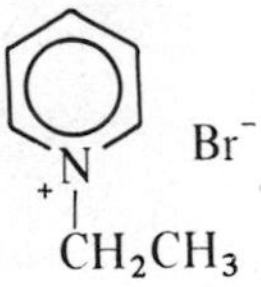

N-ethylpyridinium bromide structure

(h)

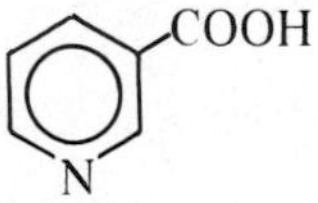

pyridine-3-carboxylic acid structure

(i)

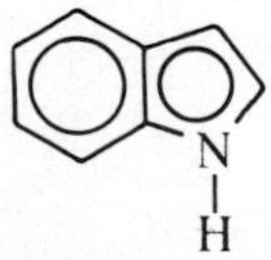

indole structure

(j)

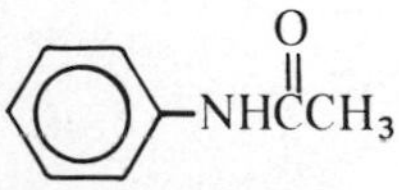

acetanilide structure

(k)

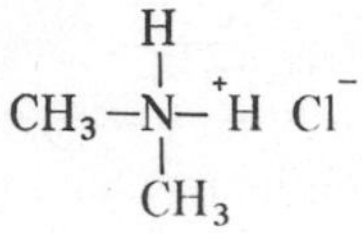

$CH_3-\overset{\overset{H}{|}}{N}-\overset{+}{H}$ Cl^-
      |
    CH_3

(l)

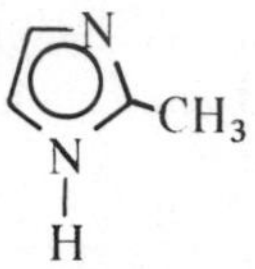

2-methylbenzimidazole structure

(m)

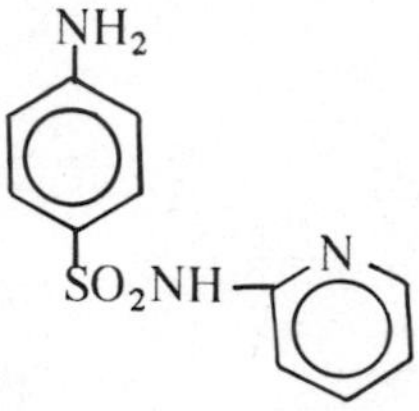

sulfapyridine structure

(n) $(CH_3CH_2CH_2)_4N^+$ Cl^-

(o)

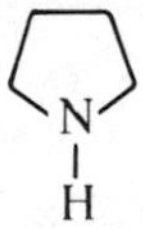

pyrrolidine structure

(p)

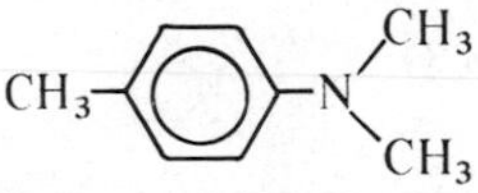

N,N-dimethyl-p-toluidine structure

(q) 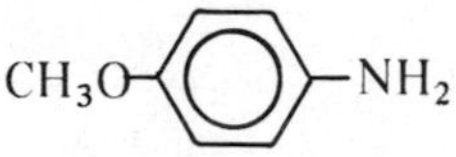 CH_3O-

$-NH_2$

(r) 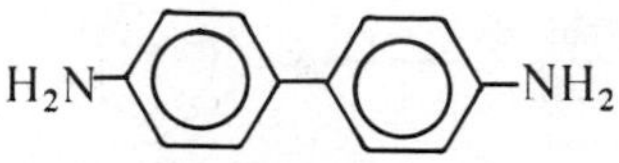 H_2N-

$-NH_2$

(s)

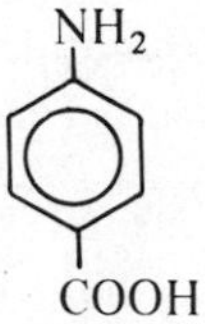

4-aminobenzoic acid structure

(t)

histidine structure

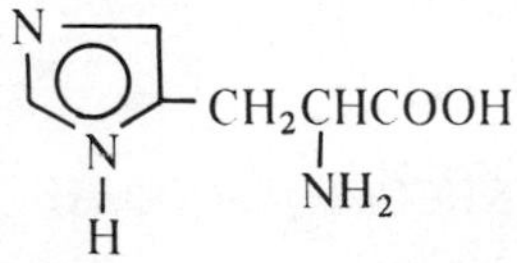

21.25
(a) Propylamine

(b) N-methylaniline

(c) Trimethylisopropylammonium iodide

(d) *o*-toluidine

(e) *o*-Anisidine (or *o*-methoxyaniline)

(f) Pyrazole

(g) 2-Aminopyrimidine

(h) Benzylammonium chloride

(i) N,N-Dipropylaniline

(j) Benzenesulfonamide

(k) Methylammonium acetate

(l) 3-Aminopropanol

(m) Purine

(n) N-Methylpyrrole

21.26

(a)

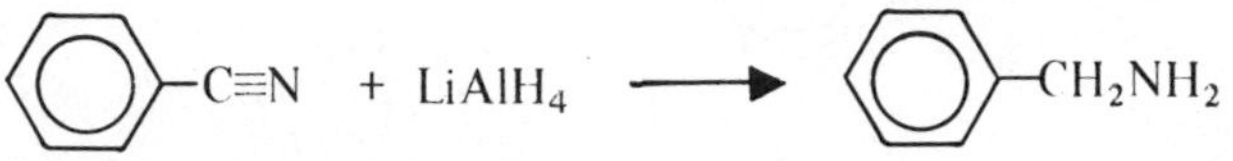

(b)

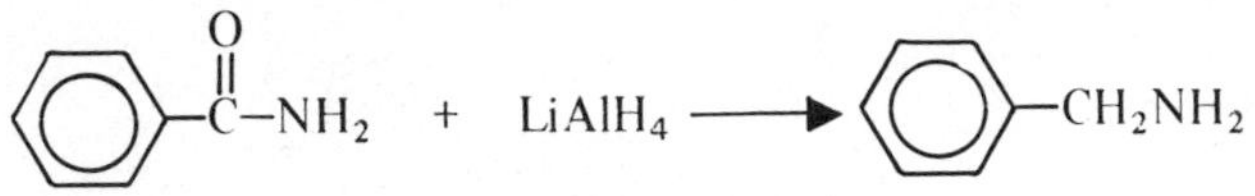

(c)

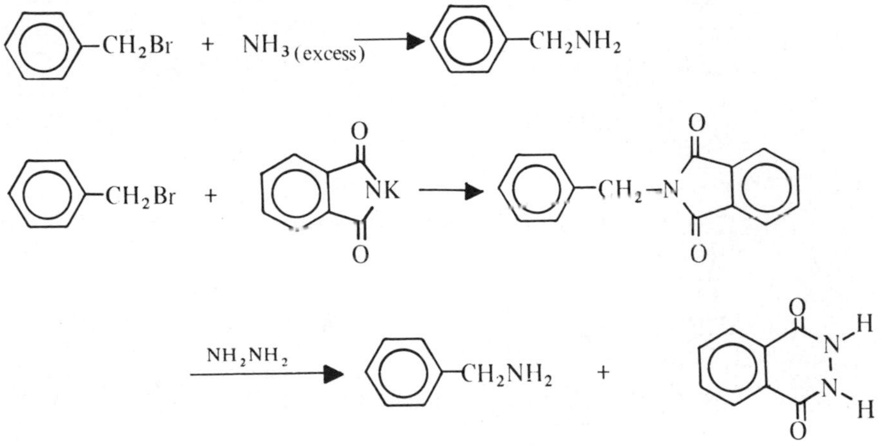

(d)

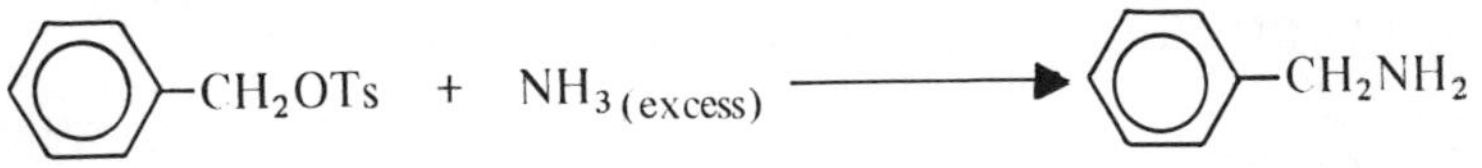

(e)

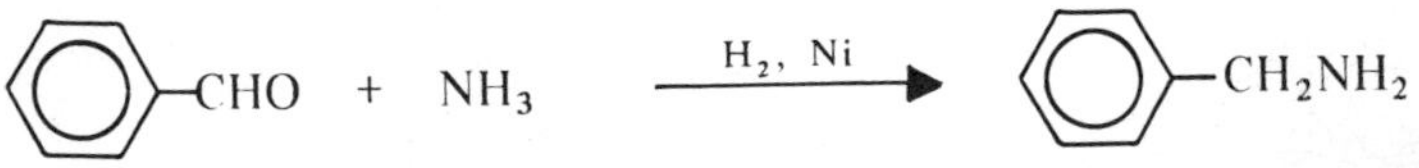

(f)

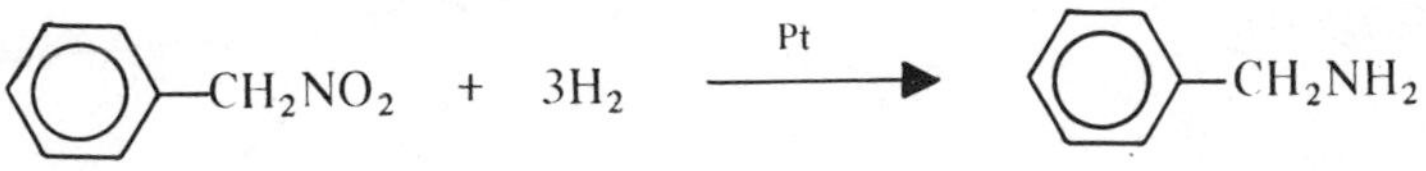

(g)

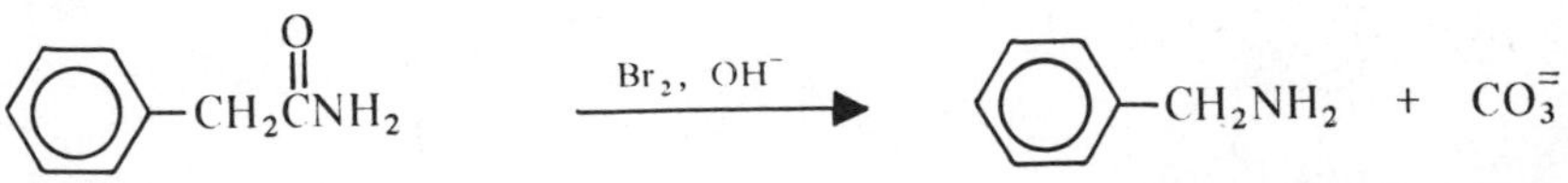

21.27

(a)

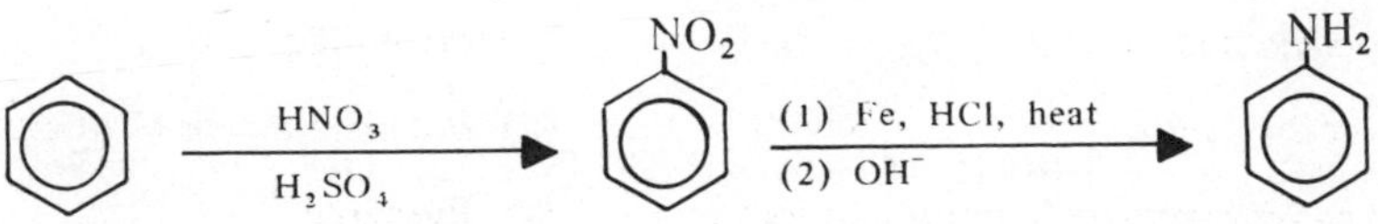

(b)

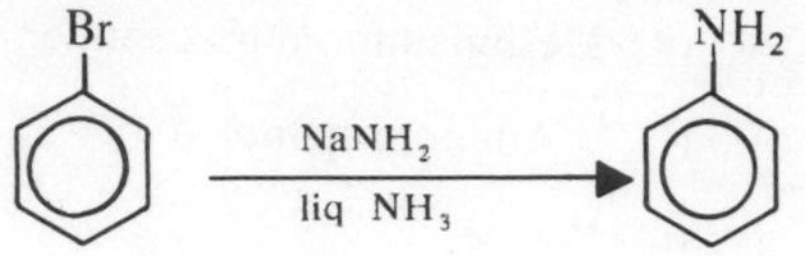

(c)

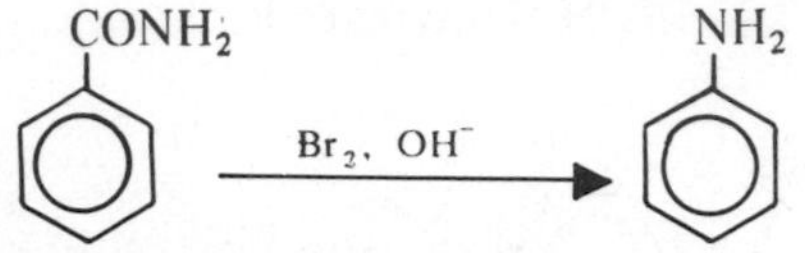

21.28

(a) 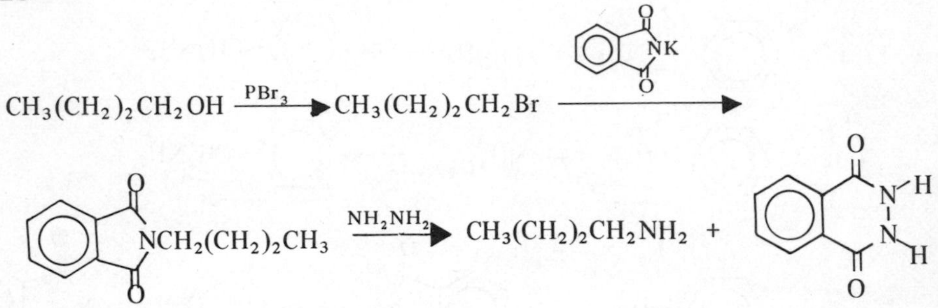 $CH_3(CH_2)_2CH_2OH \xrightarrow{PBr_3} CH_3(CH_2)_2CH_2Br \longrightarrow$

$\xrightarrow{NH_2NH_2} CH_3(CH_2)_2CH_2NH_2 +$

(b) $CH_3(CH_2)_2CH_2Br \xrightarrow{NaCN} CH_3(CH_2)_3CN \xrightarrow{LiAlH_4} CH_3(CH_2)_3CH_2NH_2$
    (from part a)

(c) $CH_3(CH_2)_2CH_2OH \xrightarrow[\text{(2) } H_3O^+]{\text{(1) } KMnO_4, \ OH^-} CH_3CH_2CH_2COOH$

$\xrightarrow[\text{(2) } NH_3]{\text{(1) } SOCl_2} CH_3CH_2CH_2CONH_2 \xrightarrow{Br_2, \ OH^-} CH_3CH_2CH_2NH_2$

(d) $CH_3CH_2CH_2CH_2OH \xrightarrow[C_5H_5N]{CrO_3} CH_3CH_2CH_2CHO \xrightarrow[H_2, \ Ni]{CH_3NH_2}$

$CH_3CH_2CH_2CH_2NHCH_3$

21.29

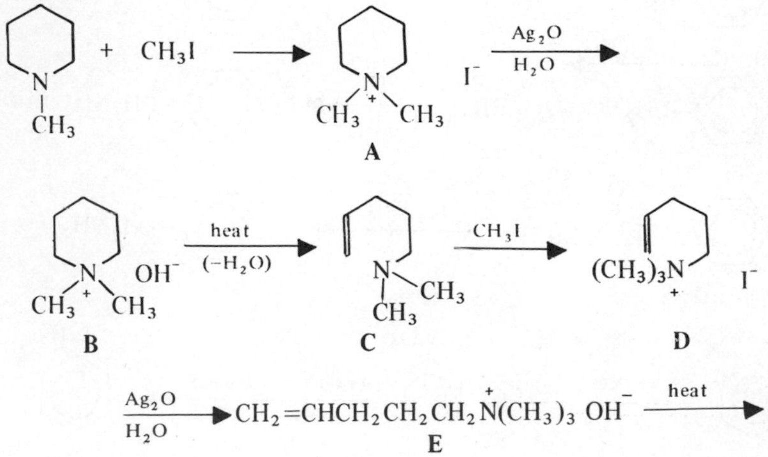

$\xrightarrow[H_2O]{Ag_2O} CH_2{=}CHCH_2CH_2CH_2\overset{+}{N}(CH_3)_3 \ OH^- \xrightarrow{heat}$
E

$$CH_2=CHCH_2CH=CH_2 + H_2O + (CH_3)_3N$$
$$\mathbf{F}$$

21.30

(a)

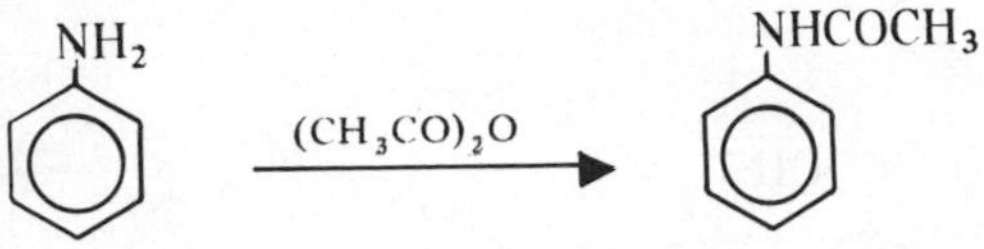

$$\xrightarrow{(CH_3CO)_2O}$$

(b)

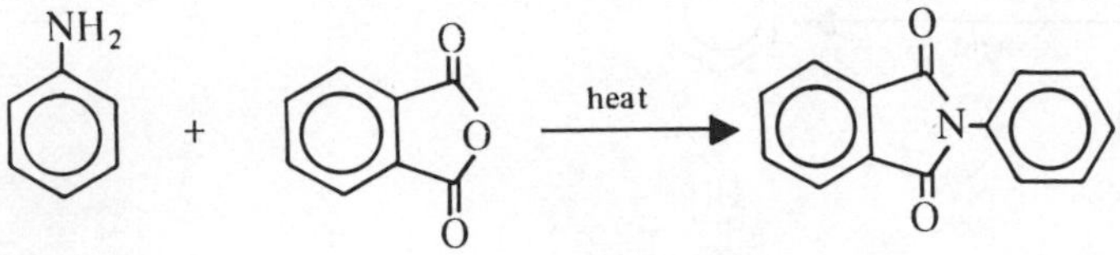

$$+ \qquad \xrightarrow{heat}$$

(c)

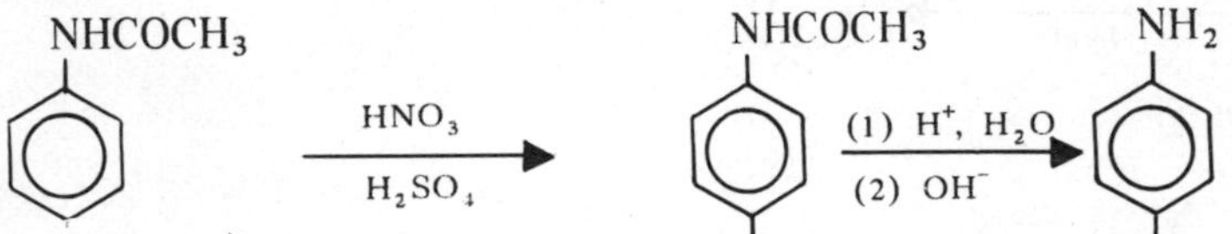

$$\xrightarrow[H_2SO_4]{HNO_3}$$

(from part a)

$$\xrightarrow[(2)\ OH^-]{(1)\ H^+,\ H_2O}$$

(d)

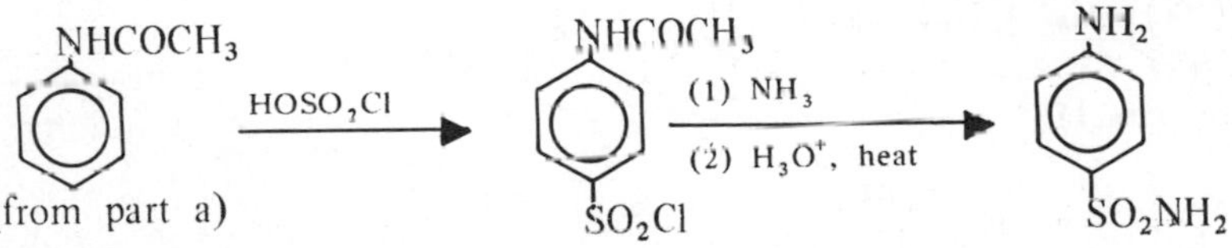

$$\xrightarrow{HOSO_2Cl}$$

(from part a)

$$\xrightarrow[(2)\ H_3O^+,\ heat]{(1)\ NH_3}$$

(e)

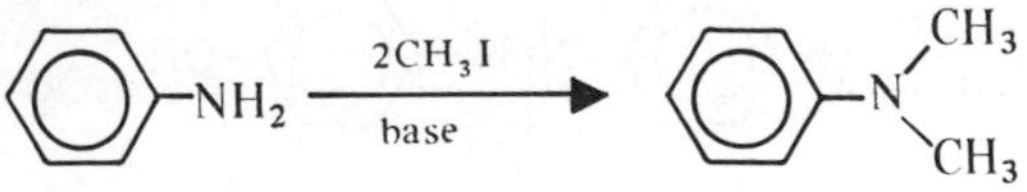

$$-NH_2 \xrightarrow[base]{2CH_3I}$$

(f)

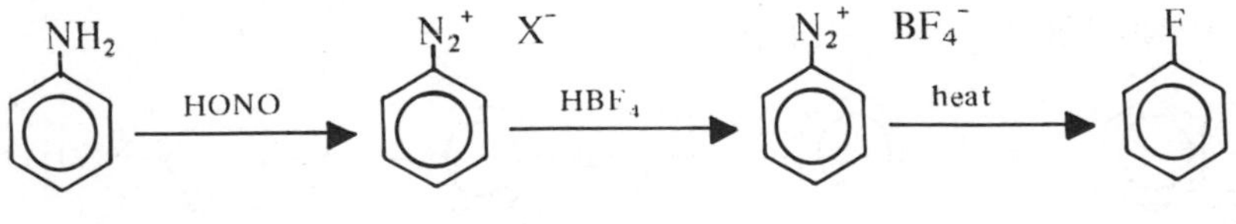

$$\xrightarrow{HONO} \qquad \xrightarrow{HBF_4} \qquad \xrightarrow{heat}$$

(g)

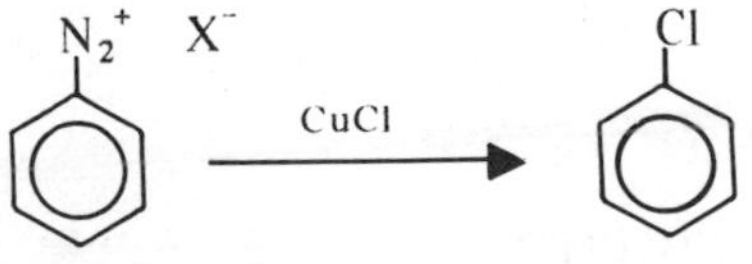

$$\xrightarrow{CuCl}$$

(from part f)

(h)

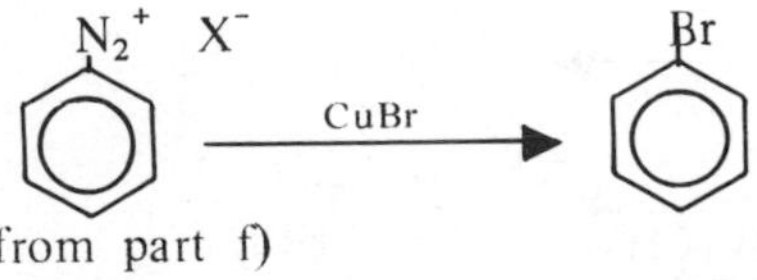

$$\xrightarrow{CuBr}$$

(from part f)

(i)

$$\xrightarrow{KI}$$

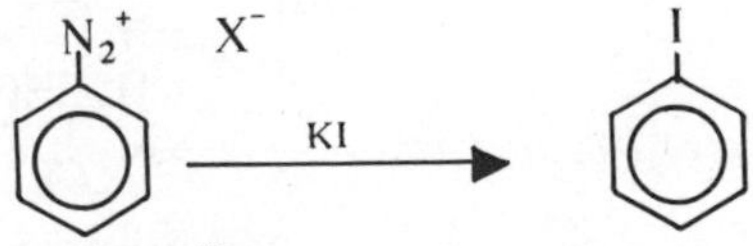

(from part f)

(j) 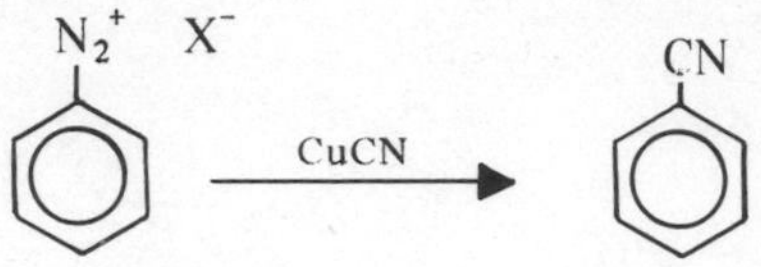
N_2^+ X^-
CuCN
CN

(from part f)

(k) 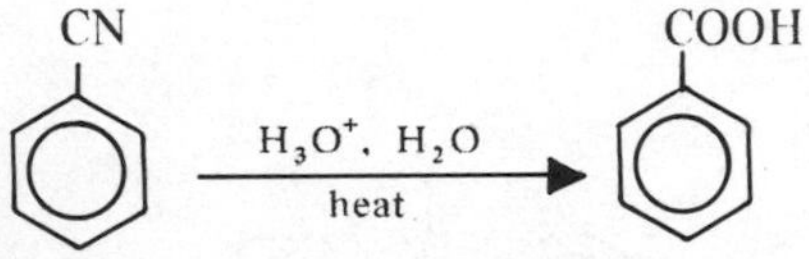
CN
H_3O^+, H_2O
heat
COOH

(from part j)

(l) 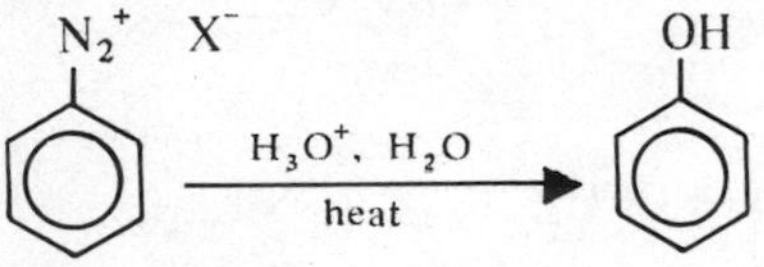
N_2^+ X^-
H_3O^+, H_2O
heat
OH

(from part f)

(m) 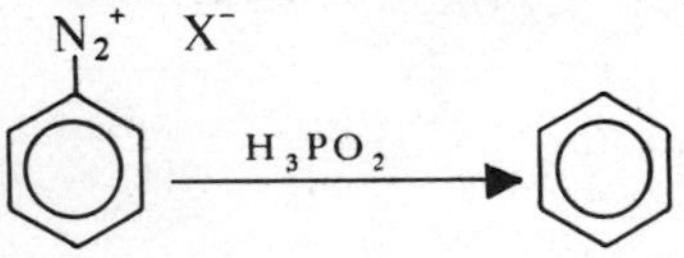
N_2^+ X^-
H_3PO_2

(from part f)

(n) 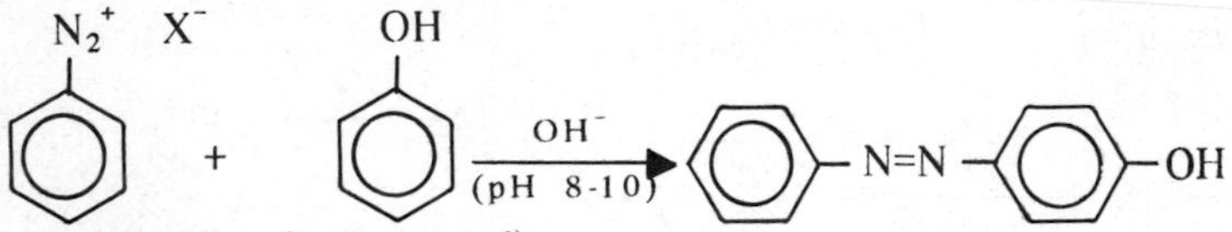
N_2^+ X^-
+
OH
OH^-
(pH 8-10)
N=N
OH

(from part f) (from part l)

(o) 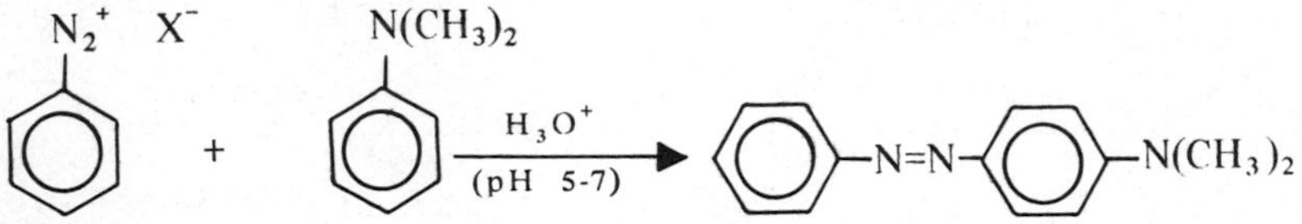
N_2^+ X^-
+
$N(CH_3)_2$
H_3O^+
(pH 5-7)
N=N
$N(CH_3)_2$

(from part f) (from part e)

21.31

(a) 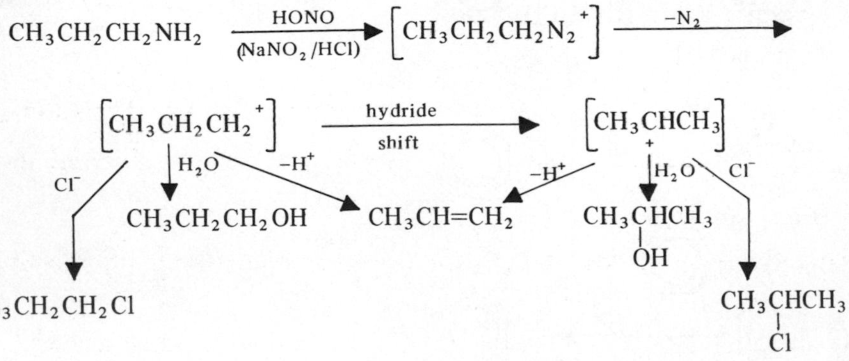
$CH_3CH_2CH_2NH_2$
HONO
($NaNO_2$/HCl)
$[CH_3CH_2CH_2N_2^+]$
$-N_2$
$[CH_3CH_2CH_2^+]$
hydride
shift
$[CH_3\overset{+}{C}HCH_3]$
Cl^-
H_2O
$-H^+$
$CH_3CH_2CH_2OH$
$CH_3CH{=}CH_2$
$CH_3CH(OH)CH_3$
$CH_3CH_2CH_2Cl$
$CH_3CH(Cl)CH_3$

(b) $(CH_3CH_2)_2NH \xrightarrow[\text{(NaNO}_2\text{/HCl)}]{\text{HONO}} (CH_3CH_2)_2N-N=O$

(c)

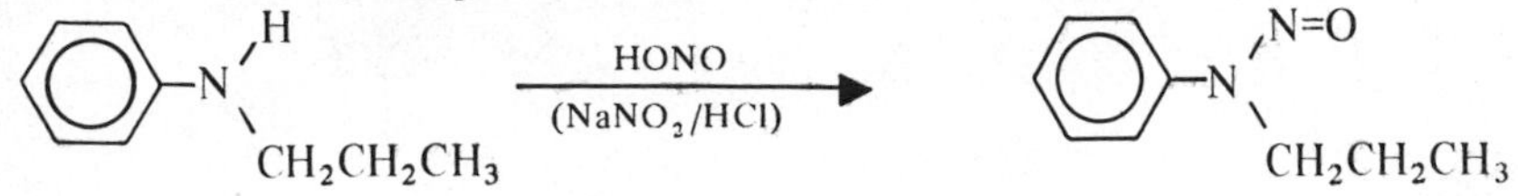

$\xrightarrow[\text{(NaNO}_2\text{/HCl)}]{\text{HONO}}$

(d)

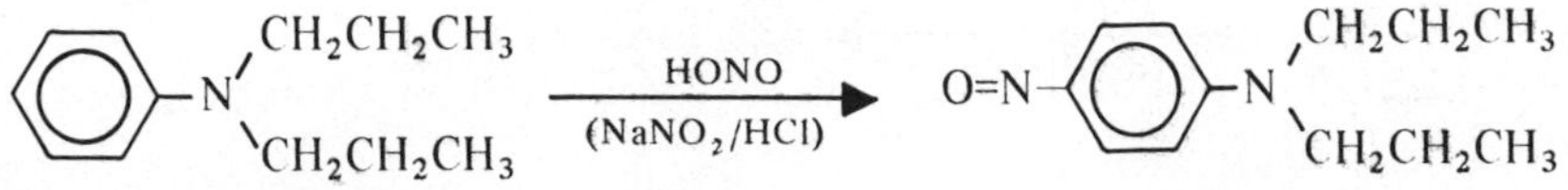

$\xrightarrow[\text{(NaNO}_2\text{/HCl)}]{\text{HONO}}$

(e) 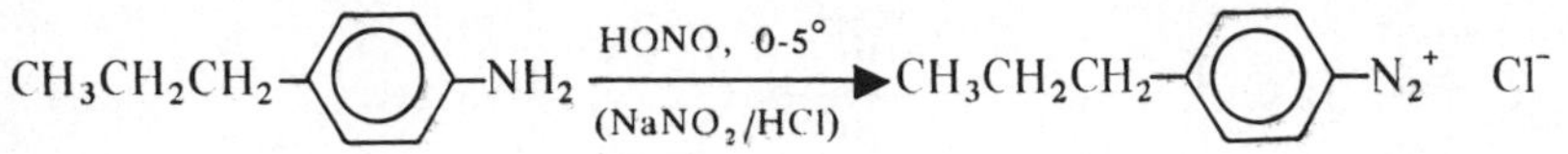 $CH_3CH_2CH_2\!-\!\langle\text{ring}\rangle\!-\!NH_2 \xrightarrow[\text{(NaNO}_2\text{/HCl)}]{\text{HONO, 0-5}°} CH_3CH_2CH_2\!-\!\langle\text{ring}\rangle\!-\!N_2^+ \quad Cl^-$

21.32

(a) $CH_3CH_2CH_2NH_2 + C_6H_5SO_2Cl \xrightarrow[H_2O]{KOH} CH_3CH_2CH_2\overset{-}{N}SO_2C_6H_5$

K^+

clear solution

$\xrightarrow{H_3O^+} CH_3CH_2CH_2NHSO_2C_6H_5$

precipitate

(b) $(CH_3CH_2CH_2)_2NH + C_6H_5SO_2Cl \xrightarrow[H_2O]{KOH} (CH_3CH_2CH_2)_2NSO_2C_6H_5$

precipitate

$\xrightarrow{H_3O^+}$ No reaction (precipitate remains)

(c)

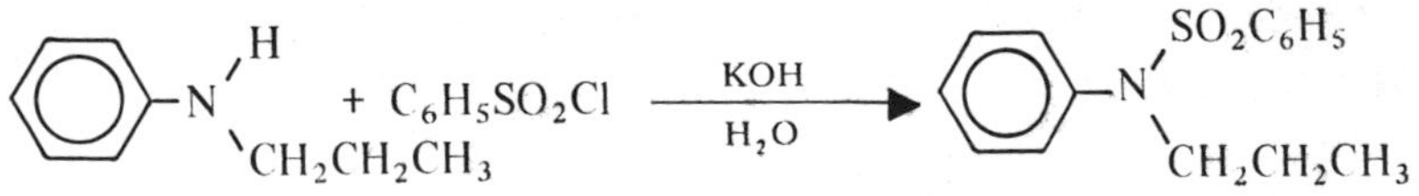

$+ C_6H_5SO_2Cl \xrightarrow[H_2O]{KOH}$

precipitate

$\xrightarrow{H_3O^+}$ No reaction (precipitate remains)

(d)

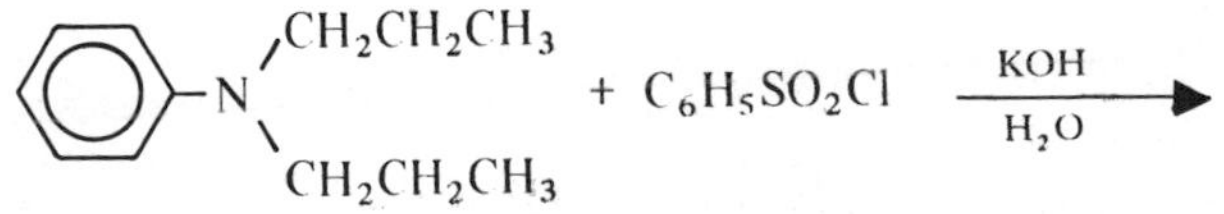

$+ C_6H_5SO_2Cl \xrightarrow[H_2O]{KOH}$ No reaction
(3° amine is insoluble)

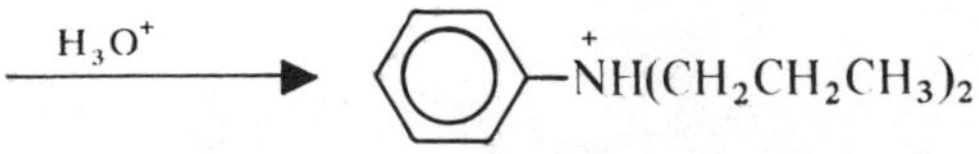

$\xrightarrow{H_3O^+}$

$-\overset{+}{N}H(CH_2CH_2CH_3)_2$

3° amine dissolves

(e) 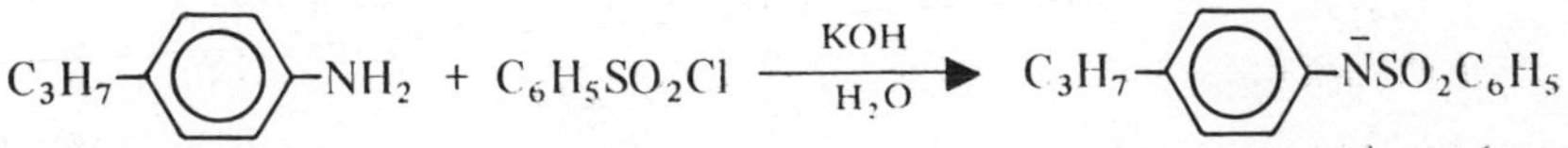 $C_3H_7\!-\!\langle\text{ring}\rangle\!-\!NH_2 + C_6H_5SO_2Cl \xrightarrow[H_2O]{KOH} C_3H_7\!-\!\langle\text{ring}\rangle\!-\!\overset{-}{N}SO_2C_6H_5$

K^+ (cont. on p. 244)

clear solution

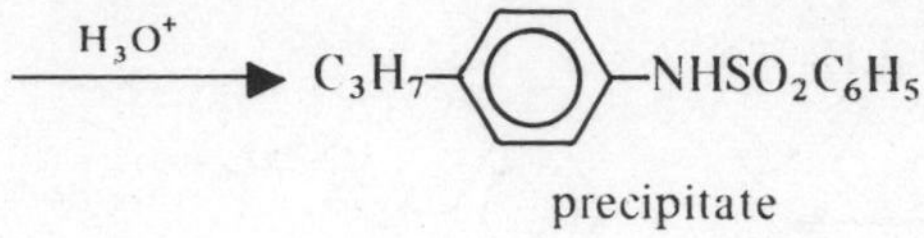

$$\xrightarrow{\text{H}_3\text{O}^+} \quad \text{C}_3\text{H}_7\text{—}\boxed{\bigcirc}\text{—NHSO}_2\text{C}_6\text{H}_5$$

precipitate

21.33

(a)

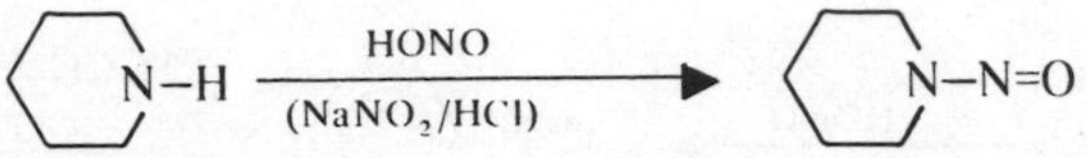

$$\boxed{\text{N—H}} \xrightarrow[\text{(NaNO}_2/\text{HCl)}]{\text{HONO}} \boxed{\text{N—N=O}}$$

(b)

$$\boxed{\text{N—H}} + \text{C}_6\text{H}_5\text{SO}_2\text{Cl} \xrightarrow[\text{H}_2\text{O}]{\text{KOH}} \boxed{\text{N—SO}_2\text{C}_6\text{H}_5}$$

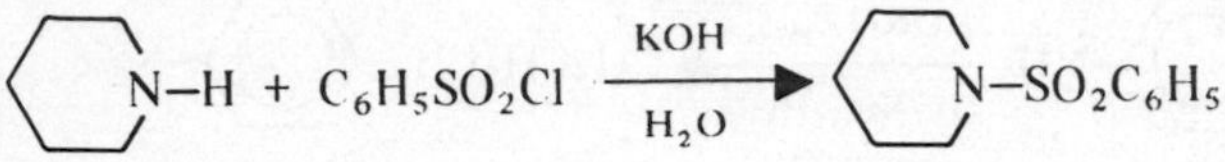

21.34

(a) $2\text{CH}_3\text{CH}_2\text{NH}_2 + \text{C}_6\text{H}_5\text{COCl} \longrightarrow \text{CH}_3\text{CH}_2\text{NHCOC}_6\text{H}_5 + \text{CH}_3\text{CH}_2\overset{+}{\text{NH}}_3\,\text{Cl}^-$

(b) $2\text{CH}_3\text{NH}_2 + (\text{CH}_3\overset{\text{O}}{\overset{\|}{\text{C}}})_2\text{O} \longrightarrow \text{CH}_3\text{NH}\overset{\text{O}}{\overset{\|}{\text{C}}}\text{CH}_3 + \text{CH}_3\overset{+}{\text{NH}}_3\ \text{CH}_3\overset{\text{O}}{\overset{\|}{\text{C}}}\text{O}^-$

(c)

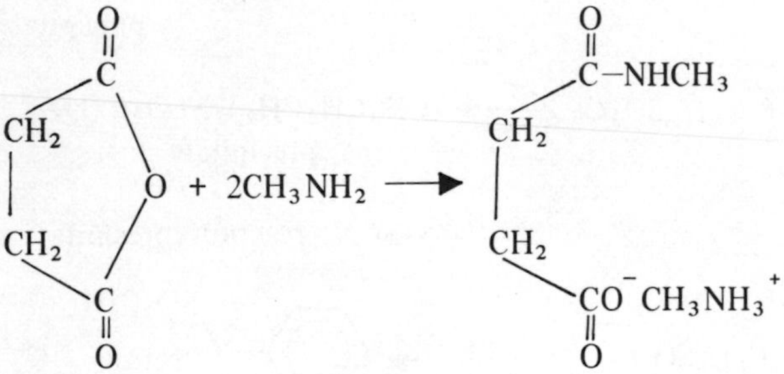

(d) (product of c) 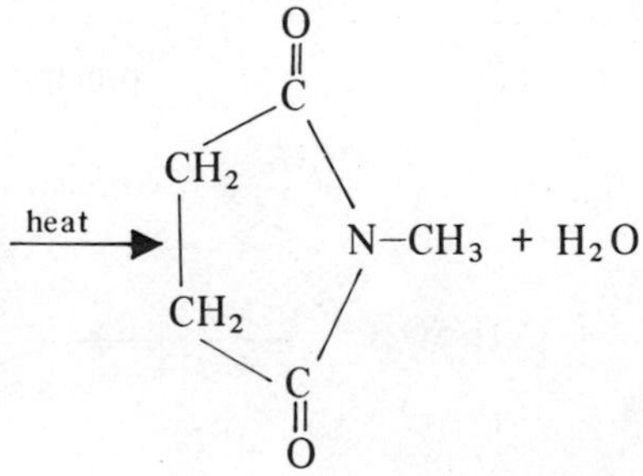 $\xrightarrow{\text{heat}}$

$+ \text{H}_2\text{O}$

(e)

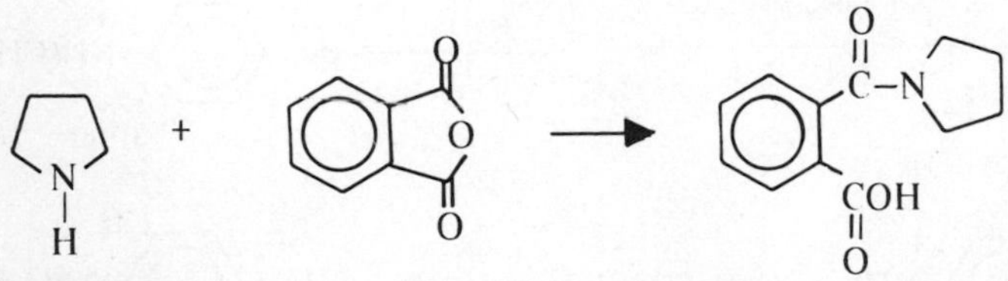

(f)

$+ \text{CH}_3\text{COOH}$

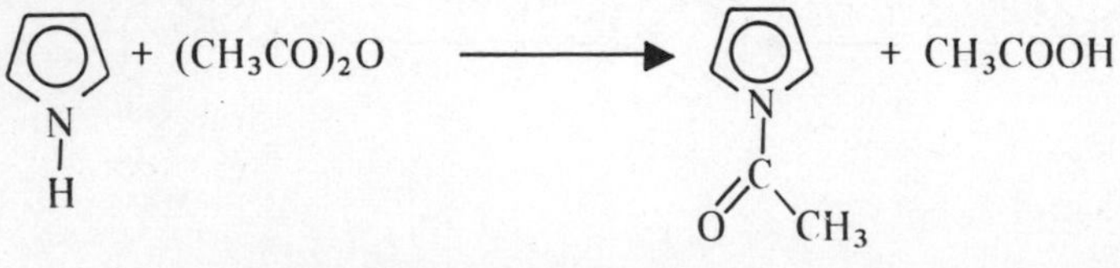

(g) 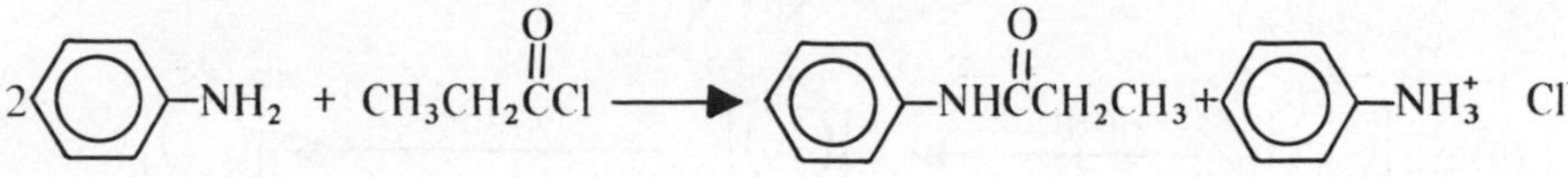 2 ⬡—NH_2 + $CH_3CH_2\overset{\overset{\displaystyle O}{\|}}{C}Cl$ ⟶ ⬡—$NH\overset{\overset{\displaystyle O}{\|}}{C}CH_2CH_3$ + ⬡—NH_3^+ Cl^-

(h) 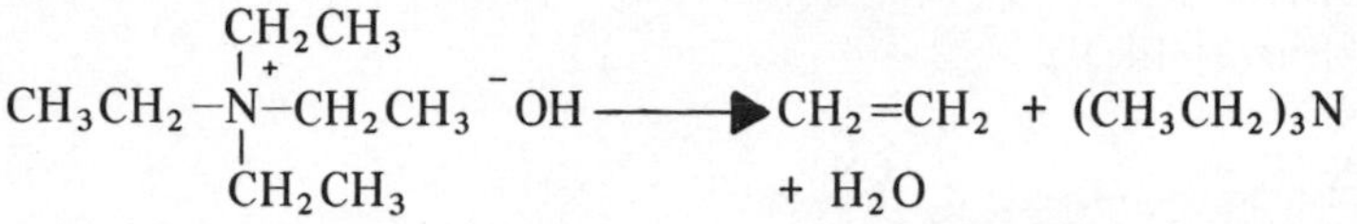 $CH_3CH_2 \overset{\overset{\displaystyle CH_2CH_3}{\underset{}{|}}}{\underset{\underset{\displaystyle CH_2CH_3}{|}}{\overset{+}{N}}} CH_2CH_3 \ ^-OH$ ⟶ $CH_2{=}CH_2$ + $(CH_3CH_2)_3N$
   $+ H_2O$

(i) 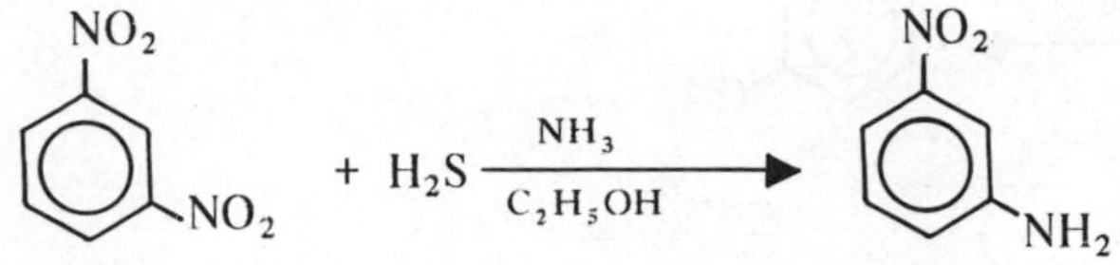 [3,5-dinitrobenzene] + H_2S $\xrightarrow[C_2H_5OH]{NH_3}$ [3-nitroaniline]

(j) 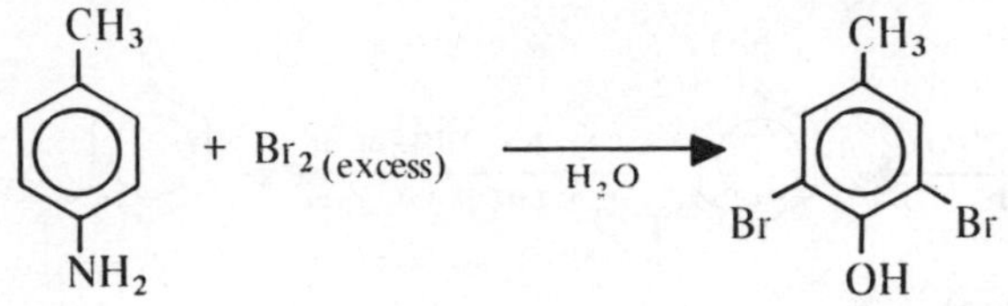 [4-aminotoluene] + $Br_{2\,(excess)}$ $\xrightarrow{H_2O}$ [2,6-dibromo-4-methylphenol]

21.35

(a) 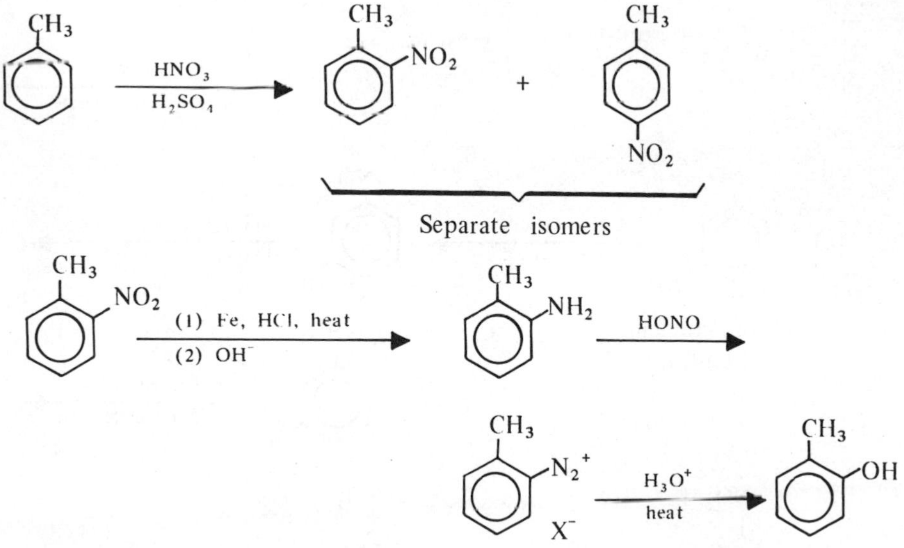 [toluene] $\xrightarrow[H_2SO_4]{HNO_3}$ [o-nitrotoluene] + [p-nitrotoluene]

$\underbrace{\hspace{8cm}}$
Separate isomers

[o-nitrotoluene] $\xrightarrow[(2)\ OH^-]{(1)\ Fe,\ HCl,\ heat}$ [o-aminotoluene] $\xrightarrow{HONO}$

[o-methylbenzenediazonium] $\xrightarrow[heat]{H_3O^+}$ [o-cresol]

(b) 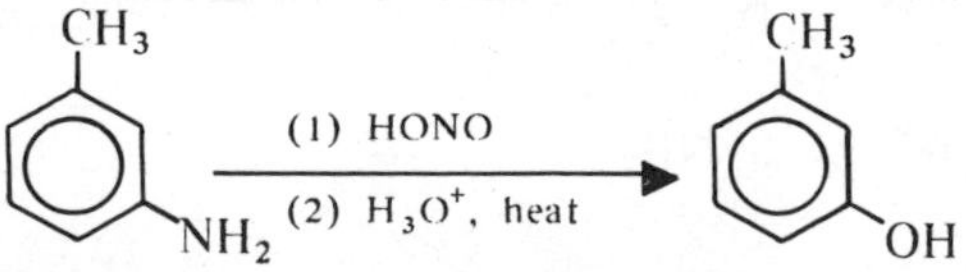 [m-aminotoluene] $\xrightarrow[(2)\ H_3O^+,\ heat]{(1)\ HONO}$ [m-cresol]

(from Problem 21.17a)

(c) 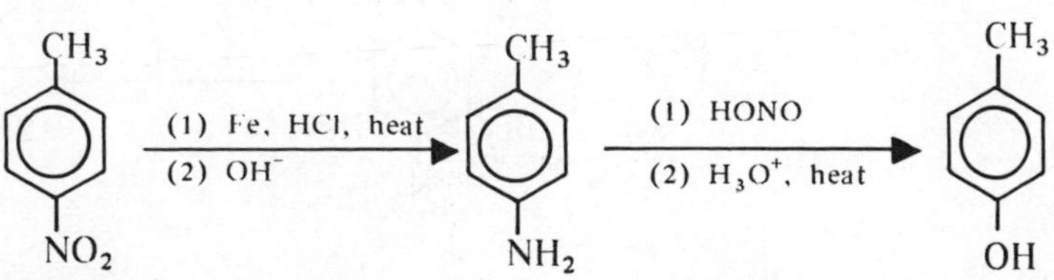 [p-nitrotoluene] $\xrightarrow[(2)\ OH^-]{(1)\ Fe,\ HCl,\ heat}$ [p-aminotoluene] $\xrightarrow[(2)\ H_3O^+,\ heat]{(1)\ HONO}$ [p-cresol]

(from part a)

(d)

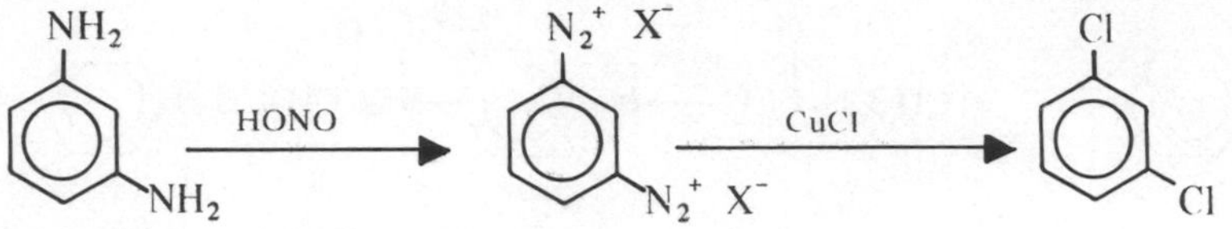

(from Problem 21.14a)

(e)

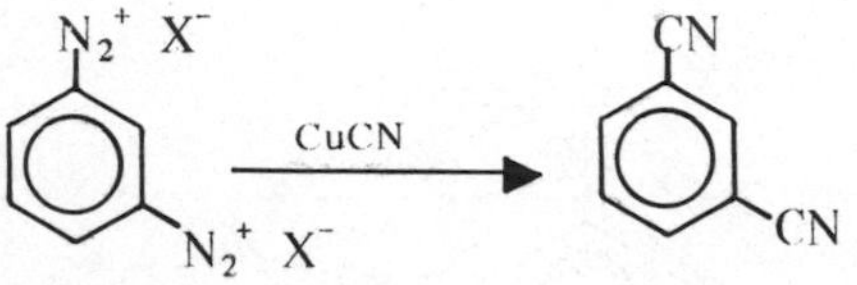

(cf. part d)

(f)

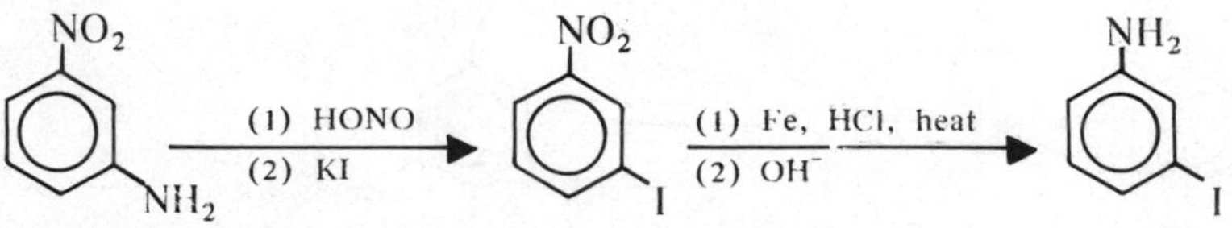

(from Problem 21.14 a)

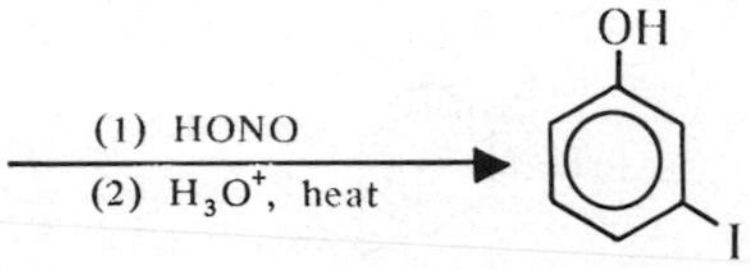

(g)

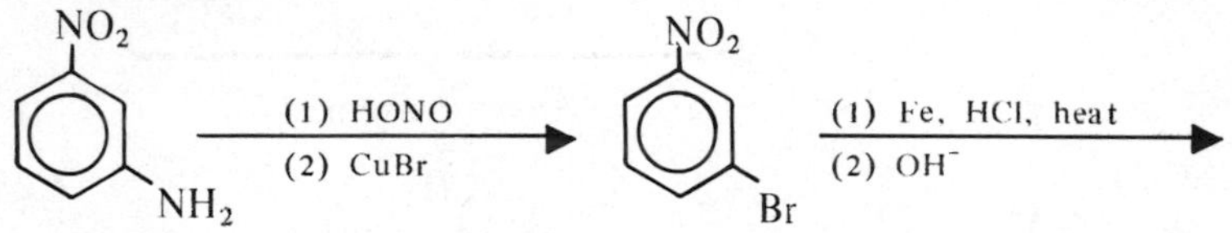

(from Problem
21.14a)

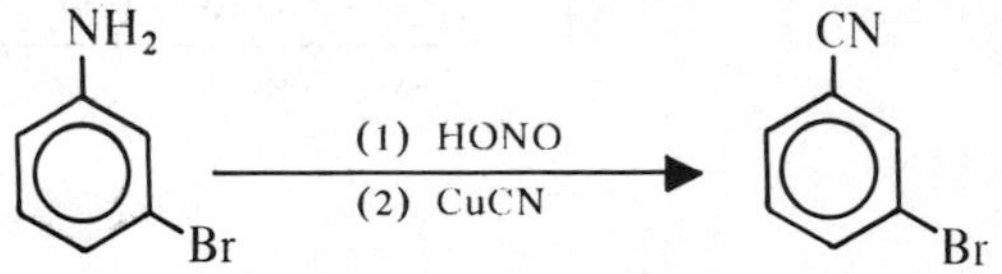

(h)

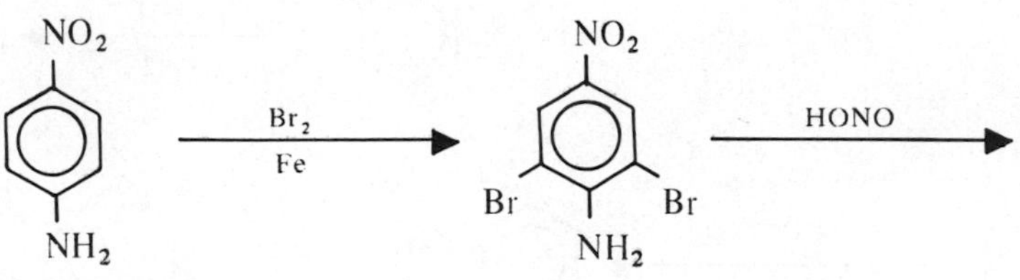

(from Problem 21.14f)

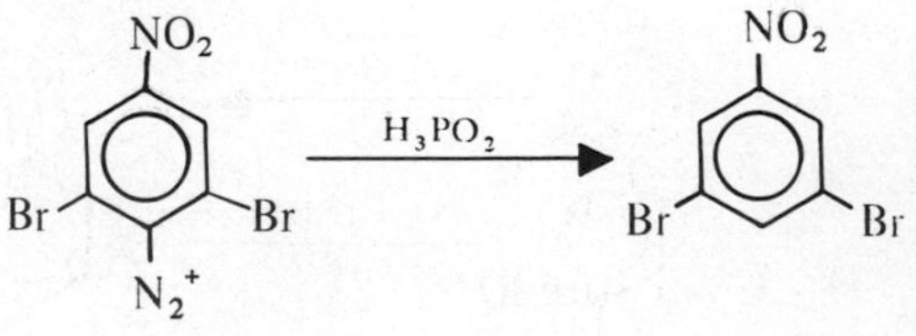

(i)

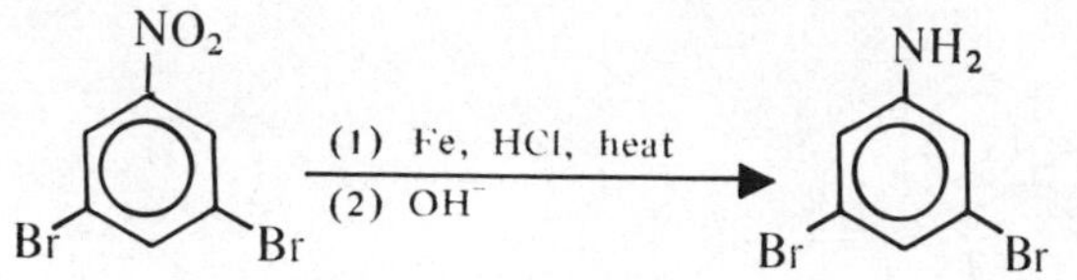

NO$_2$ / Br / Br (from part h)

$\xrightarrow[\text{(2) OH}^-]{\text{(1) Fe, HCl, heat}}$

NH$_2$ / Br / Br

(j)

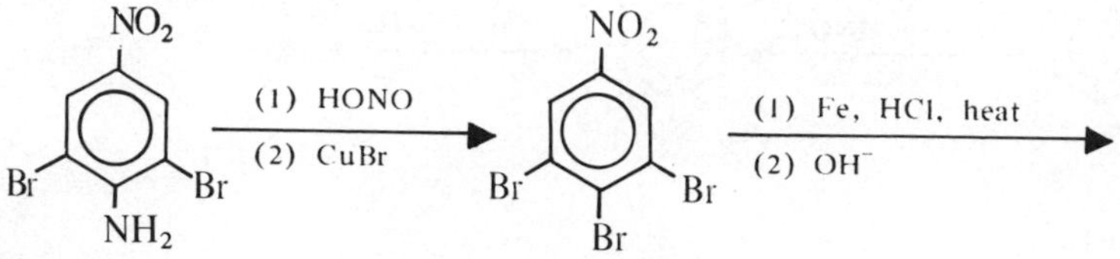

NO$_2$ / Br / Br / NH$_2$ (from part h)

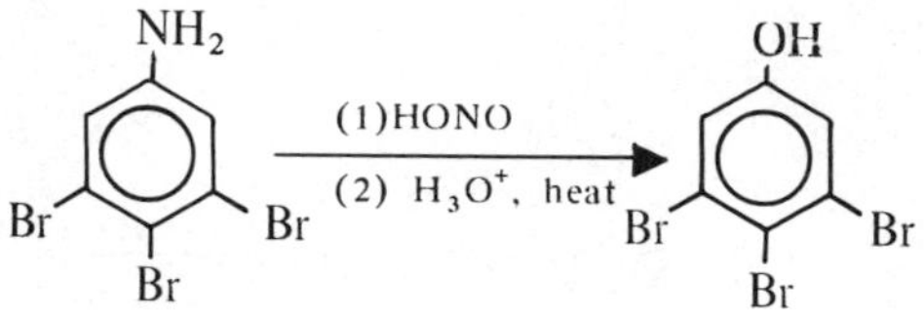

$\xrightarrow[\text{(2) CuBr}]{\text{(1) HONO}}$

NO$_2$ / Br / Br / Br

$\xrightarrow[\text{(2) OH}^-]{\text{(1) Fe, HCl, heat}}$

NH$_2$ / Br / Br / Br

$\xrightarrow[\text{(2) H}_3\text{O}^+,\text{ heat}]{\text{(1)HONO}}$

OH / Br / Br / Br

(k)

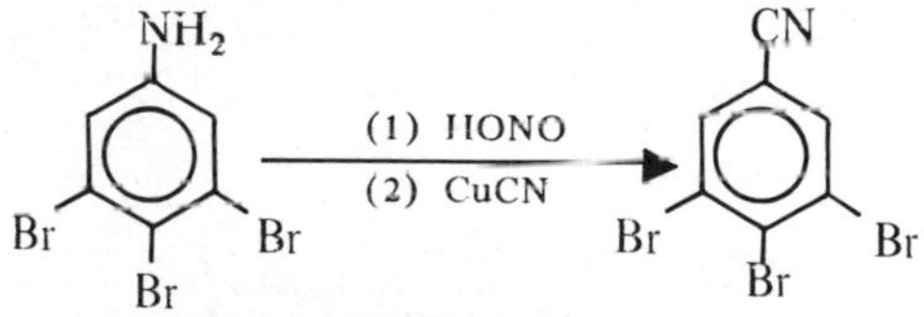

NH$_2$ / Br / Br / Br (from part j)

$\xrightarrow[\text{(2) CuCN}]{\text{(1) HONO}}$

CN / Br / Br / Br

(l)

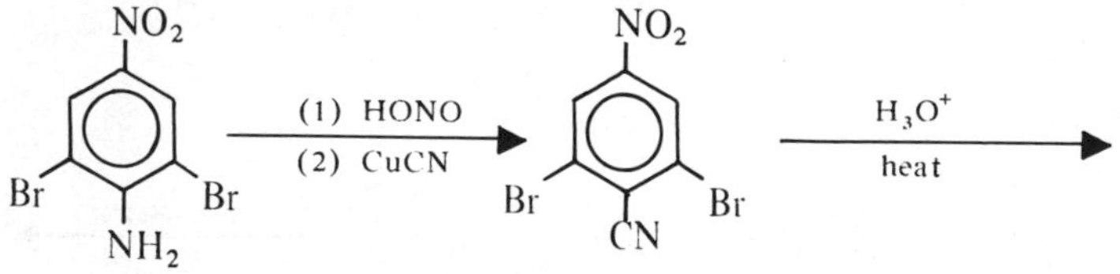

NO$_2$ / Br / Br / NH$_2$ (from part h)

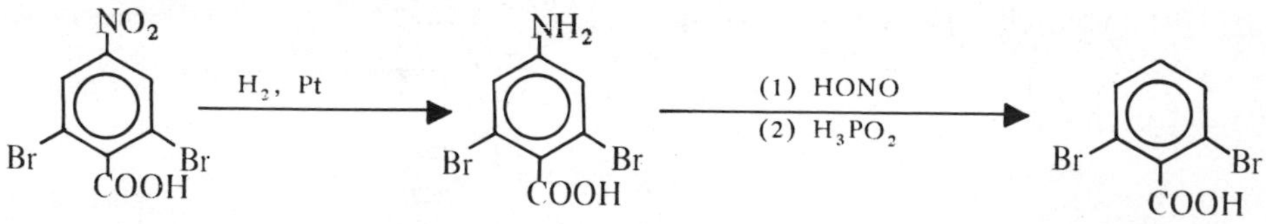

$\xrightarrow[\text{(2) CuCN}]{\text{(1) HONO}}$

NO$_2$ / Br / Br / CN

$\xrightarrow[\text{heat}]{\text{H}_3\text{O}^+}$

NO$_2$ / Br / Br / COOH

$\xrightarrow{\text{H}_2\text{, Pt}}$

NH$_2$ / Br / Br / COOH

$\xrightarrow[\text{(2) H}_3\text{PO}_2]{\text{(1) HONO}}$

Br / Br / COOH

(m)

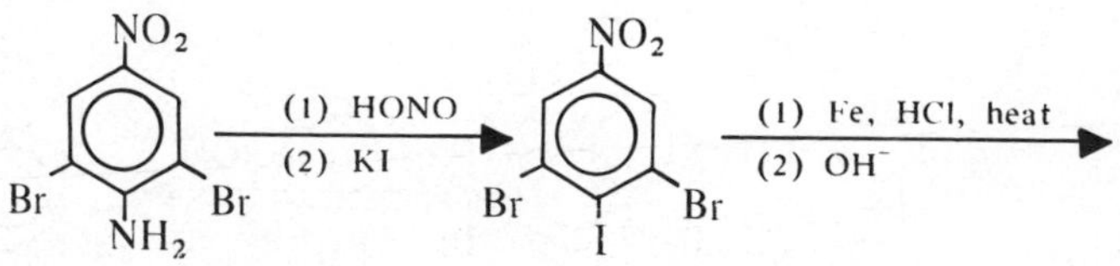

NO$_2$ / Br / Br / NH$_2$ (from part h)

$\xrightarrow[\text{(2) KI}]{\text{(1) HONO}}$

NO$_2$ / Br / Br / I

$\xrightarrow[\text{(2) OH}^-]{\text{(1) Fe, HCl, heat}}$

(cont. on p. 248)

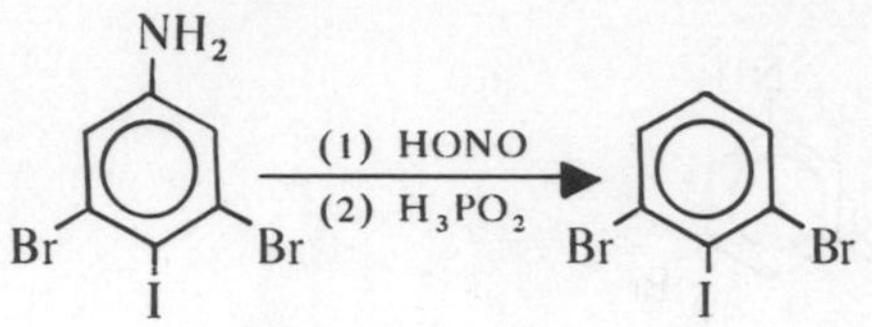

(n)

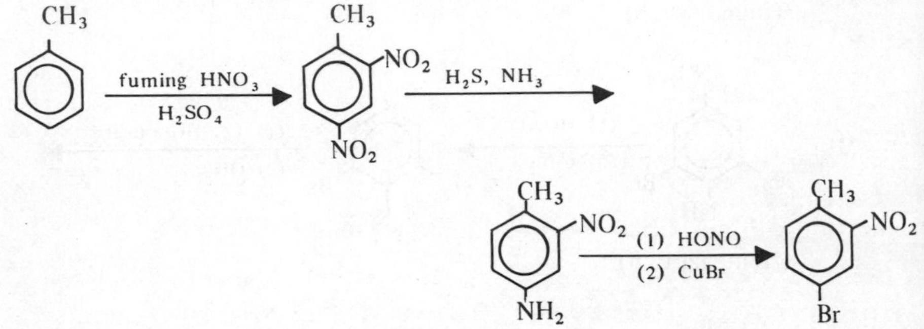

(o) 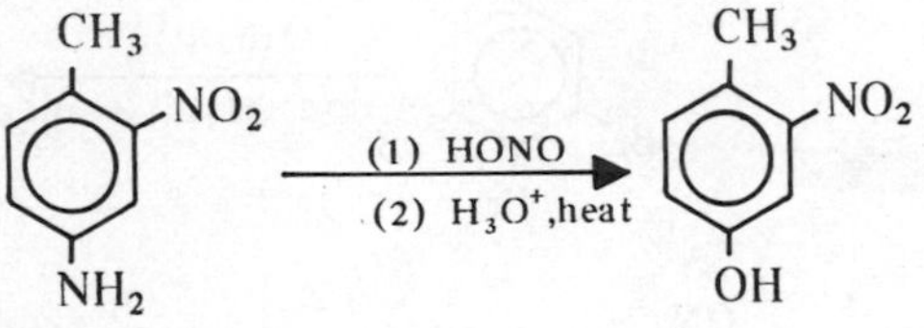 (from part n)

(p) 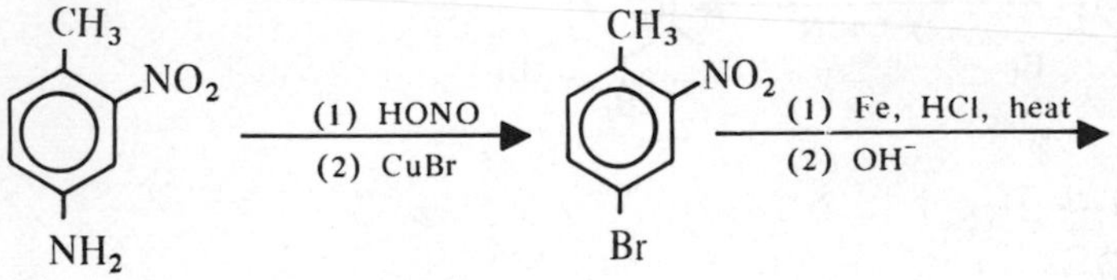 (from part n)

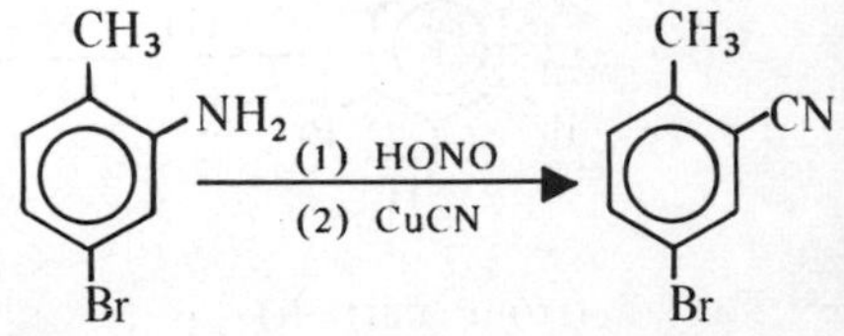

(q) 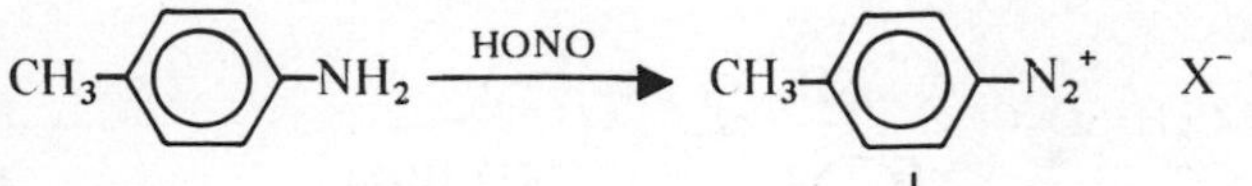 CH$_3$—⬡—NH$_2$ $\xrightarrow{\text{HONO}}$ CH$_3$—⬡—N$_2^+$ X$^-$

(from part c)

⬡—OH, pH 8-10

CH$_3$—⬡—N=N—⬡—OH

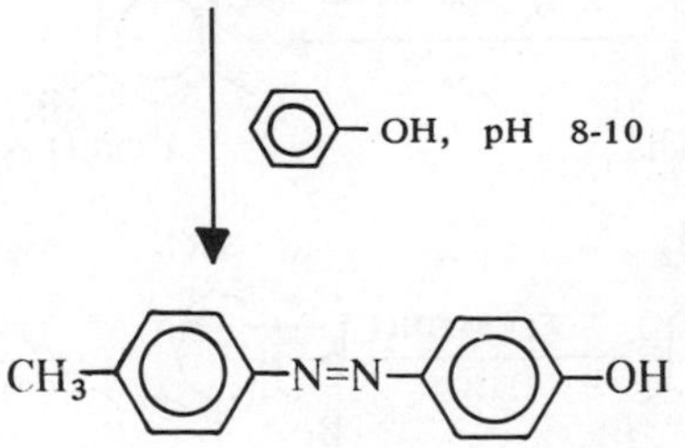

(r)

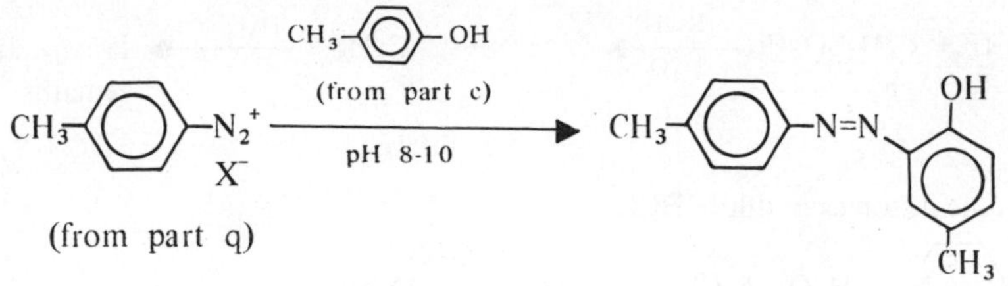

(from part q)

21.36

(a) Benzylamine dissolves in dilute HCl at room temperature,

$$C_6H_5CH_2NH_2 + H_3O^+ + Cl^- \xrightarrow{25°} C_6H_5CH_2\overset{+}{N}H_3\overset{-}{Cl}$$

benzamide does not dissolve:

$$C_6H_5CONH_2 + H_3O^+ + Cl^- \xrightarrow{25°} \text{No reaction}$$

(b) Allylamine reacts with (and decolorizes) bromine in carbon tetrachloride instantly,

$$CH_2=CHCH_2NH_2 + Br_2 \xrightarrow{CCl_4} \underset{\underset{Br}{|}}{CH_2}\underset{\underset{Br}{|}}{CHCH_2}NH_2$$

propylamine does not:

$$CH_3CH_2CH_2NH_2 + Br_2 \xrightarrow{CCl_4} \text{No reaction if the mixture is not heated or irradiated.}$$

(c) The Hinsberg test:

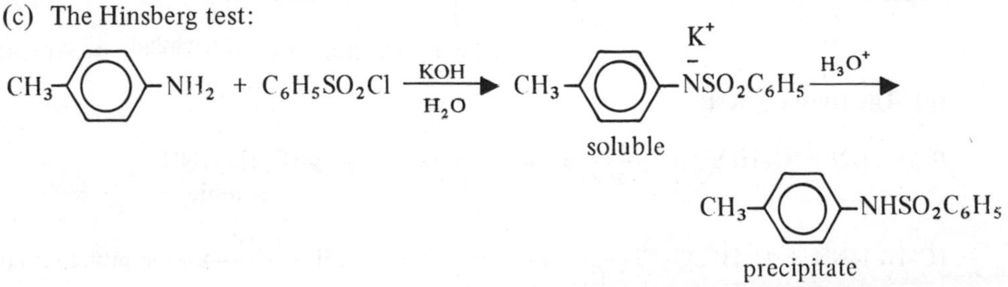

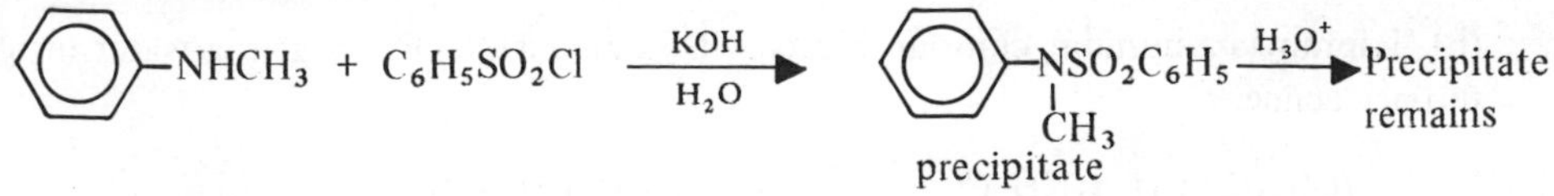

(d) The Hinsberg test:

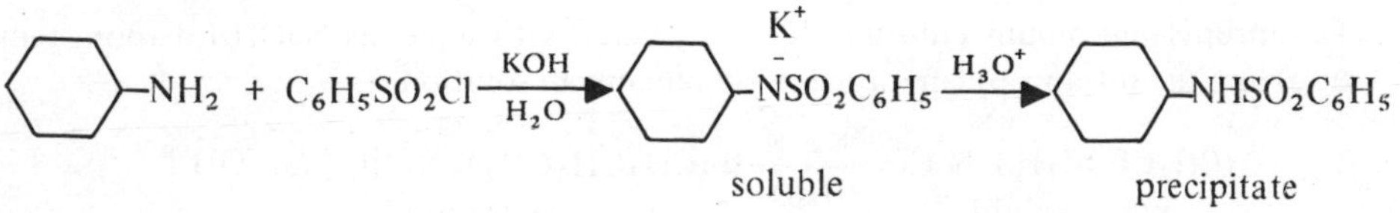

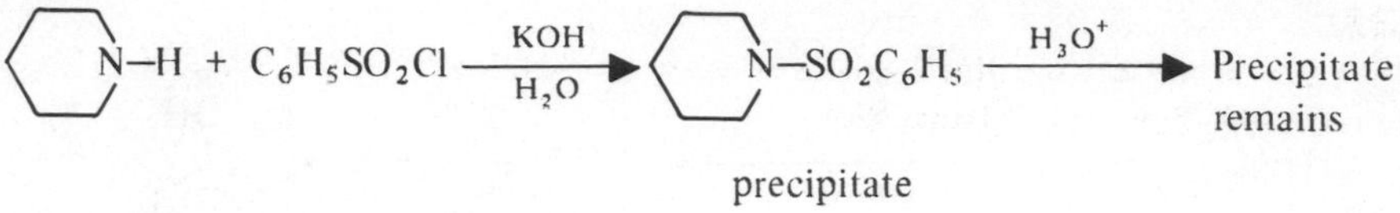

precipitate

(e) Pyridine dissolves in dilute HCl,

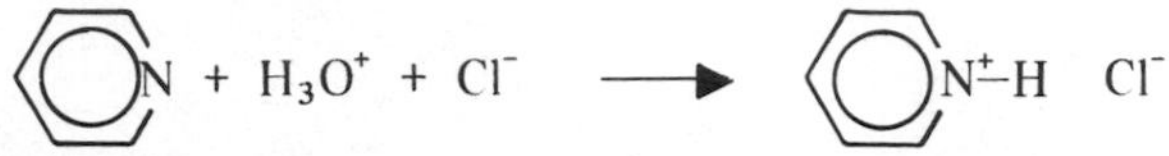

benzene does not:

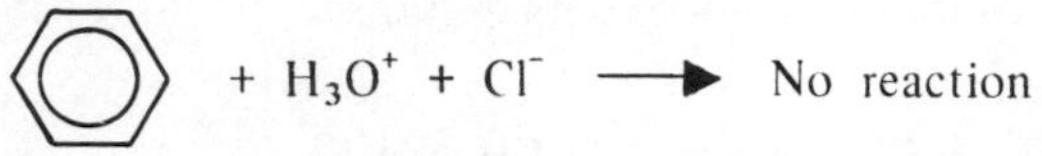

(f) Aniline reacts with nitrous acid at 0-5° to give a stable diazonium salt that couples with 2-naphthol yielding an intensely colored azo compound.

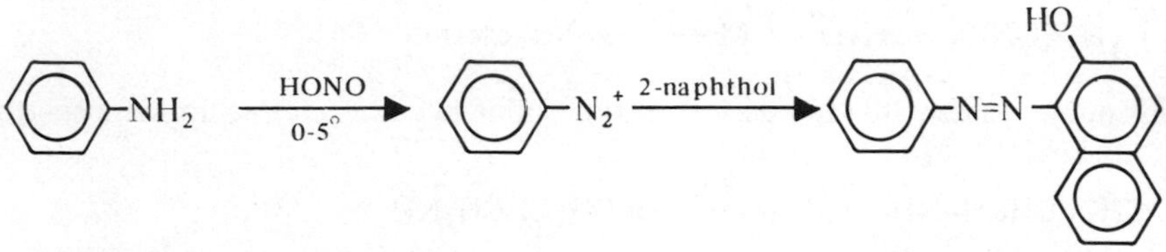

Cyclohexylamine reacts with nitrous acid at 0-5° to yield a highly unstable diazonium salt—one that decomposes so rapidly that the addition of 2-naphthol gives no azo compound.

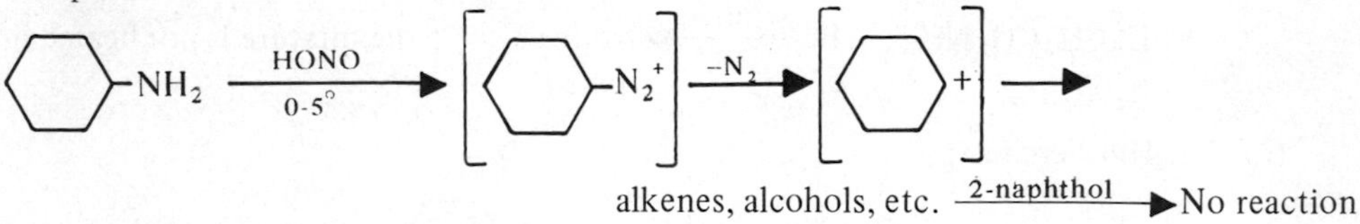

alkenes, alcohols, etc. $\xrightarrow{\text{2-naphthol}}$ No reaction

(g) The Hinsberg test:

$(C_2H_5)_3N + C_6H_5SO_2Cl \xrightarrow[H_2O]{KOH}$ No reaction $\xrightarrow{H_3O^+} (C_2H_5)_3\overset{+}{N}H$
soluble

$(C_2H_5)_2NH + C_6H_5SO_2Cl \xrightarrow[H_2O]{KOH} (C_2H_5)_2NSO_2C_6H_5 \xrightarrow{H_3O^+}$ Precipitate remains
precipitate

(h) Tripropylammonium chloride reacts with aqueous NaOH to give a water insoluble tertiary amine.

$(CH_3CH_2CH_2)_3\overset{+}{N}H\ Cl^- \xrightarrow[H_2O]{NaOH} (CH_3CH_2CH_2)_3N$
water soluble $\qquad\qquad$ water insoluble

Tetrapropylammonium chloride does not react with aqueous NaOH (at room temperature) and the tetrapropylammonium ion remains in solution.

$(CH_3CH_2CH_2)_4\overset{+}{N}\overset{-}{Cl} \xrightarrow[H_2O]{NaOH} (CH_3CH_2CH_2)_4\overset{+}{N}\ [Cl^-\ or\ OH^-]$
water soluble $\qquad\qquad$ water soluble

(i) Tetrapropylammonium chloride dissolves in water to give a neutral solution. Tetrapropylammonium hydroxide dissolves in water to give a strongly basic solution.

21.37

Follow the procedure outlined in the answer to Problem 21.4. Toluene will show the same solubility behavior as benzene.

21.38

(a)

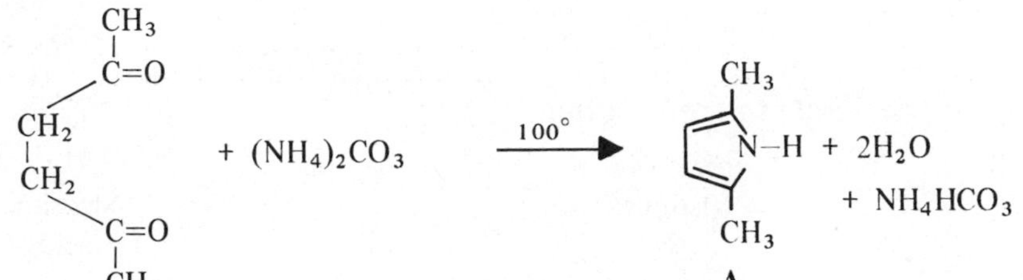

$$\underset{\substack{\text{CH}_2 \\ \text{CH}_2 \\ \text{C}=\text{O} \\ \text{CH}_3}}{\overset{\substack{\text{CH}_3 \\ \text{C}=\text{O}}}{}} + (\text{NH}_4)_2\text{CO}_3 \xrightarrow{100°} \text{A} \quad + 2\text{H}_2\text{O} + \text{NH}_4\text{HCO}_3$$

(b)

$$\xrightarrow{\text{base}} \text{B} \quad + 2\text{H}_2\text{O}$$

(c)

$$\underset{\substack{\text{CH(OCH}_3)_2 \\ \text{CH}_2 \\ \text{CH(OCH}_3)_2}}{} + \underset{\substack{\text{NH}_2 \\ \text{NH} \\ \text{CH}_3}}{} \xrightarrow[\text{H}_2\text{O}]{\text{H}^+} \text{C} \quad + 4\text{CH}_3\text{OH}$$

(d)

$$\xrightarrow{} \text{D} \xrightarrow{\text{O}_2} \text{E}$$

(e)

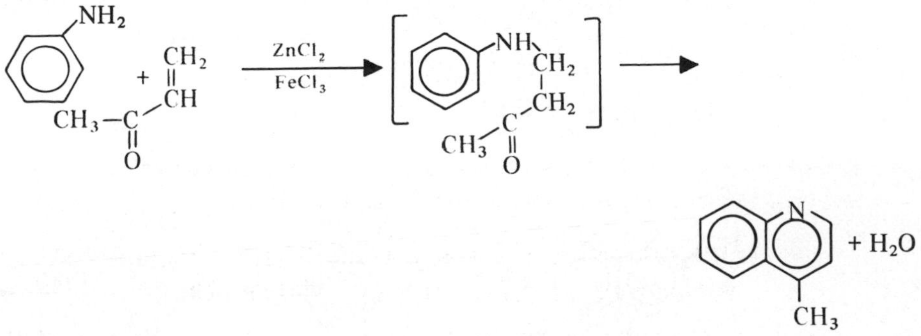

$$\text{(aniline)} + \underset{\substack{\text{CH}_2 \\ \text{CH} \\ \text{CH}_3-\text{C} \\ \text{O}}}{} \xrightarrow[\text{FeCl}_3]{\text{ZnCl}_2} \left[\text{intermediate} \right] \longrightarrow$$

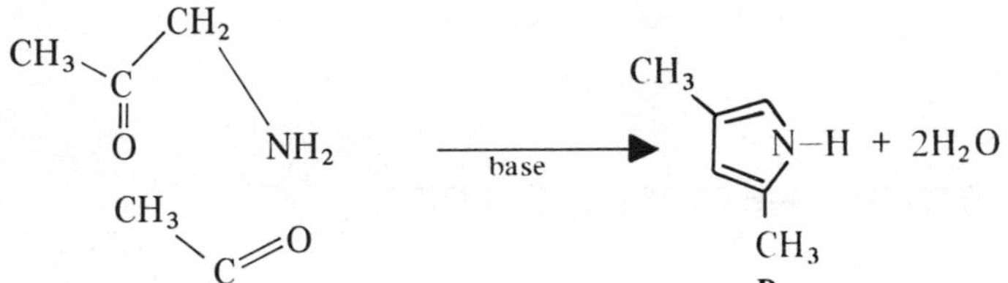

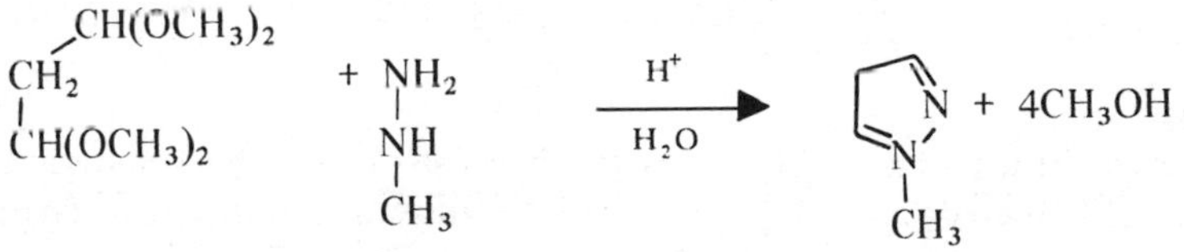

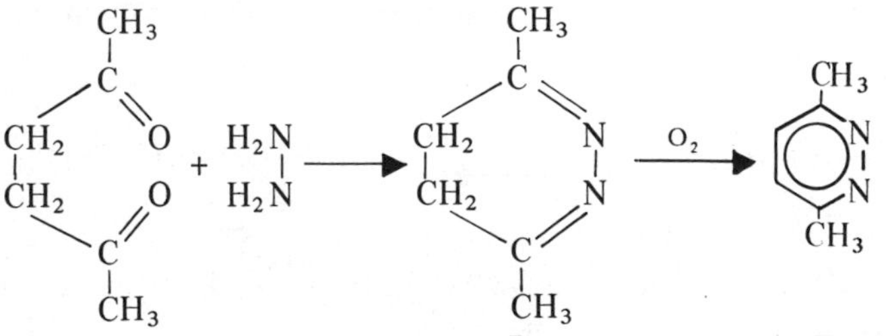

$$\text{F} \quad + \text{H}_2\text{O}$$

(f)

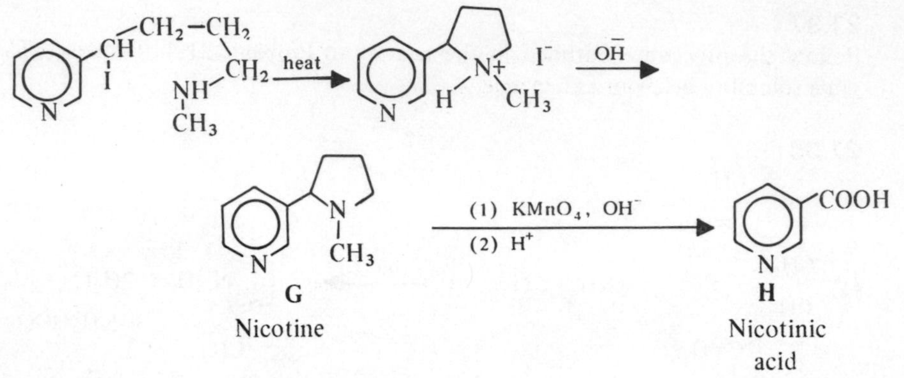

$$\text{(1) KMnO}_4,\ \text{OH}^- \quad \text{(2) H}^+$$

G
Nicotine

H
Nicotinic acid

21.39

$$+ \ NH_3 \longrightarrow H_2N\overset{O}{\overset{\|}{C}}CH_2CH_2\overset{O}{\overset{\|}{C}}OH$$

$$\xrightarrow[(-CO_3^=)]{Br_2,\ OH^-} H_2NCH_2CH_2COO^- \xrightarrow{H^+} H_2NCH_2CH_2COOH$$

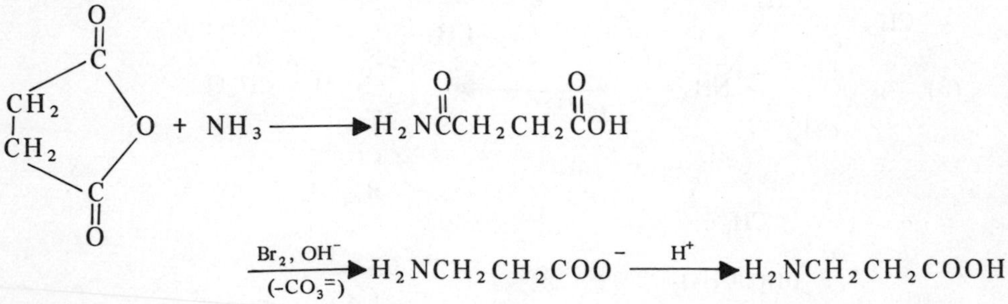

21.40
Compound (b) would probably be inactive because of the meta orientation of the groups. Compound (d) would probably be inactive because the distance that separates the $-NH_2$ from the $-SO_2NH-$ group is too large.

21.41

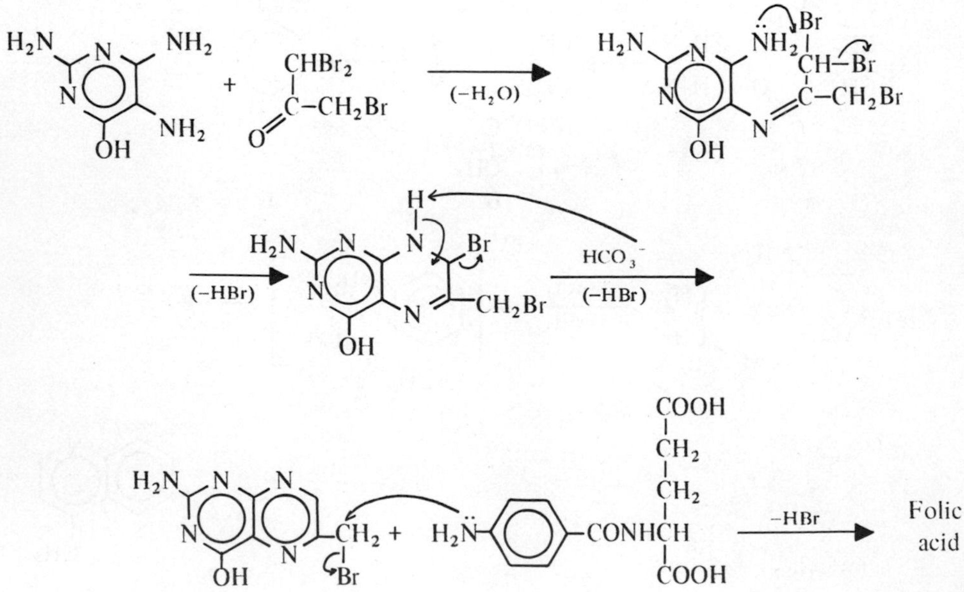

21.42

The results of the Hinsberg test indicate that compound W is a tertiary amine. The pmr spectrum provides evidence for the following:

(1) Two different C_6H_5- groups (one absorbing at $\delta 7.2$ and one at $\delta 6.7$.)
(2) A CH_3CH_2- group (the quartet at $\delta 3.3$ and the triplet at $\delta 1.2$).
(3) An unsplit $-CH_2-$ group (the singlet at $\delta 4.4$).

There is only one reasonable way to put all of this together,

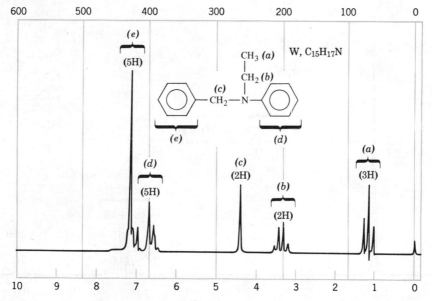

Thus W is N-benzyl-N-ethylaniline.

FIG. 21.3. The pmr spectrum of W (problem 21.42). (Spectrum courtesy of Aldrich Chemical Co.)

21.43

Compound X is benzyl bromide, $C_6H_5CH_2Br$. This is the only structure consistent with the pmr and infrared data. (The mono-substituted benzene ring is strongly indicated by the (5H), $\delta 7.3$ pmr absorption and is confirmed by the peaks at 690 cm^{-1} and 770 cm^{-1} in the infrared spectrum.)

Compound Y, therefore must be phenylacetonitrile, $C_6H_5CH_2CN$, and Z must be 2-phenylethylamine, $C_6H_5CH_2CH_2NH_2$.

$$\underset{\substack{X \\ (C_7H_7Br)}}{\text{⬡}-CH_2Br} \xrightarrow{\text{NaCN}} \underset{\substack{Y \\ (C_8H_7N)}}{\text{⬡}-CH_2CN} \xrightarrow{\text{LiAlH}_4} \underset{\substack{Z \\ (C_8H_{11}N)}}{\text{⬡}-CH_2CH_2NH_2}$$

Interpretations of the infrared and pmr spectra of Z are given in Fig. 21.4.

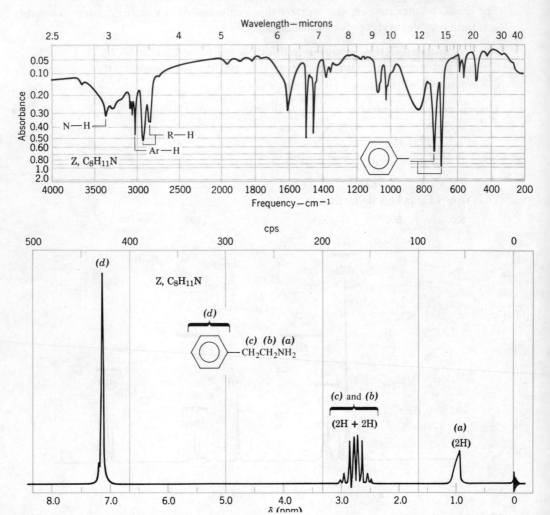

FIG. 21.4. Infrared and pmr spectra for compound Z, problem 21.43. (Spectra courtesy of Sadtler Inc.)

22 SPECIAL TOPICS III

22.1

(a) 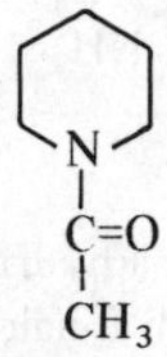 (b) 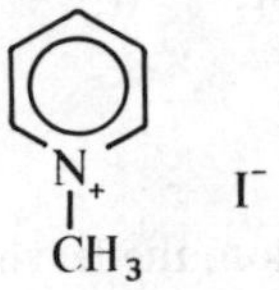

(c) 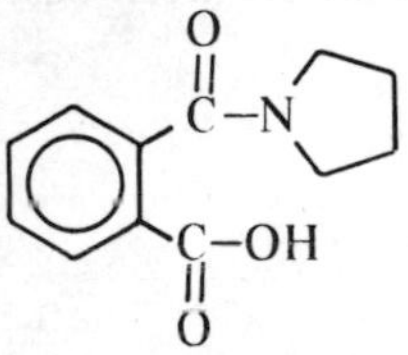 (d) 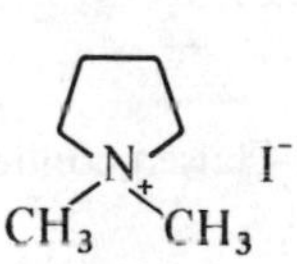

(e) $CH_2{=}CHCH_2CH_2\underset{\displaystyle CH_3}{\underset{|}{N}}{-}CH_3$

22.2

(a) The cyclopentadienide ion (cf. p. 424).

(b) The pyrrole anion is a resonance hybrid of the following structures:

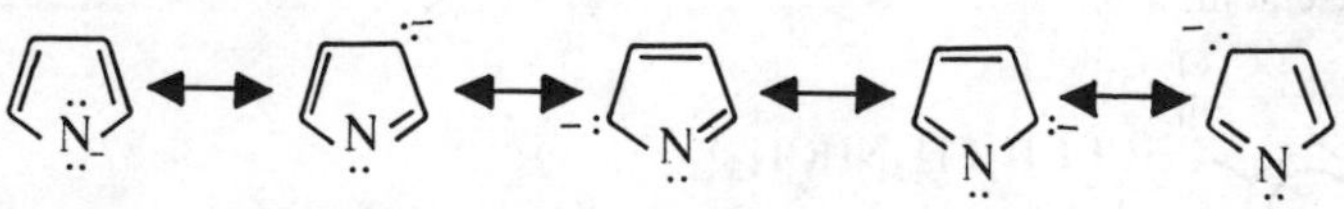

The imidazole anion is a hybrid of these:

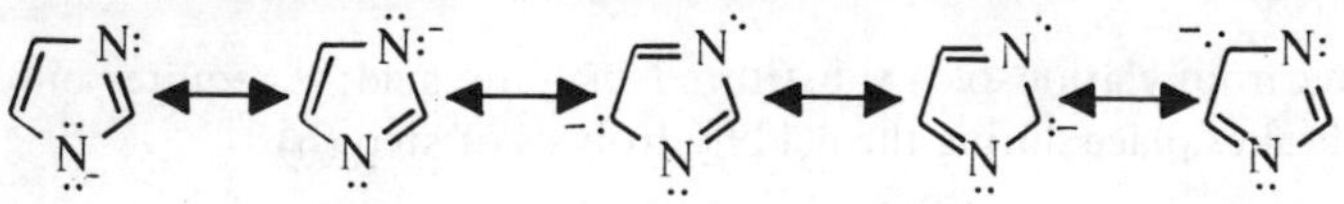

22.3

A mechanism involving a "pyridyne" intermediate would involve a net loss (of 50%) of the deuterium label.

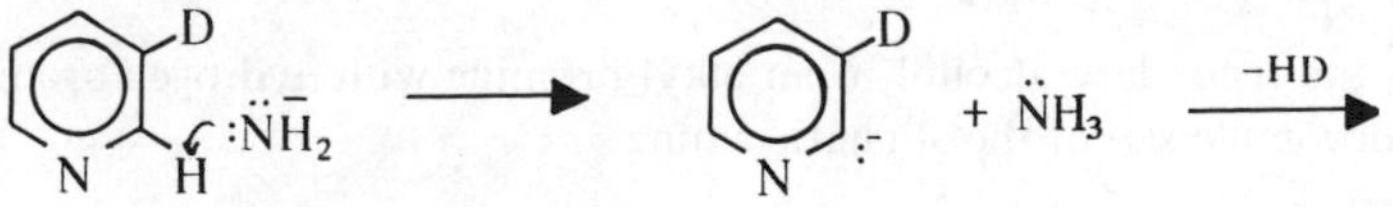

2-Pyridyne

Since in the actual experiment there was no loss of deuterium this mechanism was disallowed.

The mechanism given on page 861 would not be expected to result in a loss of deuterium, thus it is consistent with the labeling experiment.

22.4

When pyridine undergoes nucleophilic substitution, the leaving group is a hydride ion—an ion that is a strong base and, consequently, a poor leaving group. With 2-halopyridines, on the other hand, the leaving groups are halide ions—ions that are weak bases and thus good leaving groups.

22.5

(a) The first step is similar to a crossed-Claisen condensation:

(b) This step involves hydrolysis of an amide (lactam) and can be carried out with either acid or base. Here we use acid.

$$\xrightarrow[H_2O]{H_3O^+}$$

with product containing $\overset{O}{\overset{\|}{C}}-CHCH_2CH_2NHCH_3$ and $\overset{|}{C}OOH$

(c) This step is the decarboxylation of a substituted malonic acid; it requires only the application of heat and takes place during the acid hydrolysis of step (b).

(d) This is the reduction of a ketone to a secondary alcohol. A variety of reducing agents can be used, sodium borohydride, for example.

$$\xrightarrow{NaBH_4}$$

with product containing $\overset{OH}{\overset{|}{C}}HCH_2CH_2CH_2NHCH_3$

(e) Here we convert the secondary alcohol to an alkyl bromide with hydrogen bromide; this also gives a hydrobromide salt of the aliphatic amine.

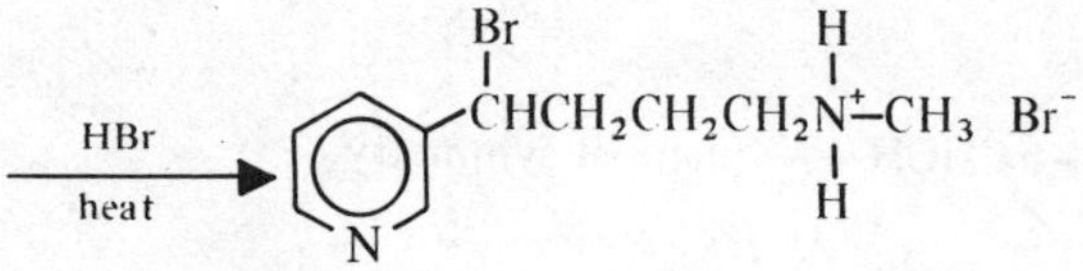

$$\xrightarrow[\text{heat}]{\text{HBr}}$$

(f) Treating the salt with base produces the secondary amine; it then acts as a nucleophile and attacks the carbon bearing the bromine. This leads to the formation of a five-membered ring and (±) nicotine.

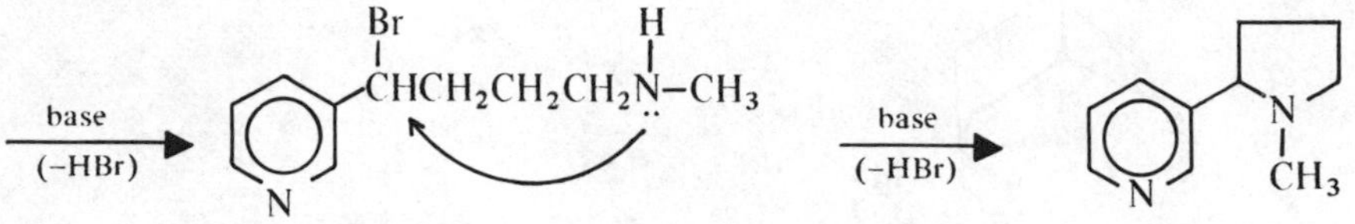

(±) nicotine

22.6

(a) The chiral carbon adjacent to the ester carbonyl group is racemized by base (probably through the formation of an anion that can undergo inversion of configuration, cf. Problem 18.24).

(b)

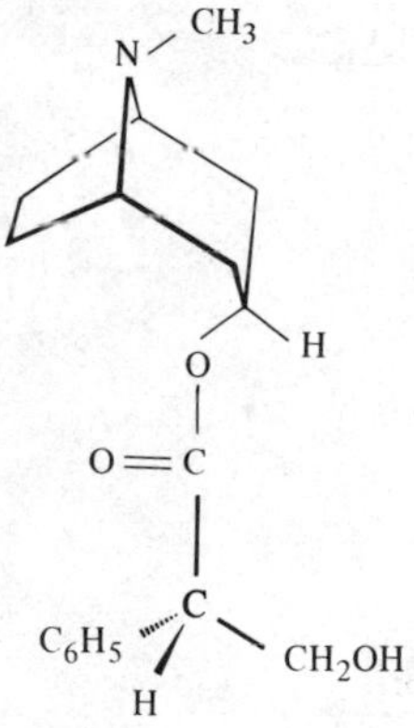

22.7

(a)

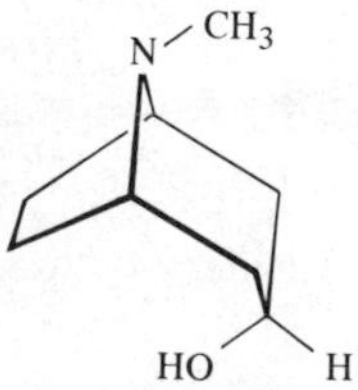

tropine

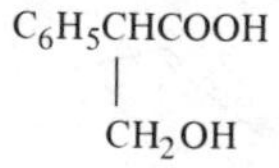

$$C_6H_5CHCOOH$$
$$\mid$$
$$CH_2OH$$

(±) tropic acid

(b) Tropine is a meso compound; it has a plane of symmetry that passes through the $>$CHOH group, the $>$NCH$_3$ group, and between the two $-$CH$_2-$ groups of the five-membered ring.

CH———CH₂

CH₂
⎸– – – NCH₃ – – – CHOH – – plane of symmetry
CH₂

CH———CH₂

(c)

ψ – tropine

22.8

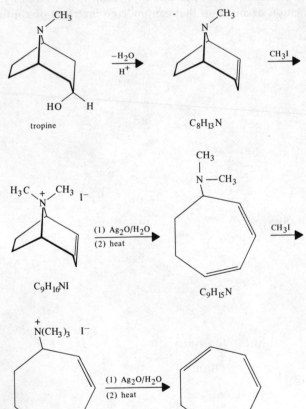

tropine

$C_8H_{13}N$

$C_9H_{16}NI$

$C_9H_{15}N$

$C_{10}H_{18}NI$

22.9

One possible sequence of steps is the following:

$$\begin{array}{c} \text{CHO} \\ \text{CH}_2 \\ \text{CH}_2 \\ \text{CHO} \end{array} + CH_3NH_2 \quad \underset{+H_2O, -H^+}{\overset{-H_2O, +H^+}{\rightleftharpoons}} \quad \begin{array}{c} \text{CHO} \\ \text{CH}_2 \\ \text{CH}_2 \\ \text{CH}=\overset{+}{N}HCH_3 \end{array}$$

$$\begin{array}{c} \text{CO}_2\text{H} \\ \text{CH}_2 \\ \text{C}=O \\ \text{CH}_2 \\ \text{CO}_2\text{H} \end{array} \quad \overset{\text{enolization}}{\rightleftharpoons} \quad \begin{array}{c} \text{CO}_2\text{H} \\ \text{CH} \\ \text{C}-O \\ \text{CH}_2 \\ \text{CO}_2\text{H} \end{array} \text{H}$$

$$\xrightarrow[\text{Mannich reaction}]{-H^+}$$

$$\begin{array}{c} \text{CHO} \\ \text{CH}_2 \\ \text{CH}_2 \\ \text{CH} \quad \text{NHCH}_3 \\ \text{CH}-\overset{O}{C}-CH_2CO_2H \\ \text{CO}_2\text{H} \end{array} \quad \underset{\substack{+H_2O \\ -H^+}}{\overset{\substack{+H^+ \\ -H_2O}}{\rightleftharpoons}} \quad \begin{array}{c} \overset{+}{N}-CH_3 \\ \text{CHCOCH}_2CO_2H \\ \text{CO}_2\text{H} \end{array} \quad \overset{\text{enolization}}{\rightleftharpoons}$$

$$\begin{array}{c} \text{CO}_2\text{H} \\ \text{CH} \\ \overset{+}{N}-CH_3 \quad \text{C}-\overset{..}{O}H \\ \text{CH} \\ \text{CO}_2\text{H} \end{array} \quad \xrightarrow[\text{Mannich reaction}]{-H^+} \quad \begin{array}{c} \text{CO}_2\text{H} \\ N-CH_3 \quad =O \\ \text{CO}_2\text{H} \end{array}$$

$$\xrightarrow{-2CO_2} \quad \begin{array}{c} N-CH_3 =O \end{array} \quad \equiv \quad \begin{array}{c} N-CH_3 \\ \\ O \end{array}$$

tropinone

22.10

$$\begin{array}{c} CH_3O \\ CH_3O \end{array} \text{—} \underset{NH_2}{\text{CH}_2CH_2} \quad + \quad \begin{array}{c} CH_3O \\ CH_3O \end{array} \text{—} \underset{COCl}{} \quad \xrightarrow{OH^-} \quad \begin{array}{c} CH_3O \\ CH_3O \end{array} \text{—} \overset{O}{\underset{}{}} N \text{—} H \quad \xrightarrow[\substack{\text{heat} \\ (-H_2O)}]{P_2O_5}$$

$$\begin{array}{c} CH_3O \\ CH_3O \end{array}$$

$$C_{20}H_{25}NO_5$$

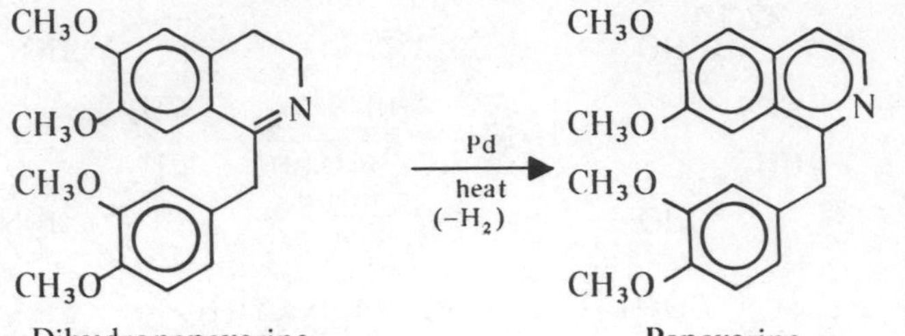

Dihydropapaverine → (Pd, heat, $-H_2$) → Papaverine

22.11

A Diels-Alder reaction was carried out using 1,3-butadiene as the diene component.

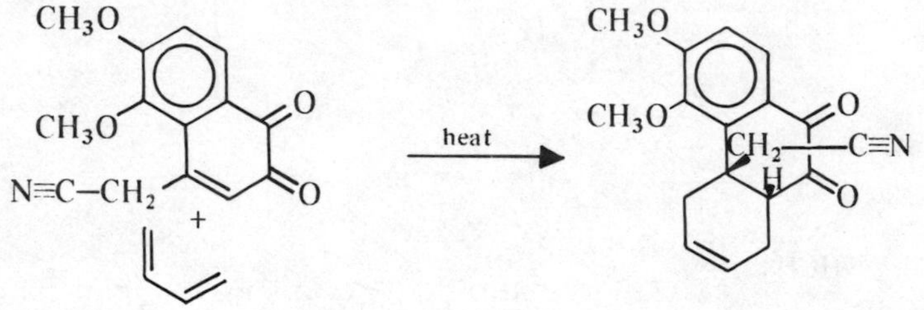

22.12

Acetic anhydride acetylates both —OH groups.

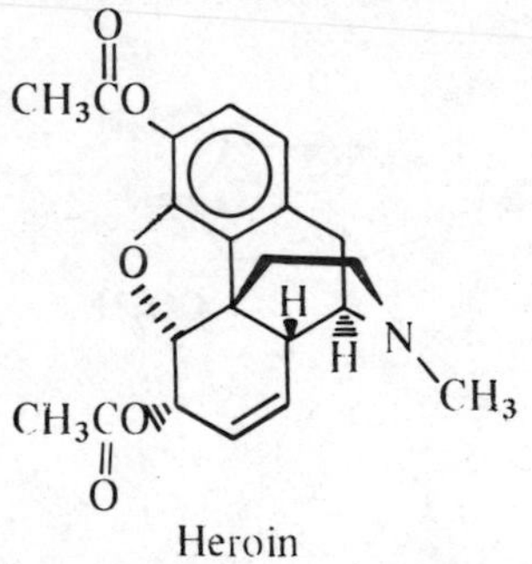

Heroin

22.13

(a) A Mannich reaction.

(b) CH_2O + $HN(CH_3)_2$ $\underset{-H_2O}{\overset{+H^+}{\rightleftharpoons}}$ $CH_2\overset{+}{=}N(CH_3)_2$

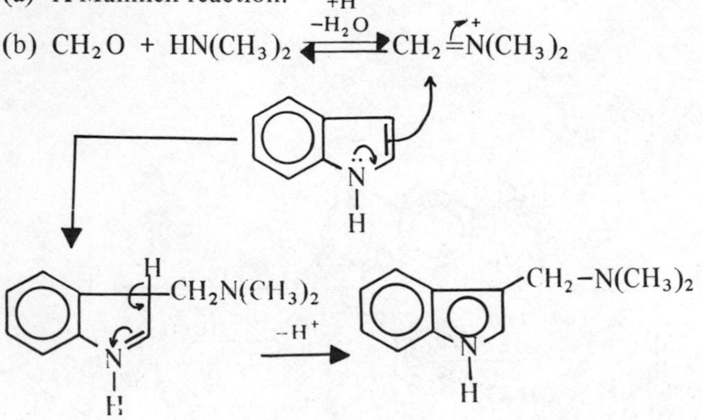

Gramine

22.14

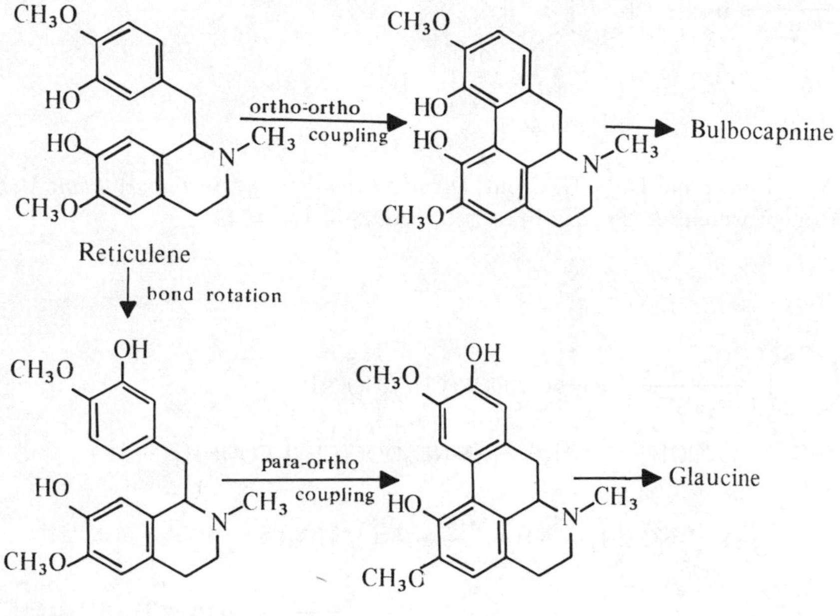

CH₃O ... HO ... HO ... CH₃O ... Reticulene

ortho-ortho coupling → Bulbocapnine

bond rotation ↓

para-ortho coupling → Glaucine

22.15

Yes, because according to the pathway given on pages 871-872, carbons 1 and 3 of papaverine arise from the α-carbons of two molecules of tyrosine.

(b) Yes.

(c) Methylation of the four phenolic hydroxyl groups and dehydrogenation of the nitrogen-containing ring.

22.16

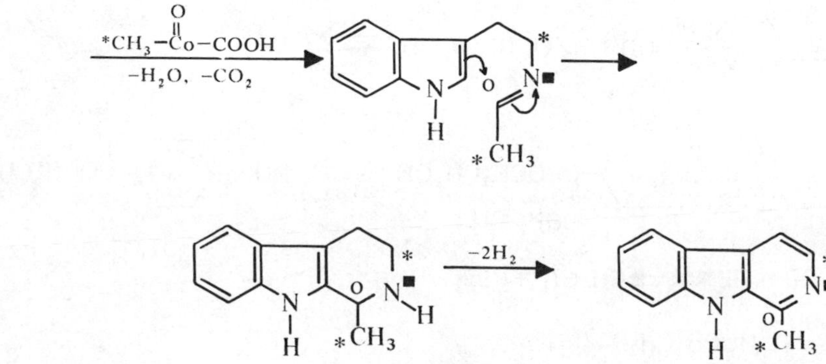

Tryptophan $\xrightarrow{-CO_2}$ Tryptamine

$$*CH_3-\overset{O}{\overset{\|}{C}}-COOH \xrightarrow{-H_2O,\ -CO_2}$$

$$\xrightarrow{-2H_2}$$

(cont. on p. 262)

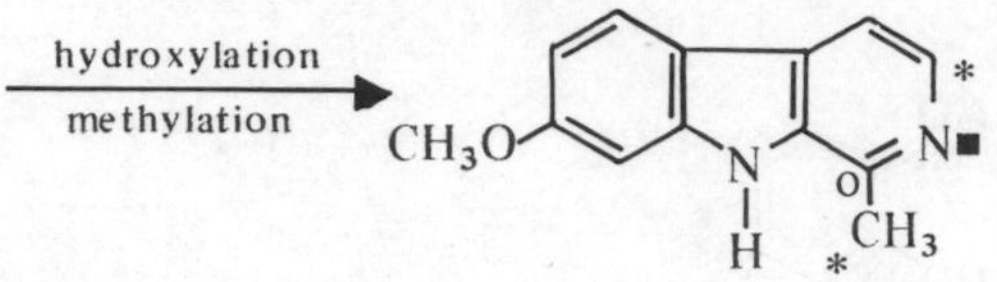

Harmine

Cf. T. A. Geissman and D. H. G. Crout, *Organic Chemistry of Secondary Plant Metabolism*, Freeman, Cooper & Co., San Francisco, 1969, pp. 473-474.

22.17

(a)

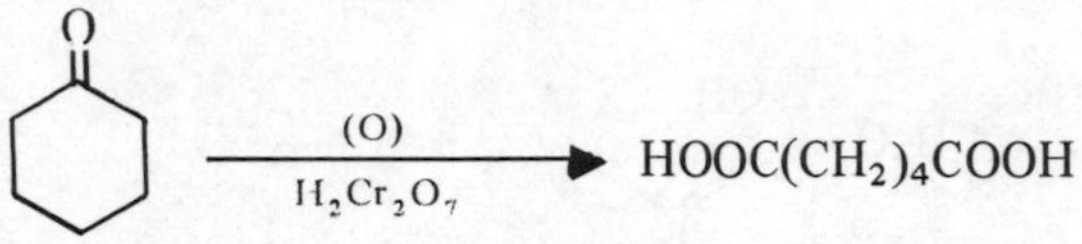

$\xrightarrow[\text{H}_2\text{Cr}_2\text{O}_7]{(O)}$ $HOOC(CH_2)_4COOH$

(b) $HOOC(CH_2)_4COOH + 2NH_3 \longrightarrow NH_4OOC(CH_2)_4COONH_4$

$\xrightarrow{\text{heat}} H_2N\overset{\displaystyle O}{\overset{\displaystyle \|}{C}}(CH_2)_4\overset{\displaystyle O}{\overset{\displaystyle \|}{C}}NH_2 \xrightarrow[\text{catalyst}]{350°} N\equiv C(CH_2)_4C\equiv N$

$\xrightarrow[\text{catalyst}]{4H_2} H_2NCH_2(CH_2)_4CH_2NH_2$

(c) $CH_2{=}CH{-}CH{=}CH_2 \xrightarrow{Cl_2} ClCH_2CH{=}CHCH_2Cl \xrightarrow{2NaCN}$

$N\equiv CCH_2CH{=}CHCH_2C\equiv N \xrightarrow[\text{Ni}]{H_2} N\equiv C(CH_2)_4C\equiv N$

$\xrightarrow[\text{catalyst}]{4H_2} H_2NCH_2(CH_2)_4CH_2NH_2$

(d) 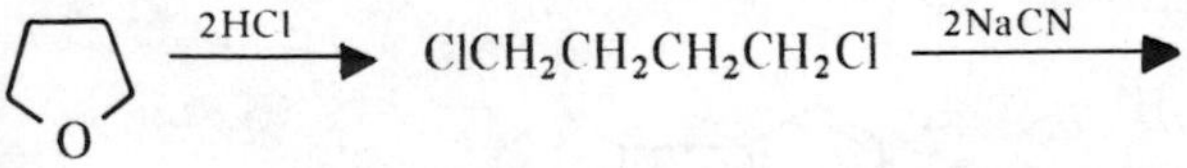 $\xrightarrow{2HCl}$ $ClCH_2CH_2CH_2CH_2Cl \xrightarrow{2NaCN}$

$N\equiv C(CH_2)_4C\equiv N \xrightarrow[\text{catalyst}]{4H_2} H_2NCH_2(CH_2)_4CH_2NH_2$

22.18

(a) $HOCH_2CH_2OH + {:}B \rightleftharpoons HOCH_2CH_2O^- + HB$

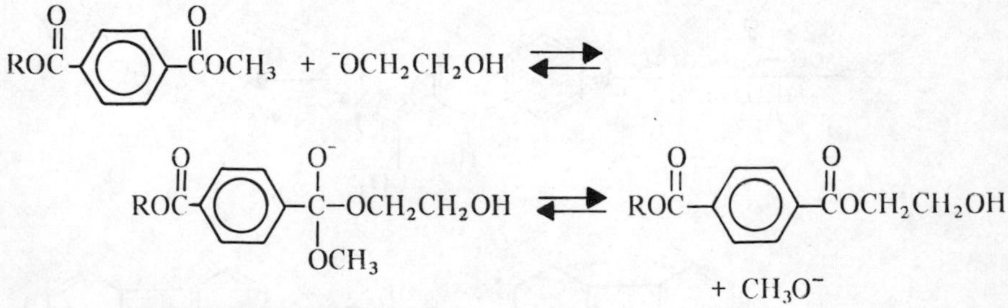

$CH_3O^- + HB \rightleftharpoons CH_3OH + {:}B^-$

$R = CH_3{-}$ or $HOCH_2CH_2{-}$

(b) $\underset{\text{ROC}}{\overset{\text{O}}{\parallel}}$—⬡—$\underset{\text{C—OCH}_3}{\overset{\text{O}}{\parallel}}$ $\underset{-H^+}{\overset{+H^+}{\rightleftharpoons}}$ $\underset{\text{ROC}}{\overset{\text{O}}{\parallel}}$—⬡—$\underset{\text{C—OCH}_3}{\overset{\overset{+}{\text{OH}}}{\parallel}}$

$\xrightarrow[-\text{HOCH}_2\text{CH}_2\text{OH}]{+\text{HOCH}_2\text{CH}_2\text{OH}}$ $\underset{\text{ROC}}{\overset{\text{O}}{\parallel}}$—⬡—$\underset{\underset{\underset{\text{CH}_3}{|}}{\overset{|}{\text{O}}\ \text{H}}}{\overset{\overset{\text{OH}}{|}}{\underset{|}{\text{C}}\text{—O}^+\text{—CH}_2\text{CH}_2\text{OH}}}$ $\rightleftharpoons$

$\underset{\text{ROC}}{\overset{\text{O}}{\parallel}}$—⬡—$\underset{\underset{\underset{\text{CH}_3}{|}}{\overset{|}{+\text{O—H}}}}{\overset{\overset{\text{OH}}{|}}{\underset{|}{\text{C}}\text{—OCH}_2\text{CH}_2\text{OH}}}$ $\underset{+\text{CH}_3\text{OH}}{\overset{-\text{CH}_3\text{OH}}{\longrightarrow}}$ $\underset{\text{ROC}}{\overset{\text{O}}{\parallel}}$—⬡—$\underset{\text{COCH}_2\text{CH}_2\text{OH}}{\overset{\overset{+}{\text{OH}}}{\parallel}}$

$\underset{+H^+}{\overset{-H^+}{\rightleftharpoons}}$ $\underset{\text{ROC}}{\overset{\text{O}}{\parallel}}$—⬡—$\underset{\text{C—OCH}_2\text{CH}_2\text{OH}}{\overset{\text{O}}{\parallel}}$

R = CH₃— or HOCH₂CH₂—

22.19

(a) $\text{CH}_3\underset{\overset{\text{O}}{\parallel}}{\text{OC}}$—⬡—$\underset{\overset{\text{O}}{\parallel}}{\text{COCH}_3}$ + HOCH₂—⬡—CH₂OH

(b) By high-pressure catalytic hydrogenation

23

SPECIAL TOPICS IV
LIPIDS

23.1

(a) $CH_3(CH_2)_{16}COOH + C_2H_5OH \overset{H^+}{\rightleftharpoons} CH_3(CH_2)_{16}COOC_2H_5 + H_2O$

$CH_3(CH_2)_{16}COOH \xrightarrow{SOCl_2} CH_3(CH_2)_{16}COCl \xrightarrow{C_2H_5OH} CH_3(CH_2)_{16}COOC_2H_5$

(b) $CH_3(CH_2)_{16}COCl \xrightarrow{(CH_3)_3COH} CH_3(CH_2)_{16}COOC(CH_3)_3$

(c) $CH_3(CH_2)_{16}COCl \xrightarrow{NH_3} CH_3(CH_2)_{16}CONH_2$

(d) $CH_3(CH_2)_{16}COCl \xrightarrow{(CH_3)_2NH} CH_3(CH_2)_{16}CON(CH_3)_2$

(e) $CH_3(CH_2)_{16}CONH_2 \xrightarrow{LiAlH_4} CH_3(CH_2)_{16}CH_2NH_2$

(f) $CH_3(CH_2)_{16}CONH_2 \xrightarrow{NaOBr} CH_3(CH_2)_{15}CH_2NH_2$

(g) $CH_3(CH_2)_{16}COCl \xrightarrow{LiAlH[OC(CH_3)_3]_3} CH_3(CH_2)_{16}CHO$

(h) $CH_3(CH_2)_{16}COOC_2H_5 \xrightarrow{H_2,\ Ni} CH_3(CH_2)_{16}CH_2OH$

$CH_3(CH_2)_{16}COCl$

$CH_3(CH_2)_{16}COOCH_2(CH_2)_{16}CH_3$

(i) $CH_3(CH_2)_{16}COOH \xrightarrow[(2)\ H_2O]{(1)\ LiAlH_4} CH_3(CH_2)_{16}CH_2OH$

$CH_3(CH_2)_{16}COOC_2H_5 \xrightarrow{H_2,\ Ni} CH_3(CH_2)_{16}CH_2OH$

(j) $CH_3(CH_2)_{16}COCl + (CH_3)_2Cd \longrightarrow CH_3(CH_2)_{16}COCH_3$

$\qquad\qquad$ or $(CH_3)_2CuLi$

(k) $CH_3(CH_2)_{16}CH_2OH \xrightarrow{PBr_3} CH_3(CH_2)_{16}CH_2Br$

(l) $CH_3(CH_2)_{16}CH_2Br \xrightarrow[(2)\ H^+,\ H_2O,\ heat]{(1)\ NaCN} CH_3(CH_2)_{16}CH_2COOH$

23.2

(a) $CH_3(CH_2)_{11}CH_2COOH \xrightarrow{Br_2,\ P} CH_3(CH_2)_{11}\underset{\underset{Br}{|}}{C}HCOOH$

(b) $CH_3(CH_2)_{11}\underset{\displaystyle Br}{\underset{|}{C}}HCOOH \xrightarrow[(2)\ H^+]{(1)\ OH^-,\ heat} CH_3(CH_2)_{11}\underset{\displaystyle OH}{\underset{|}{C}}HCOOH$

(c) $CH_3(CH_2)_{11}\underset{\displaystyle Br}{\underset{|}{C}}HCOOH \xrightarrow[(2)\ H^+]{NaCN} CH_3(CH_2)_{11}\underset{\displaystyle CN}{\underset{|}{C}}HCOOH$

(d) $CH_3(CH_2)_{11}\underset{\displaystyle Br}{\underset{|}{C}}HCOOH \xrightarrow[(2)\ H^+]{NH_3\,(excess)} CH_3(CH_2)_{11}\underset{\displaystyle NH_2}{\underset{|}{C}}HCOOH$ or $CH_3(CH_2)_{11}\underset{\displaystyle {}^+NH_3}{\underset{|}{C}}HCOO^-$

(cf. p. 952)

23.3

(a) $CH_3(CH_2)_5CH{=}CH(CH_2)_7COOH \xrightarrow{I_2} CH_3(CH_2)_5CHICHI(CH_2)_7COOH$

(b) $CH_3(CH_2)_5CH{=}CH(CH_2)_7COOH \xrightarrow{H_2,\ Ni} CH_3(CH_2)_{14}COOH$

(c) $CH_3(CH_2)_5CH{=}CH(CH_2)_7COOH \xrightarrow{KMnO_4} CH_3(CH_2)_5CHOHCHOH(CH_2)_7COOH$

(d) $CH_3(CH_2)_5CH{=}CH(CH_2)_7COOH \xrightarrow{HCl} CH_3(CH_2)_7CH_2CHCl(CH_2)_7COOH$
$+$
$CH_3(CH_2)_7CHClCH_2(CH_2)_7COOH$

23.4

(a) There are two sets of enantiomers, giving a total of four stereoisomers

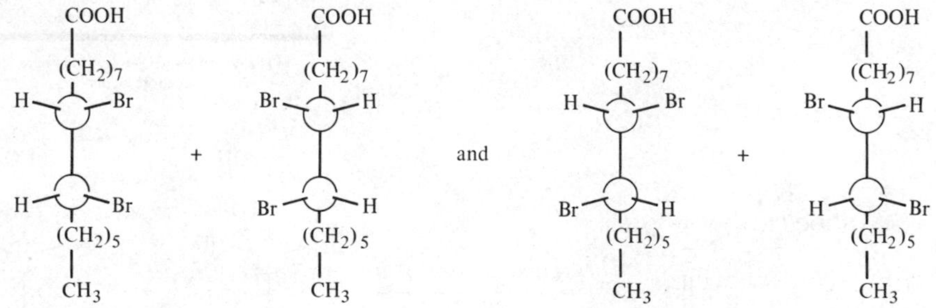

(b)

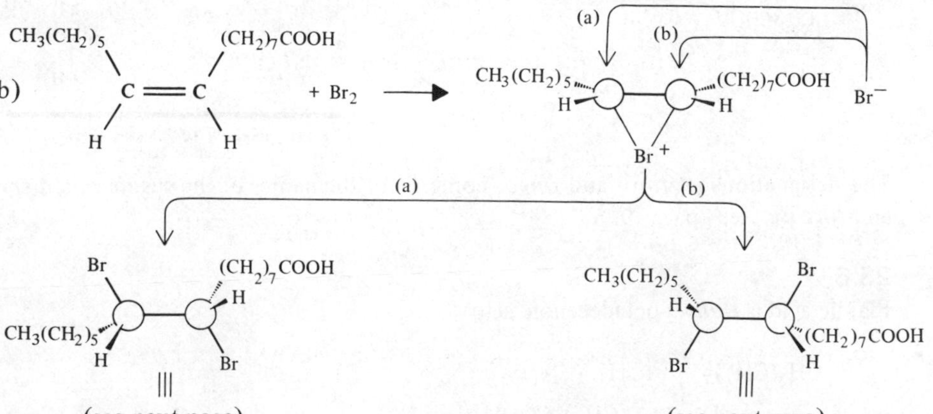

(see next page) (see next page)

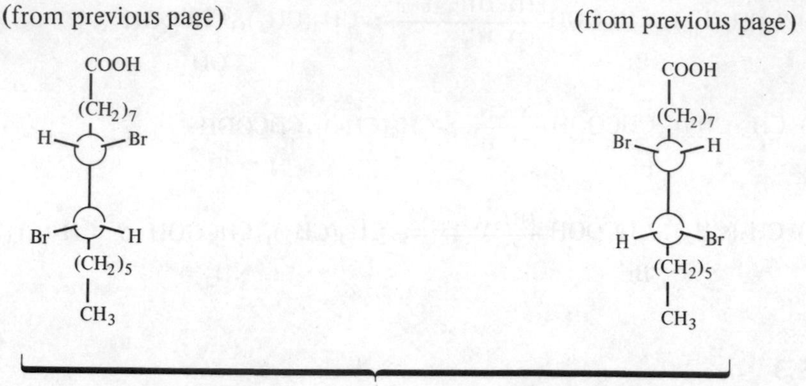

(from previous page)

COOH
|
$(CH_2)_7$
H——Br

Br——H
|
$(CH_2)_5$
|
CH_3

(from previous page)

COOH
|
$(CH_2)_7$
Br——H

H——Br
|
$(CH_2)_5$
|
CH_3

(±)–*threo*–9, 10–dibromohexadecanoic acids

Formation of a bromonium ion at the other face of palmitoleic acid gives a result such that the *threo*-enantiomers are obtained as a racemic modification.

23.5

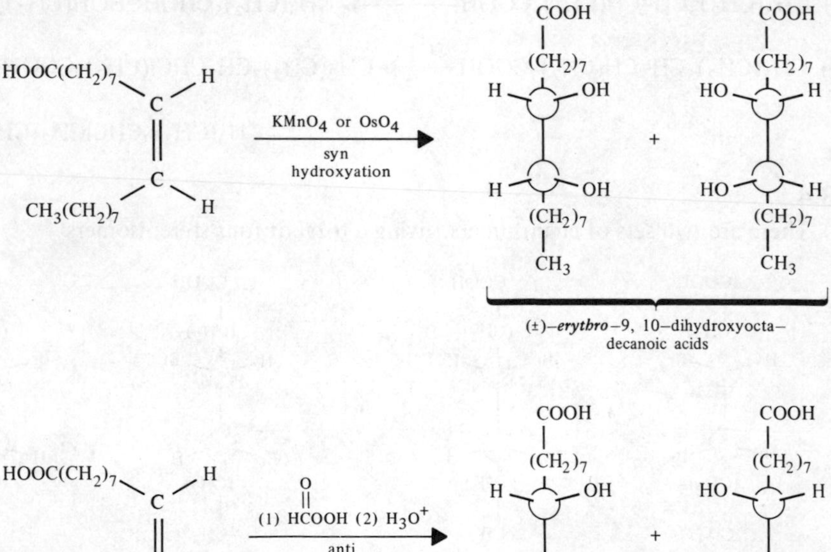

$HOOC(CH_2)_7$ H
     C
    ‖
    C
$CH_3(CH_2)_7$ H

$\xrightarrow[\text{hydroxyation}]{\text{KMnO}_4 \text{ or OsO}_4 \atop \text{syn}}$

COOH
|
$(CH_2)_7$
H——OH

H——OH
|
$(CH_2)_7$
|
CH_3

+

COOH
|
$(CH_2)_7$
HO——H

HO——H
|
$(CH_2)_7$
|
CH_3

(±)–*erythro*–9, 10–dihydroxyocta– decanoic acids

$HOOC(CH_2)_7$ H
     C
    ‖
    C
$CH_3(CH_2)_7$ H

$\xrightarrow[\text{hydroxylation}]{\text{(1) HCOOH (2) H}_3\text{O}^+ \atop \text{anti}}$

COOH
|
$(CH_2)_7$
H——OH

HO——H
|
$(CH_2)_7$
|
CH_3

+

COOH
|
$(CH_2)_7$
HO——H

H——OH
|
$(CH_2)_7$
|
CH_3

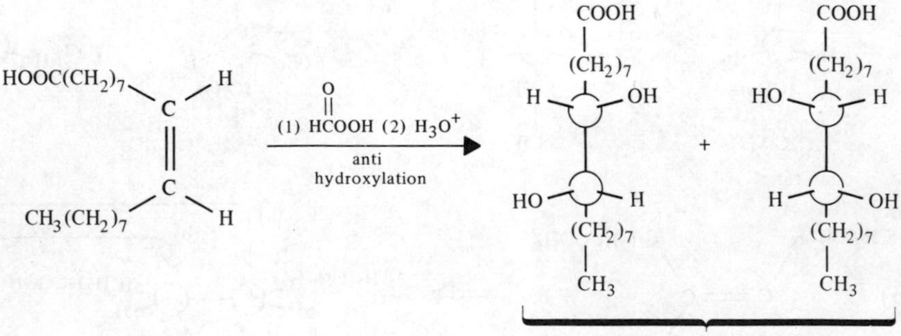

(±)–*threo*–9, 10–dihydroxyocta– decanoic acids

The designations *erythro*- and *threo*- come from the names of the sugars called *erythrose* and *threose;* see page 935.

23.6

Elaidic acid is *trans*-9-octadecenoic acid:

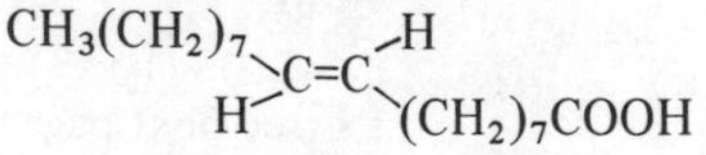

$CH_3(CH_2)_7$ H
    C=C
H $(CH_2)_7COOH$

It is formed by the isomerization of oleic acid.

23.7

(a)
$$CH_3(CH_2)_9\diagdown C=C\diagup (CH_2)_7COOH \atop H\diagup \diagdown H \quad \text{and} \quad CH_3(CH_2)_9\diagdown C=C\diagup H \atop H\diagup \diagdown (CH_2)_7COOH$$

(b) Infrared spectroscopy

(c) A peak in the 675-730 cm^{-1} region would indicate that the double bond is *cis*; a peak in the 960-975 cm^{-1} region would indicate that it is *trans*.

23.8

$$CH_3(CH_2)_5C{\equiv}CH + NaNH_2 \xrightarrow[NH_3]{} CH_3(CH_2)_5C{\equiv}CNa$$
$$\textbf{A}$$

$$\xrightarrow{ICH_2(CH_2)_7CH_2Cl} CH_3(CH_2)_5C{\equiv}CCH_2(CH_2)_7CH_2Cl \xrightarrow{NaCN}$$
$$\textbf{B}$$

$$CH_3(CH_2)_5C{\equiv}CCH_2(CH_2)_7CH_2CN \xrightarrow{KOH, H_2O} CH_3(CH_2)_5C{\equiv}CCH_2(CH_2)_7CH_2COOK$$
$$\textbf{C} \qquad\qquad\qquad\qquad\qquad\qquad\qquad\qquad \textbf{D}$$

$$\xrightarrow{H_3O^+} CH_3(CH_2)_5C{\equiv}CCH_2(CH_2)_7CH_2COOH \xrightarrow{H_2, \ Pd\text{-}BaSO_4}$$
$$\textbf{E}$$

$$CH_3(CH_2)_5\diagdown C=C\diagup CH_2(CH_2)_7CH_2COOH \atop H\diagup \diagdown H$$

Vaccenic acid

23.9

$$FCH_2(CH_2)_6CH_2Br + HC{\equiv}CNa \longrightarrow FCH_2(CH_2)_6CH_2C{\equiv}CH$$
$$\textbf{F}$$

$$\xrightarrow[(2)\ I(CH_2)_7Cl]{(1)\ NaNH_2} FCH_2(CH_2)_6CH_2C{\equiv}C(CH_2)_7Cl \xrightarrow[DMSO]{NaCN}$$
$$\textbf{G}$$

$$FCH_2(CH_2)_6CH_2C{\equiv}C(CH_2)_7CN \xrightarrow[(2)\ H^+]{(1)\ KOH} FCH_2(CH_2)_6CH_2C{\equiv}C(CH_2)_7COOH$$
$$\textbf{H} \qquad\qquad\qquad\qquad\qquad\qquad\qquad \textbf{I}$$

$$\xrightarrow[Pd\text{-}BaSO_4]{H_2} FCH_2(CH_2)_6CH_2\diagdown C=C\diagup (CH_2)_7COOH \atop H\diagup \diagdown H$$

23.10

(a)
$$\begin{array}{l} CH_2OH \\ | \\ CHOH \\ | \\ CH_2OH \end{array} + R\overset{O}{\overset{||}{C}}OH + R'\overset{O}{\overset{||}{C}}OH + H_3PO_4 + HOCH_2CH_2\overset{+}{N}(CH_3)_3 \ \ X^-$$

(b)
$$\begin{array}{l} CH_2OH \\ | \\ CHOH \\ | \\ CH_2OH \end{array} + R\overset{O}{\overset{||}{C}}OH + R'\overset{O}{\overset{||}{C}}OH + H_3PO_4 + HOCH_2CH_2NH_2$$

(c)
$$CH_2OH$$
$$|$$
$$CHOH + CH_3(CH_2)_nCH_2\overset{O}{\overset{||}{C}}H + R^1\overset{O}{\overset{||}{C}}OH + H_3PO_4$$
$$|$$
$$CH_2OH$$

$$+ HOCH_2CH_2\overset{+}{N}(CH_3)_3 \ X^-$$

23.11

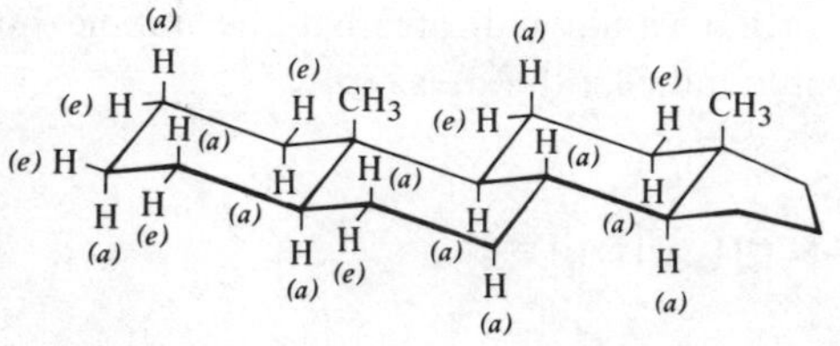

5α—series

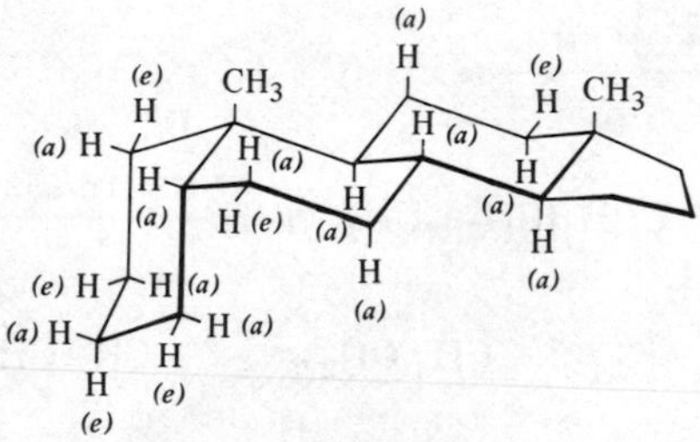

5β—series

23.12

(a)

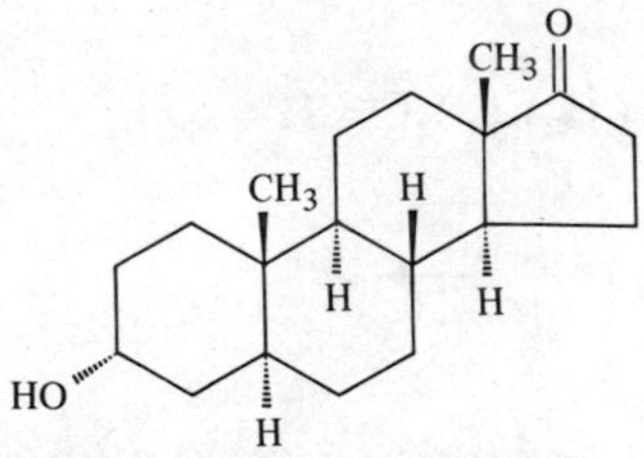

androstan−3α−ol−17−one
(androsterone)

(b)

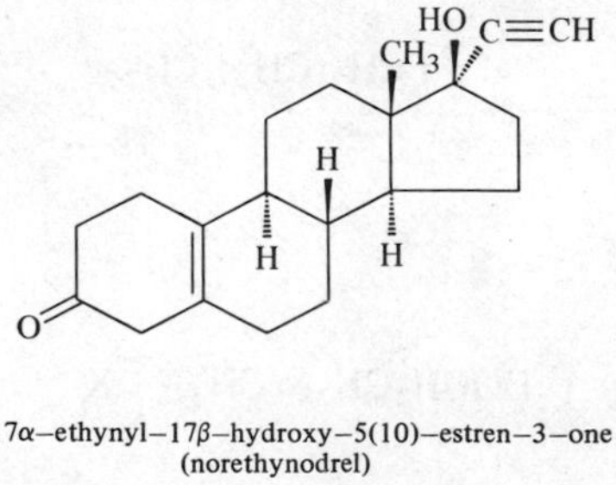

17α−ethynyl−17β−hydroxy−5(10)−estren−3−one
(norethynodrel)

23.13

Estrone and estradiol are *phenols* and thus are soluble in aqueous sodium hydroxide.
Extraction with aqueous sodium hydroxide separates the estrogens from the androgens.

23.14

(a)

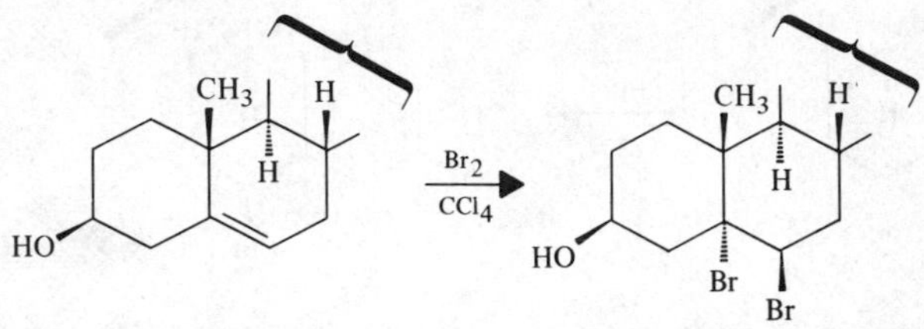

cholesterol

5α, 6β–dibromocholestan–3β–ol

(b)

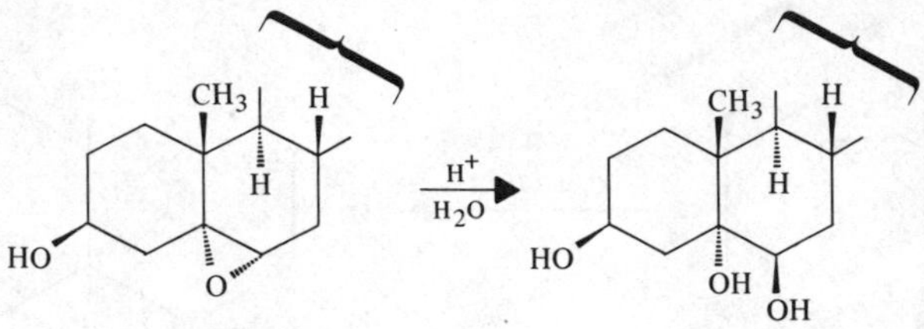

5α, 6α–oxidocholestan–3β–ol
(prepared by epoxidation of
cholesterol, cf. p.902)

cholestan–3β, 5α, 6β–triol

(c)

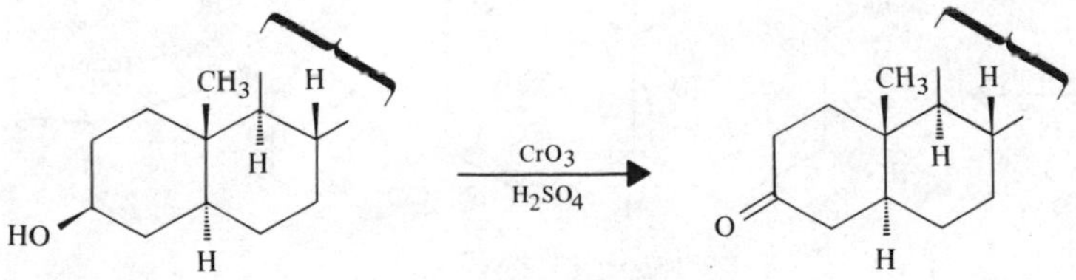

5α–cholestan–3β–ol
(prepared by hydrogenation
of cholesterol, cf. p. 902)

5α–cholestan–3–one

(d)

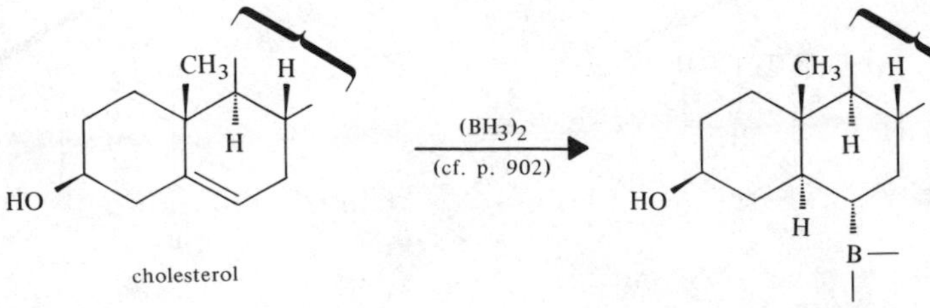

cholesterol

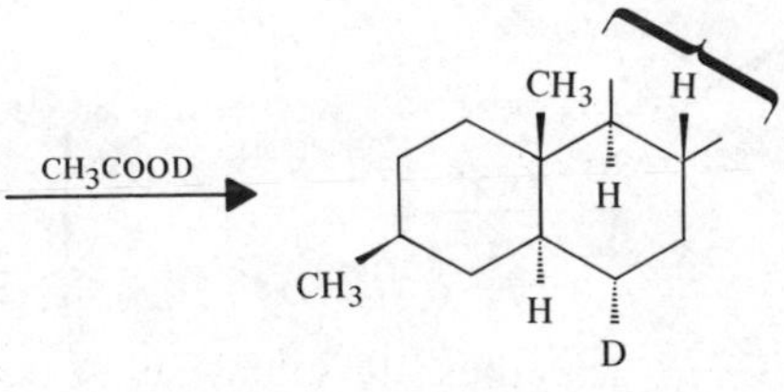

6α–deuterio–5α–cholestan–3β–ol

(e)

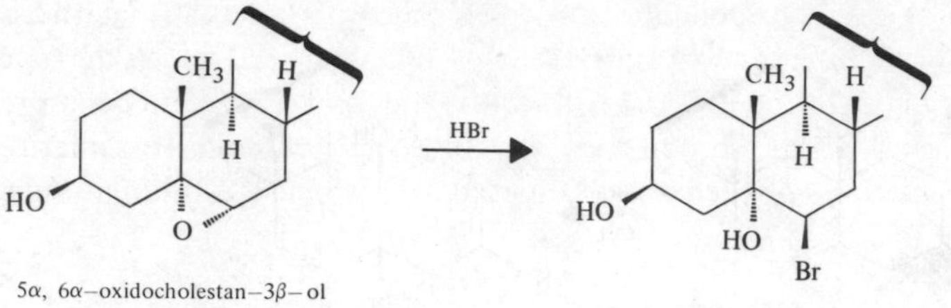

5α, 6α–oxidocholestan–3β–ol

6β–bromocholestan–3β, 5α–diol

23.15

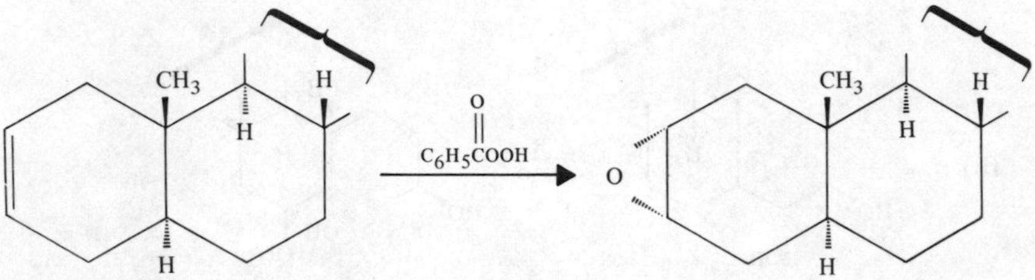

5α–cholest–2–ene

A

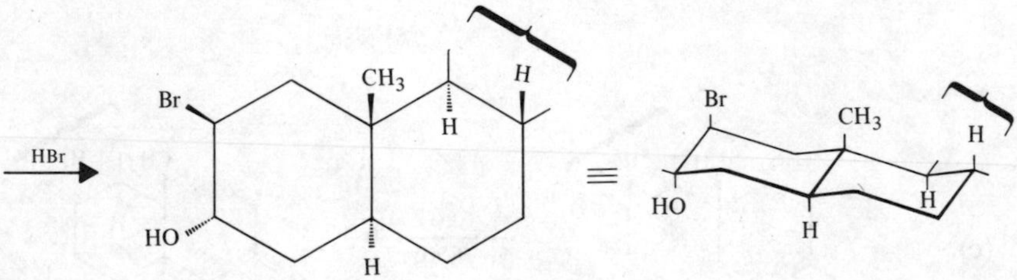

B

Here we find that epoxidation takes place at the less hindered α face (cf. p. 902). Ring opening by HBr takes place in an *anti* fashion to give a product with diaxial substituents.

23.16

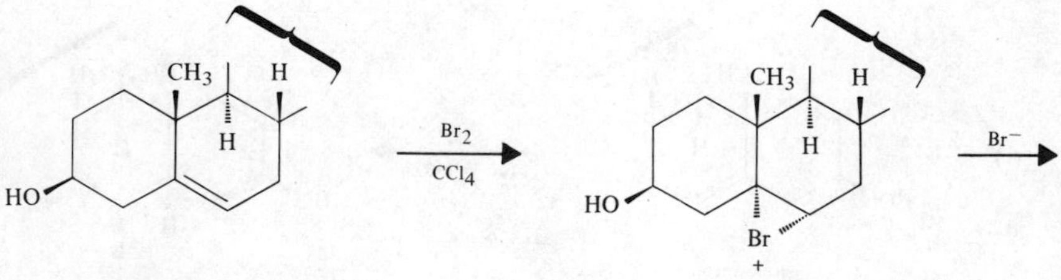

cholesterol

5α, 6α–bromonium ion

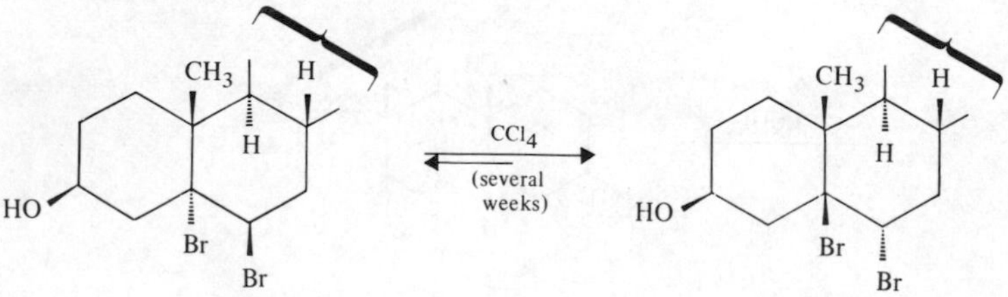

+

5α, 6β–dibromocholestan–3β–ol

5β, 6α–dibromocholestan–3β–ol

Here formation of the bromonium ion takes place preferentially at the α face. Ring opening takes place in an *anti* manner (with a bromide ion attacking the 6-position from above) to give, initially, the $5\alpha,6\beta$-dibromo compound. The $5\alpha,6\beta$-dibromocholestan-3β-ol, however is a *diaxial* dibromide and is, therefore, unstable. It isomerizes to the 5β, 6α-dibromocholestan-3β-ol (below)—a compound in which the bromine substituents are both equatorial.

$5\alpha, 6\beta$—dibromide
(bromines are diaxial)

$5\beta, 6\alpha$—dibromide
(bromines are diequatorial)

This isomerization does not result from a simple "flipping" of the cylohexane rings; it requires an inversion of configuration at carbons 5 and 6. One mechanism that has been proposed for the isomerization involves the formation of a "bromonium-bromide" ion pair:

diaxial   ion pair   diequatorial

24

24.1

(a) Two,   CHO
　　　　　 |
　　　　　 *CHOH
　　　　　 |
　　　　　 *CHOH
　　　　　 |
　　　　　 CH₂OH

(b) Two,   CH₂OH
　　　　　 |
　　　　　 C=O
　　　　　 |
　　　　　 *CHOH
　　　　　 |
　　　　　 *CHOH
　　　　　 |
　　　　　 CH₂OH

(c) There would be four stereoisomers (two sets of enantiomers) with each general structure: $2^2 = 4$.

24.2

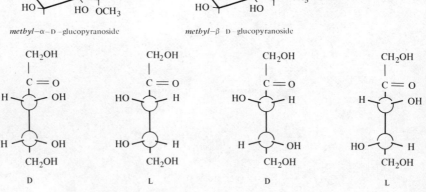

24.3

Since glycosides are acetals they undergo hydrolysis in aqueous acid to form cyclic hemiacetals that then undergo mutarotation.

24.4

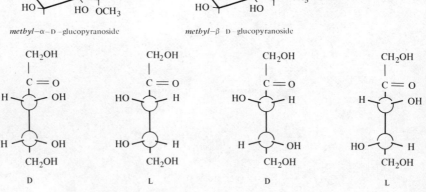

methyl–α–D–glucopyranoside   methyl–β–D–glucopyranoside

24.5

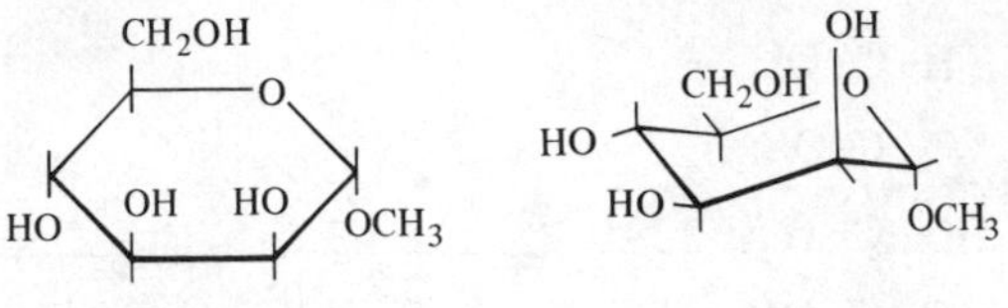

Haworth formula   conformational formula

methyl–α–D–mannopyranoside

24.6

α-D-glucopyranose will give a positive test with Benedict's or Tollens' solution because it is a cyclic hemiacetal. Methyl-α-D-glucopyranoside, because it is a cyclic acetal, will not.

24.7

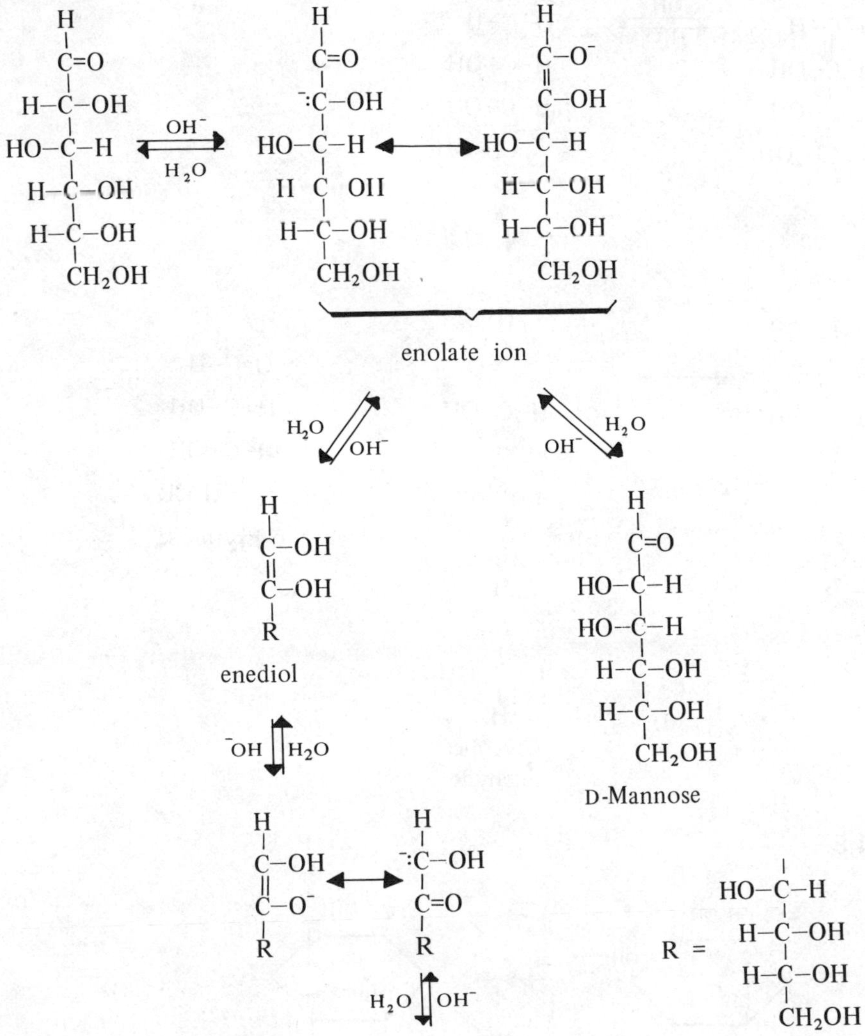

enolate ion

enediol

D-Mannose

(see next page)

$$
\begin{array}{c}
\text{H} \\
| \\
\text{H--C--OH} \\
| \\
\text{C=O} \\
| \\
\text{HO--C--H} \\
| \\
\text{H--C--OH} \\
| \\
\text{H--C--OH} \\
| \\
\text{CH}_2\text{OH}
\end{array}
$$

D-Fructose

$$
\begin{array}{c}
\text{H} \\
| \\
\text{C=O} \\
| \\
\text{H--C--OH} \\
| \\
\text{HO--C--H} \\
| \\
\text{H--C--OH} \\
| \\
\text{H--C--OH} \\
| \\
\text{CH}_2\text{OH}
\end{array}
\quad \underset{\text{H}_2\text{O}}{\overset{\text{OH}^-}{\rightleftarrows}} \quad
\begin{array}{c}
\text{H} \\
| \\
\text{C=O} \\
| \\
\text{H--C--OH} \\
| \\
{}^-\text{O--C--H} \\
| \\
\text{H--C--OH} \\
| \\
\text{H--C--OH} \\
| \\
\text{CH}_2\text{OH}
\end{array}
$$

$$
\begin{array}{c}
\text{H} \\
| \\
\text{C--O}^- \\
| \\
\text{H--C--OH}
\end{array}
\quad \longleftrightarrow \quad
\begin{array}{c}
\text{H} \\
| \\
\text{C=O} \\
| \\
\text{H--C--OH} \\
\ddot{}
\end{array}
\quad + \quad
\begin{array}{c}
\text{O=C--H} \\
| \\
\text{H--C--OH} \\
| \\
\text{H--C--OH} \\
| \\
\text{CH}_2\text{OH}
\end{array}
$$

D-Erythrose

$$
\text{H}_2\text{O} \updownarrow \text{OH}
$$

$$
\begin{array}{c}
\text{H} \\
| \\
\text{C=O} \\
| \\
\text{H--C--OH} \\
| \\
\text{H}
\end{array}
$$

Glycolic
aldehyde

24.8

β–D–mannopyranose δ–D–mannolactone

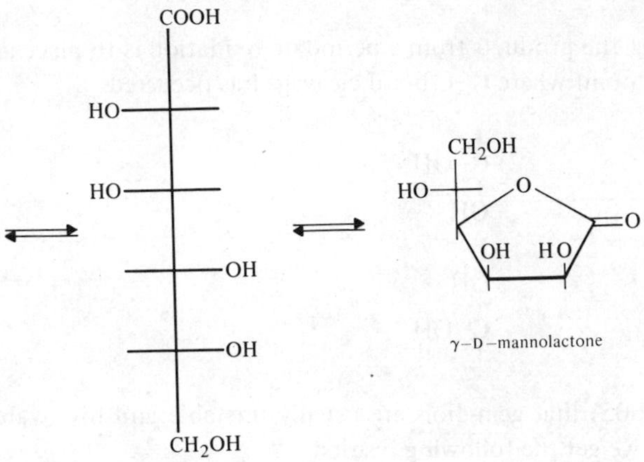
COOH
HO
HO
OH
OH
CH2OH
CH2OH
HO
O
O
OH
HO

D-mannonic acid

γ-D-mannolactone

24.9

(a) Yes (b)

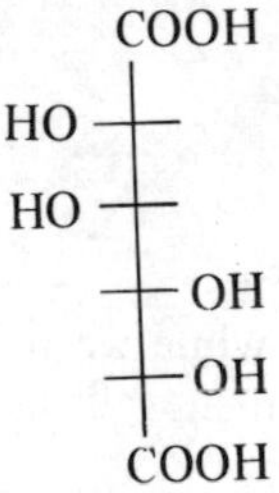
COOH
HO
HO
OH
OH
COOH

D-mannaric acid

(c) Yes

(d)

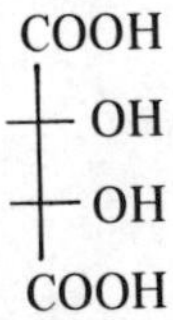
COOH
OH
OH
COOH

(e) No (f)

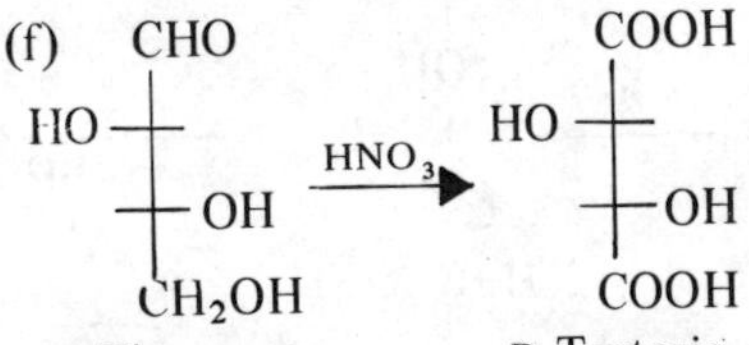
CHO
HO
OH
CH2OH
HNO3
COOH
HO
OH
COOH

D-Threose

D-Tartaric acid

(g) The aldaric acid obtained from D-erythrose is *meso*-tartaric acid; the aldaric acid obtained from D-Threose is D-tartaric acid.

24.10

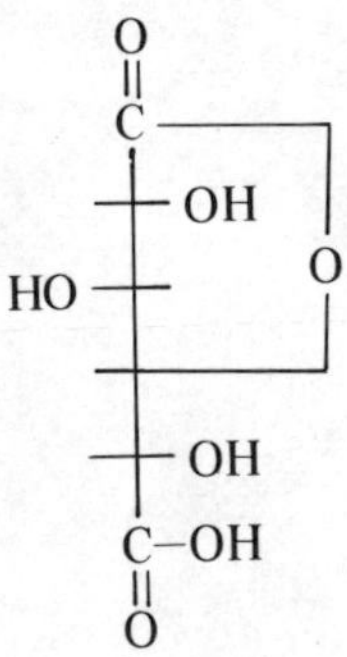
O
C
OH
HO
O
OH
C–OH
O

and

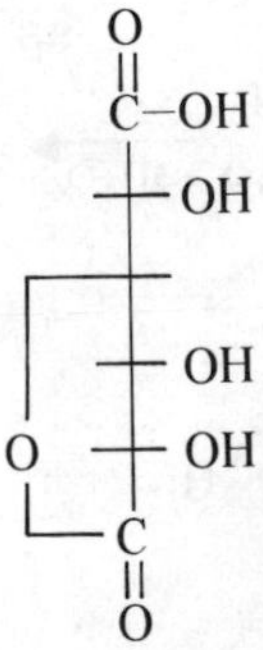
O
C–OH
OH
OH
O
OH
C
O

24.11

One way of predicting the products from a periodate oxidation is to place an −OH group
on each carbon at the point where C−C bond cleavage has occurred:

$$
\begin{array}{c}
-\overset{|}{\underset{|}{C}}-OH \\
-\overset{|}{\underset{|}{C}}-OH
\end{array}
\quad \xrightarrow{\ IO_4^-\ } \quad
\begin{array}{c}
-\overset{|}{C}-OH \\
OH \\
+ \\
OH \\
-\overset{|}{\underset{|}{C}}-OH
\end{array}
$$

Then if we recall (p. 605) that gem-diols are usually unstable and lose water to produce
carbonyl compounds, we get the following results:

$$
-\overset{|}{\underset{\underset{OH}{|}}{C}}-O-H \longrightarrow -\overset{|}{C}=O + H_2O
$$

$$
-\overset{\overset{OH}{|}}{\underset{|}{C}}-O-H \longrightarrow -\overset{|}{C}=O + H_2O
$$

Let us apply this procedure to several examples here while we remember that for every
C−C bond that is broken one mole of HIO_4 is consumed.

(a)

$$
\begin{array}{c}
CH_3 \\
H-\overset{|}{C}-OH \\
\text{-----}|\text{-----} \\
H-\overset{|}{C}-OH \\
CH_3
\end{array}
\ + \ HIO_4 \ \longrightarrow \
\begin{array}{c}
CH_3 \\
H-\overset{|}{C}-O-H \\
\overset{|}{OH} \\
+ \\
\overset{OH}{|} \\
H-\overset{|}{C}-O-H \\
CH_3
\end{array}
\ \xrightarrow{-2H_2O} \ 2CH_3\overset{\overset{O}{\|}}{C}-H
$$

(b)

$$
\begin{array}{c}
H \\
H-\overset{|}{C}-OH \\
\text{--}|\text{--} \\
H-\overset{|}{C}-OH \\
\text{--}|\text{--} \\
H-\overset{|}{C}-OH \\
CH_3
\end{array}
\ + \ 2HIO_4 \ \longrightarrow \
\begin{array}{c}
H \\
H-\overset{|}{C}-O-H \\
\overset{|}{OH} \\
+ \\
O-H \\
H-\overset{|}{C}-OH \\
\overset{|}{OH} \\
+ \\
\overset{OH}{|} \\
H-\overset{|}{C}-O-H \\
CH_3
\end{array}
\ \xrightarrow{-3H_2O} \
\begin{array}{c}
H \\
H-C=O \\
+ \\
\overset{\overset{O}{\|}}{} \\
H-C-OH \\
+ \\
H-C=O \\
CH_3
\end{array}
$$

(c)

$$
\begin{array}{l}
\text{H} \\
\text{H–C–OH} \\
\text{----|----} \\
\text{H–C–OH} \;+\; \text{HIO}_4 \\
\text{H–C–OCH}_3 \\
\text{OCH}_3
\end{array}
\longrightarrow
\begin{array}{l}
\text{H} \\
\text{H–C–OH} \\
\text{OH} \\
+ \\
\text{OH} \\
\text{H–C–OH} \\
\text{H–C–OCH}_3 \\
\text{OCH}_3
\end{array}
\xrightarrow{-2\text{H}_2\text{O}}
\begin{array}{l}
\text{H} \\
\text{H–C=O} \\
+ \\
\text{O} \\
\text{H–C} \\
\text{H–C–OCH}_3 \\
\text{OCH}_3
\end{array}
$$

(d)

$$
\begin{array}{l}
\text{H} \\
\text{H–C–OH} \\
\text{--|----} \\
\text{H–C–OH} \;+\; 2\text{HIO}_4 \\
\text{--|--} \\
\text{C=O} \\
\text{CH}_3
\end{array}
\longrightarrow
\begin{array}{l}
\text{H} \\
\text{H–C–OH} \\
\text{OH} \\
+ \\
\text{OH} \\
\text{H–C–OH} \\
\text{OH} \\
+ \\
\text{OH} \\
\text{C=O} \\
\text{CH}_3
\end{array}
\xrightarrow{-2\text{H}_2\text{O}}
\begin{array}{l}
\text{H} \\
\text{H–C=O} \\
+ \\
\text{O} \\
\text{H–C–OH} \\
+ \\
\text{O} \\
\text{CH}_3\text{COH}
\end{array}
$$

(e)

$$
\begin{array}{l}
\text{CH}_3 \\
\text{C=O} \\
\text{--|--} \\
\text{H–C–OH} \;+\; 2\text{HIO}_4 \\
\text{--|--} \\
\text{C=O} \\
\text{CH}_3
\end{array}
\longrightarrow
\begin{array}{l}
\text{CH}_3 \\
\text{H–C–OH} \\
\text{OH} \\
+ \\
\text{OH} \\
\text{H–C–OH} \\
\text{OH} \\
+ \\
\text{OH} \\
\text{C=O} \\
\text{CH}_3
\end{array}
\xrightarrow{-2\text{H}_2\text{O}}
2\text{CH}_3\overset{\text{O}}{\text{C}}\text{OH} + \text{H}\overset{\text{O}}{\text{C}}\text{OH}
$$

(f)

$$
\begin{array}{l}
\text{CH}_2\diagdown \text{C–OH} \\
\text{CH}_2 \;\text{----|----} \\
\text{C–OH} \\
\text{CH}_2 \diagup
\end{array}
\;+\; \text{HIO}_4 \longrightarrow
\begin{array}{l}
\text{CH}_2\diagdown \text{C–OH} \\
\text{CH}_2 \quad \text{OH} \\
\text{OH} \\
\text{CH}_2\diagup \text{C–OH}
\end{array}
\xrightarrow{-2\text{H}_2\text{O}}
\overset{\text{O}}{\text{H}\text{C}}\text{CH}_2\text{CH}_2\text{CH}_2\overset{\text{O}}{\text{C}}\text{H}
$$

(g)

$$
\begin{array}{c}
\overset{\displaystyle H}{\underset{\displaystyle CH_3}{\overset{\displaystyle |}{\underset{\displaystyle |}{\text{H–C–OH}}}}} \\
\text{- -}\,|\,\text{- -} \;+\; HIO_4 \;\longrightarrow \\
\text{CH}_3\text{–C–OH} \\
CH_3
\end{array}
\qquad
\begin{array}{c}
\text{H} \\
\text{H–C–OH} \\
\text{OH} \\
+ \\
\text{OH} \\
\text{CH}_3\text{–C–OH} \\
\text{CH}_3
\end{array}
\xrightarrow{-2H_2O}
\begin{array}{c}
\text{H} \\
\text{H–C=O} \\
+ \\
\text{CH}_3\text{–C=O} \\
\text{CH}_3
\end{array}
$$

(h)

$$
\begin{array}{c}
\overset{\displaystyle O}{\overset{\displaystyle \|}{\text{H–C}}} \\
\text{- -}\,|\,\text{- -} \\
\text{H–C–OH} \\
\text{- -}\,|\,\text{- -} \\
\text{H–C–OH} \;+\; 3HIO_4 \;\longrightarrow \\
\text{- -}\,|\,\text{- -} \\
\text{H–C–OH} \\
\text{H}
\end{array}
\qquad
\begin{array}{c}
\overset{\displaystyle O}{\overset{\displaystyle \|}{\text{H–C–OH}}} \\
+ \\
\text{OH} \\
\text{H–C–OH} \\
\text{OH} \\
+ \\
\text{OH} \\
\text{H–C–OH} \\
\text{OH} \\
+ \\
\text{OH} \\
\text{H–C–OH} \\
\text{H}
\end{array}
\xrightarrow{-3H_2O}
\;\;3\overset{\displaystyle O}{\overset{\displaystyle \|}{\text{HCOH}}} \;+\; \overset{\displaystyle O}{\overset{\displaystyle \|}{\text{HCH}}}
$$

D-erythrose

24.12

Oxidation of an aldohexose and a ketohexose would each require five moles of HIO_4 but would give different results

$$
\begin{array}{c}
\text{CHO} \\
\text{-}\,|\,\text{- - -} \\
\text{CHOH} \\
\text{-}\,|\,\text{- -} \\
\text{CHOH} \\
\text{-}\,|\,\text{- - -} \;+\; 5HIO_4 \;\longrightarrow \\
\text{CHOH} \\
\text{-}\,|\,\text{- - -} \\
\text{CHOH} \\
\text{-}\,|\,\text{- - -} \\
\text{CH}_2\text{OH}
\end{array}
\qquad
\begin{array}{c}
\text{HCOOH} \\
+ \\
\text{HCOOH} \\
+ \\
\text{HCOOH} \\
+ \\
\text{HCOOH} \\
+ \\
\text{HCOOH} \\
+ \\
\text{HCHO}
\end{array}
\qquad
(5\;\text{HCOOH} + \text{HCHO})
$$

Aldohexose

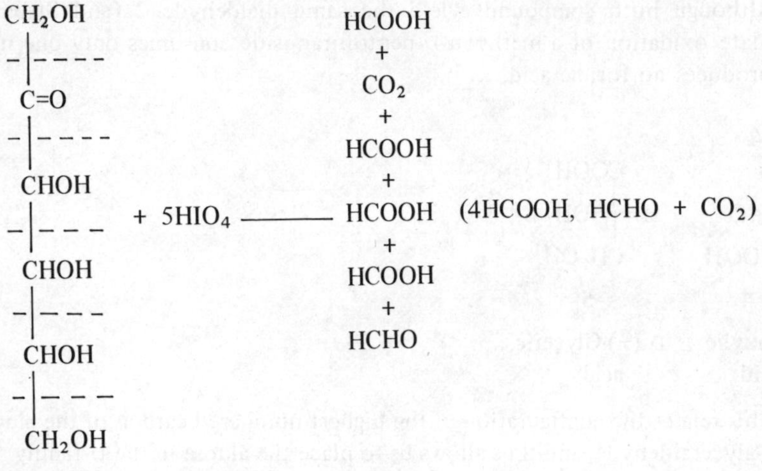

CH$_2$OH
|
C=O
|
CHOH
| + 5HIO$_4$ ————
CHOH
|
CHOH
|
CH$_2$OH

HCOOH
+
CO$_2$
+
HCOOH
+
HCOOH   (4HCOOH, HCHO + CO$_2$)
+
HCOOH
+
HCHO

Ketohexose

23.13
(a)

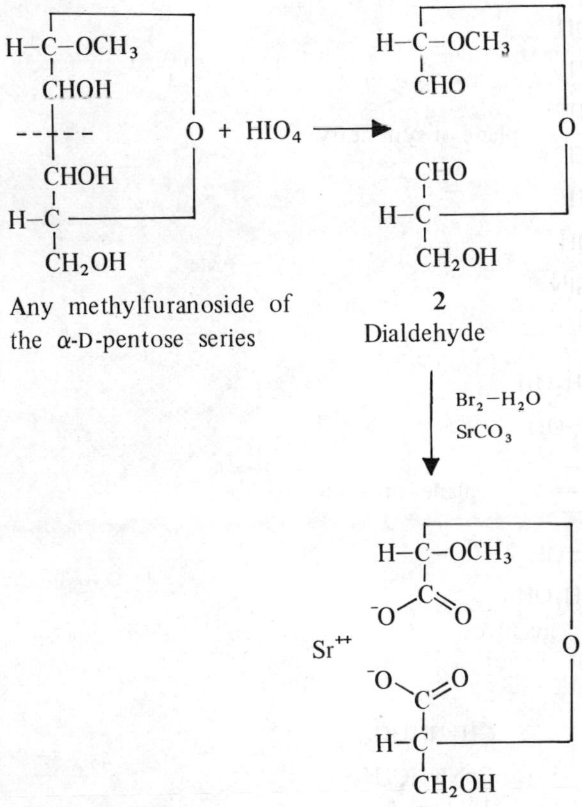

H–C–OCH$_3$
   CHOH
         O + HIO$_4$ ⟶
   CHOH
H–C
   CH$_2$OH

H–C–OCH$_3$
   CHO
        O
   CHO
H–C
   CH$_2$OH

Any methylfuranoside of
the α-D-pentose series

2
Dialdehyde

Br$_2$–H$_2$O
SrCO$_3$

H–C–OCH$_3$
     $^-$O–C=O
Sr^{++} O
     $^-$O–C=O
H–C
   CH$_2$OH

3
Same strontium salt

(b) Although both compounds yield the same dialdehyde **2** (and Strontium salt **3**), periodate oxidation of a methyl-α-D-pentofuranoside consumes only one mole of HIO_4 and produces no formic acid.

24.14

(a)

```
     H                    COOH
     |                
     C=O            H ——— OH
     |                     |
    COOH              CH₂OH

      4                    5

  Glyoxylic        D-(−)-Glyceric
    acid                acid
```

(b) This relates the configuration of the highest numbered carbon of the aldose to that of D-(+)-glyceraldehyde, and thus allows us to place the aldose in the D-family.

25.15

(a) Yes, D-glucitol would be optically active; only those alditols (below) whose molecules possess a plane of symmetry would be optically inactive.

(b)

```
    CHO                    CH₂OH
     |                       |
     ┼ OH                    ┼ OH
     |                       |
     ┼ OH   NaBH₄            ┼ OH    - - - - -  plane of symmetry
     |      ───────►         |
     ┼ OH                    ┼ OH
     |                       |
     ┼ OH                    ┼ OH
     |                       |
    CH₂OH                  CH₂OH

                          optically
                          inactive
```

```
     CHO                   CH₂OH
      |                      |
      ┼ OH                   ┼ OH
      |                      |
  HO ─┼       NaBH₄      HO ─┼     - - - - -  plane of symmetry
      |       ───────►       |
  HO ─┼                  HO ─┼
      |                      |
      ┼ OH                   ┼ OH
      |                      |
    CH₂OH                  CH₂OH

                        optically  inactive
```

24.16

(a)

```
     CH₂OH                            CH=NNHC₆H₅
      |                                |
      C=O                              C=NNHC₆H₅
      |          C₆H₅NHNH₂             |
  HO ─┼          ───────────►      HO ─┼
      |                                |
      ┼ OH                             ┼ OH
      |                                |
      ┼ OH                             ┼ OH
      |                                |
    CH₂OH                            CH₂OH
```

(b) This experiment shows that D-glucose and D-fructose have the same configurations at C-3, C-4, and C-5.

24.17

(a)

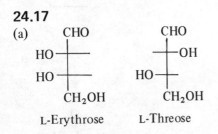

    L-Erythrose   L-Threose

(b) L-Glyceraldehyde,

$$
\begin{array}{c}
\text{CHO} \\
\text{HO} \!-\!\!\!-\!\!\!-\!\text{H} \\
\text{CH}_2\text{OH}
\end{array}
$$

24.18

(a)

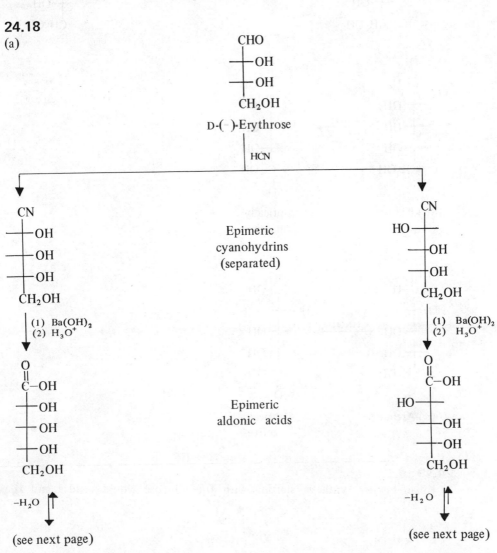

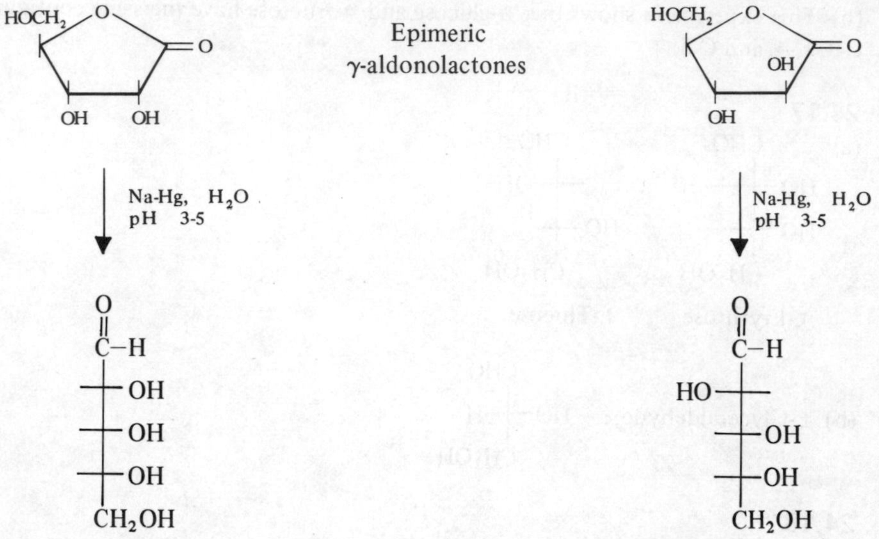

HOCH₂ ... O ... =O ... OH OH   Epimeric γ-aldonolactones   HOCH₂ ... O ... =O ... OH ... OH

Na-Hg, H₂O
pH 3-5

Na-Hg, H₂O
pH 3-5

$$\begin{array}{c} \text{O} \\ \| \\ \text{C-H} \\ \text{---OH} \\ \text{---OH} \\ \text{---OH} \\ \text{CH}_2\text{OH} \end{array}$$

$$\begin{array}{c} \text{O} \\ \| \\ \text{C-H} \\ \text{HO---} \\ \text{---OH} \\ \text{---OH} \\ \text{CH}_2\text{OH} \end{array}$$

(b)

$$\begin{array}{c} \text{O} \\ \| \\ \text{C-H} \\ \text{---OH} \\ \text{---OH} \\ \text{---OH} \\ \text{CH}_2\text{OH} \end{array} \xrightarrow{\text{HNO}_3} \begin{array}{c} \text{O} \\ \| \\ \text{C-OH} \\ \text{---OH} \\ \text{---OH} \\ \text{---OH} \\ \text{C-OH} \\ \| \\ \text{O} \end{array}$$

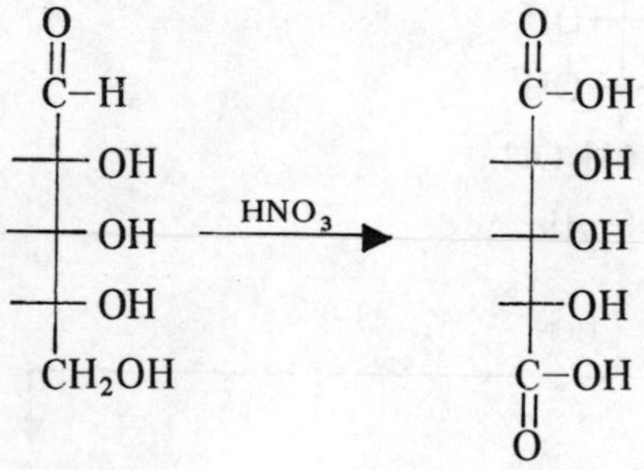

D-(−)-Ribose   optically inactive

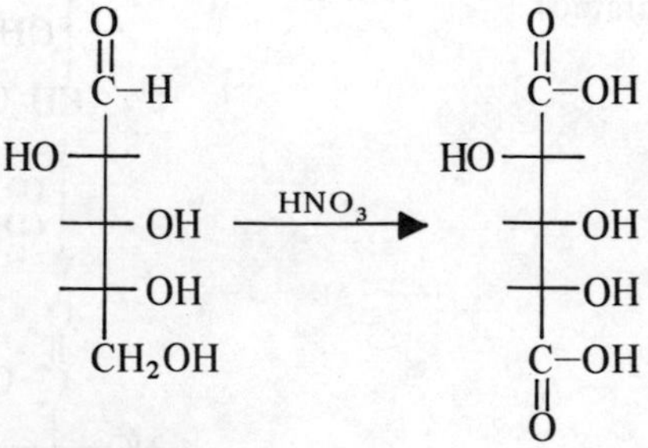

$$\begin{array}{c} \text{O} \\ \| \\ \text{C-H} \\ \text{HO---} \\ \text{---OH} \\ \text{---OH} \\ \text{CH}_2\text{OH} \end{array} \xrightarrow{\text{HNO}_3} \begin{array}{c} \text{O} \\ \| \\ \text{C-OH} \\ \text{HO---} \\ \text{---OH} \\ \text{---OH} \\ \text{C-OH} \\ \| \\ \text{O} \end{array}$$

D-(−)-Arabinose   optically active

24.19

A Kiliani-Fischer synthesis starting with D-(−)-threose would yield **I** and **II**.

$$
\begin{array}{cc}
\text{COOH} & \text{COOH} \\
\quad\;\;\vert\text{—OH} & \text{HO—}\vert \\
\text{HO—}\vert & \text{HO—}\vert \\
\quad\;\;\vert\text{—OH} & \quad\;\;\vert\text{—OH} \\
\text{CH}_2\text{OH} & \text{CH}_2\text{OH} \\
\mathbf{I} & \mathbf{II}
\end{array}
$$

(D)-(+)-Xylose)   (D-(−)-Lyxose)

I must be D-(+)-xylose because when oxidized by nitric acid, it yields an optically inactive aldaric acid:

$$
\mathbf{I} \xrightarrow{\text{HNO}_3}
\begin{array}{c}
\text{COOH} \\
\vert\text{—OH} \\
\text{HO—}\vert \\
\vert\text{—OH} \\
\text{COOH}
\end{array}
$$

optically
inactive

II must be D-(−)-lyxose because when oxidized by nitric acid it yields an optically active aldaric acid:

$$
\mathbf{II} \xrightarrow{\text{HNO}_3}
\begin{array}{c}
\text{COOH} \\
\text{HO—}\vert \\
\text{HO—}\vert \\
\vert\text{—OH} \\
\text{COOH}
\end{array}
$$

optically
active

24.20

$$
\begin{array}{cccc}
\text{CHO} & \text{CHO} & \text{CHO} & \text{CHO} \\
\text{HO—}\vert & \vert\text{—OH} & \text{HO—}\vert & \vert\text{—OH} \\
\text{HO—}\vert & \text{HO—}\vert & \quad\;\;\vert\text{—OH} & \vert\text{—OH} \\
\text{HO—}\vert & \text{HO—}\vert & \text{HO—}\vert & \text{HO—}\vert \\
\text{CH}_2\text{OH} & \text{CH}_2\text{OH} & \text{CH}_2\text{OH} & \text{CH}_2\text{OH} \\
\text{L-(+)-Ribose} & \text{L-(+)-Arabinose} & \text{L-(−)-Xylose} & \text{L-(+)-Lyxose}
\end{array}
$$

24.21

Since D-(+)-galactose yields an optically inactive aldaric acid it must have either structure

III or structure **IV** below.

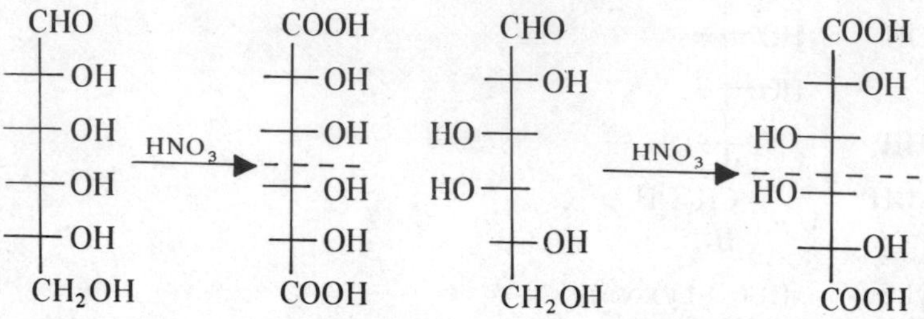

| III | optically inactive | IV | optically inactive |

A Ruff degradation beginning with **III** would yield D-(−)-ribose

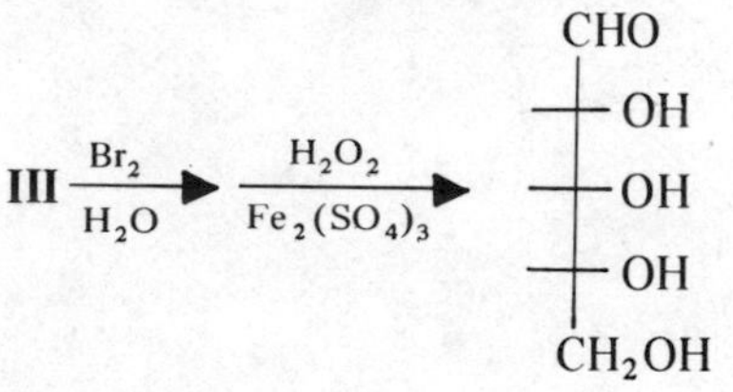

D-(−)-Ribose

A Ruff degradation beginning with **IV** would yield D-(−)-lyxose; thus D-(+)-galactose must have structure **IV**.

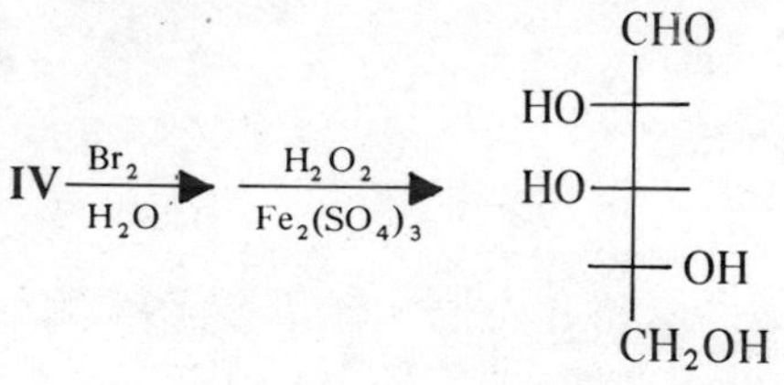

D-(−)-Lyxose

24.22

D-(+)-glucose, as shown below.

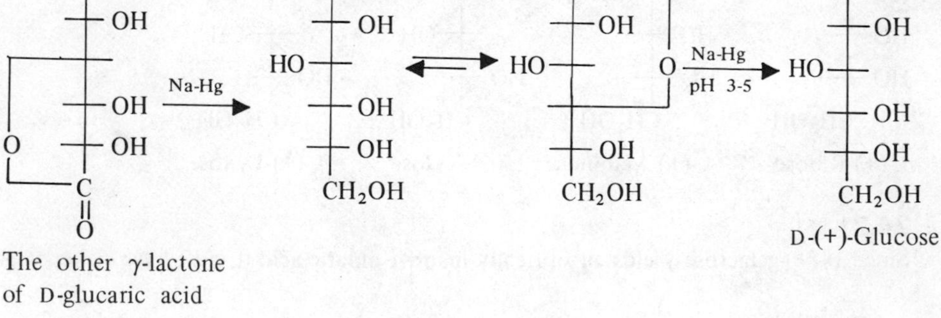

D-(+)-Glucose

The other γ-lactone of D-glucaric acid

24.23

If the methyl glucoside had been a furanoside, hydrolysis of the methylation product would have given:

```
        CHO
    ——|—— OCH₃
CH₃O——|——
    ——|—— OH
    ——|—— OCH₃
        CH₂OH
```

And, oxidation would have given:

```
        CHO                          COOH          COOH          COOH
    ——|—— OCH₃                   ——|—— OCH₃    ——|—— OCH₃ +  ——|—— OCH₃
CH₃O——|——        HNO₃    CH₃O ——|——        +   CH₂OCH₃       COOH
    — —|— —                       COOH
    — —|— — OCH₃            A dimethoxysuc-   A dimethoxy-  Methoxyma-
        CH₂OCH₃             cinic acid        propanoic     lonic acid
                                               acid
```

$$\downarrow -CO_2$$

```
        COOH
        |
        CH₂OCH₃
```

Methoxyacetic
acid

24.24

(a)
```
CHO
|
CHOH
|
CHOH
|
CHOH
|
CH₂OH
```

(b)
```
CH₂OH
|
C=O
|
CHOH
|
CHOH
|
CHOH
|
CH₂OH
```

(c)
```
    CHO
    |
    (CHOH)ₙ
    |
HO——|——H
    |
    CH₂OH
```
or
```
    CH₂OH
    |
    C=O
    |
    (CHOH)ₙ
    |
HO——|——H
    |
    CH₂OH
```

(d)
```
⌐CHOR        ⌐
 (CHOH)ₙ    O
 CH————————⌐
 |
 CH₂OH
```

(e)
```
COOH
|
(CHOH)ₙ
|
CH₂OH
```

(f)
```
COOH
|
(CHOH)ₙ
|
COOH
```

(g)

$$\begin{array}{c} \text{O} \\ \| \\ \text{C}\!-\!\!\!\rule{0pt}{0pt} \\ | \\ (\text{CHOH})_n \quad \text{O} \\ | \\ \text{CH}\!-\!\!\!\rule{0pt}{0pt} \\ | \\ \text{CH}_2\text{OH} \end{array}$$

(h)

$$\begin{array}{c} \text{OH} \\ | \\ \text{CH}\!-\!\!\!\rule{0pt}{0pt} \\ | \\ \text{CHOH} \\ | \\ \text{CHOH} \quad \text{O} \\ | \\ \text{CHOH} \\ | \\ \text{CH}\!-\!\!\!\rule{0pt}{0pt} \\ | \\ \text{CH}_2\text{OH} \end{array}$$

or

$$\begin{array}{c} \text{CH}_2\text{OH} \\ | \\ \text{CH}\!\!-\!\!\text{O} \\ / \qquad \backslash \\ \text{CHOH} \qquad \text{CHOH} \\ \backslash \qquad\quad / \\ \text{CHOH}\!-\!\text{CHOH} \end{array}$$

(i)

$$\begin{array}{c} \text{OH} \\ | \\ \text{CH}\!-\!\!\!\rule{0pt}{0pt} \\ | \\ \text{CHOH} \\ | \qquad \text{O} \\ \text{CHOH} \\ | \\ \text{CH}\!-\!\!\!\rule{0pt}{0pt} \\ | \\ \text{CHOH} \\ | \\ \text{CH}_2\text{OH} \end{array}$$

or

$$\begin{array}{c} \text{CH}_2\text{OH} \\ | \\ \text{CHOH} \\ | \qquad\;\; \text{O} \\ \text{CH} \qquad\;\; \text{CHOH} \\ \backslash \qquad\quad / \\ \text{CHOH}\!-\!\text{CHOH} \end{array}$$

(j) Any sugar that has a free aldehyde or ketone group or one that exists as a cyclic hemiacetal or hemiketal. Examples are:

$$\begin{array}{c} \text{CHO} \\ | \\ (\text{CHOH})_n \\ | \\ \text{CHOH} \\ | \\ \text{CH}_2\text{OH} \end{array} \rightleftarrows \begin{array}{c} \text{OH} \\ | \\ \text{CH}\!-\!\!\!\rule{0pt}{0pt} \\ | \\ (\text{CHOH})_n \quad \text{O} \\ | \\ \text{CH}\!-\!\!\!\rule{0pt}{0pt} \\ | \\ \text{CH}_2\text{OH} \end{array}$$

or

$$\begin{array}{c} \text{CH}_2\text{OH} \\ | \\ \text{C}\!=\!\text{O} \\ | \\ (\text{CHOH})_n \\ | \\ \text{CHOH} \\ | \\ \text{CH}_2\text{OH} \end{array} \rightleftarrows \begin{array}{c} \text{CH}_2\text{OH} \\ | \\ \text{C}\!-\!\text{OH} \\ | \\ (\text{CHOH})_n \quad \text{O} \\ | \\ \text{CH}\!-\!\!\!\rule{0pt}{0pt} \\ | \\ \text{CH}_2\text{OH} \end{array}$$

(k)

$$\begin{array}{c} \text{CH}_2\text{OH} \\ | \\ \text{CH}\!\!-\!\!\text{O} \\ / \qquad \backslash \\ \text{CHOH} \qquad \text{CHOR} \\ \backslash \qquad\quad / \\ \text{CHOH}\!-\!\text{CHOH} \end{array}$$

(l)

$$\begin{array}{c} \text{CH}_2\text{OH} \\ | \\ \text{CHOH} \\ | \qquad\; \text{O} \\ \text{CH} \qquad\;\; \text{CHOR} \\ \backslash \qquad\quad / \\ \text{CHOH}\!-\!\text{CHOH} \end{array}$$

(m) Any two aldoses that differ only in configuration at C-2. (See also page 932 for a broader definition.) D-Erythrose and D-threose are examples.

$$\begin{array}{c} \text{CHO} \\ | \\ \!\!\!-\!\!\!-\text{OH} \\ | \\ \!\!\!-\!\!\!-\text{OH} \\ | \\ \text{CH}_2\text{OH} \end{array} \qquad \begin{array}{c} \text{CHO} \\ | \\ \text{HO}\!-\!\!\!- \\ | \\ \!\!\!-\!\!\!-\text{OH} \\ | \\ \text{CH}_2\text{OH} \end{array}$$

D-Erythrose   D-Threose

(n) Cyclic sugars that differ only in the configuration of C-1. Examples are:

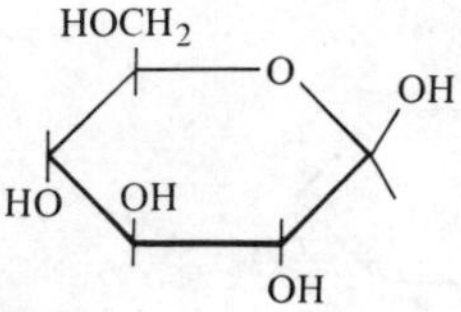

and

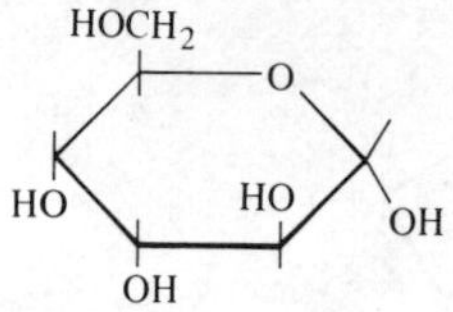

(o) CH=NNHC$_6$H$_5$
　　C=NNHC$_6$H$_5$
　　(CHOH)$_n$
　　CH$_2$OH

(p) Maltose is an example:

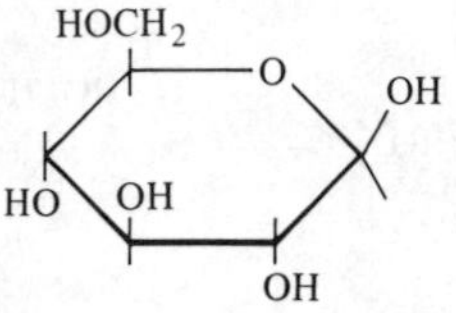

and

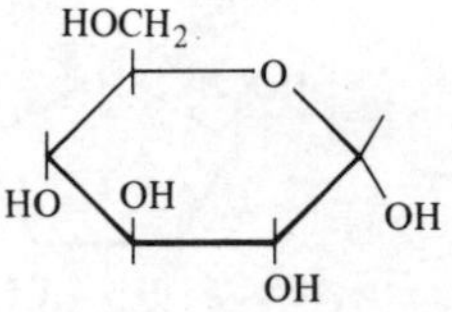

(q) Amylose is an example:

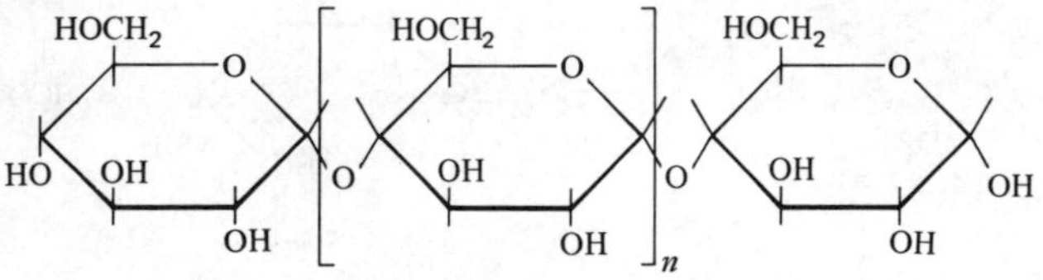

(r) Any sugar in which all potential carbonyl groups are present as acetals or ketals (i.e., as glycosides). Sucrose (p. 939) is an example of a nonreducing disaccharide; the methyl-D-glucopyranosides (p. 922) are examples of nonreducing monosaccharides.

24.25

(a)

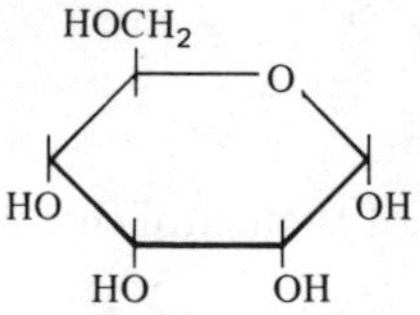

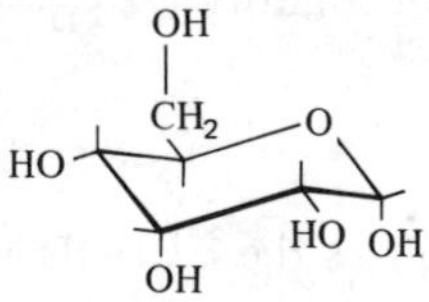

(b)

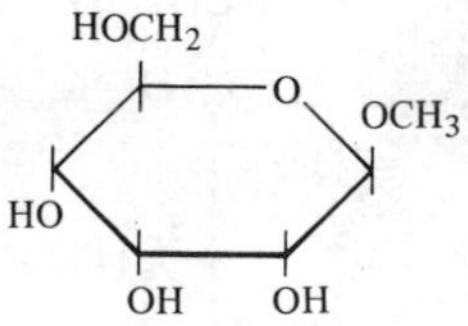

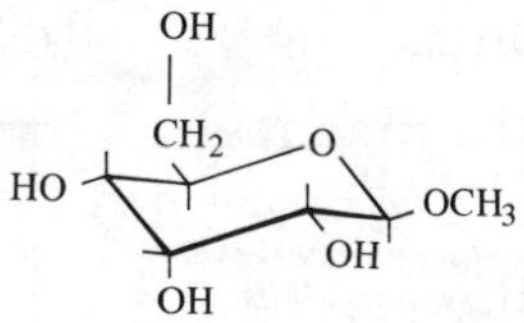

(c)

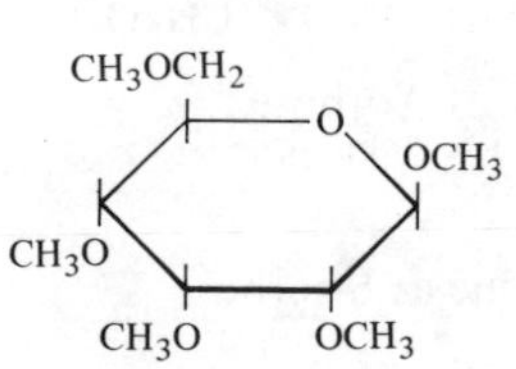

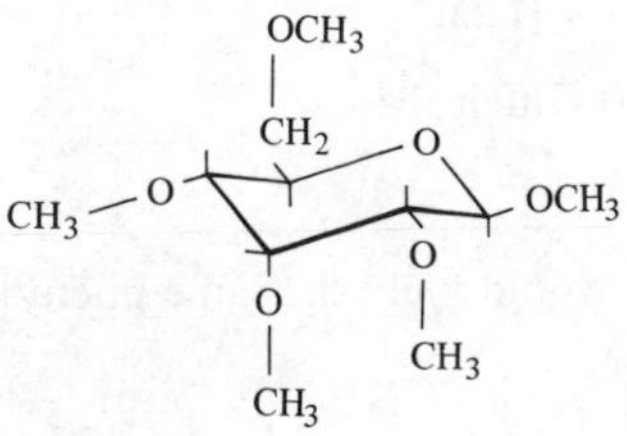

24.26

α–D–ribofuranoside

α–D–ribpyranoside

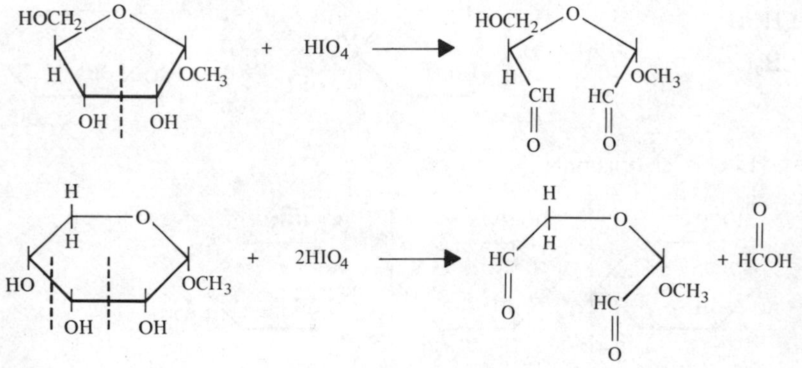

A methyl ribofuranoside would consume only one mole of HIO_4; a methyl ribopyranoside would consume two moles of HIO_4 and would also produce one mole of formic acid.

24.27
One anomer of D-mannose is dextrorotatory ($[\alpha]_D = +29.3°$), the other is levorotatory ($[\alpha]_D = -17.0°$).

24.28
The microorganism selectively oxidizes the $-CHOH$ group of D-glucitol that corresponds to C-5 of D-glucose.

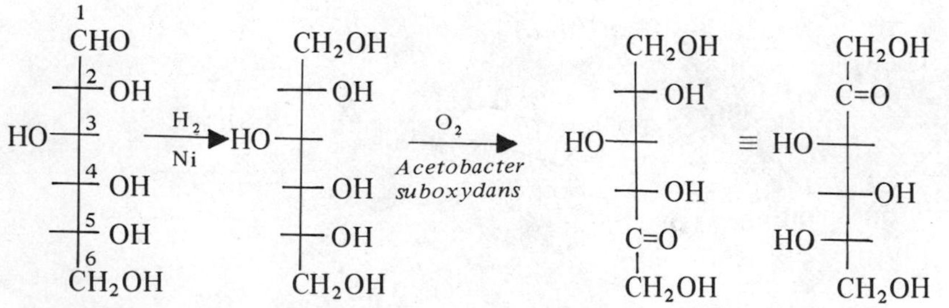

D-Glucose   D-Glucitol   L-Sorbose

24.29
L-Gulose and L-idose would yield the same phenylosazone as L-sorbose.

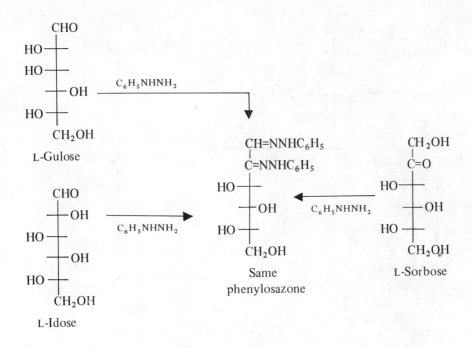

L-Gulose

L-Idose

Same phenylosazone

L-Sorbose

24.30

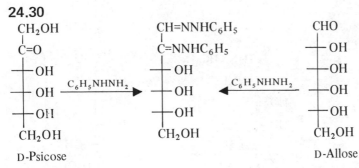

D-Psicose

D-Allose

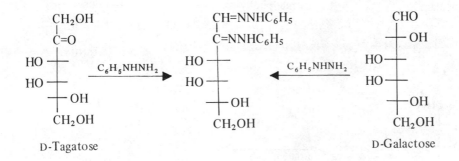

D-Tagatose

D-Galactose

24.31

A is D-altrose, **B** is D-talose, and **C** is D-galactose:

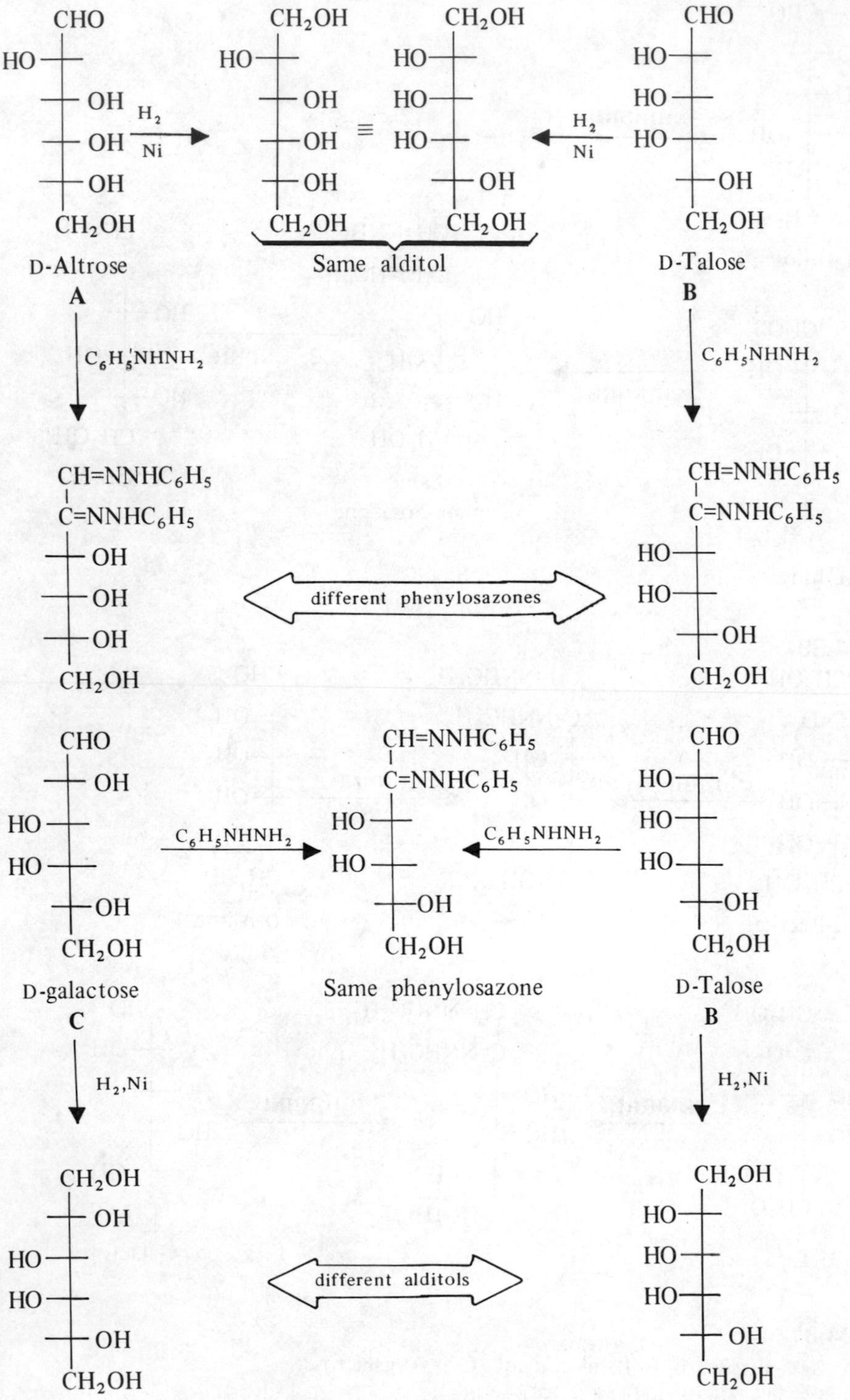

$$
\begin{array}{ccc}
\text{CHO} & & \text{CHO} \\
\text{HO}\!-\!\!\vdash & & \text{HO}\!-\!\!\vdash \\
\vdash\!-\!\text{OH} \quad \underset{\text{Ni}}{\overset{H_2}{\longrightarrow}} & & \underset{\text{Ni}}{\overset{H_2}{\longleftarrow}} \quad \text{HO}\!-\!\!\vdash \\
\end{array}
$$

D-Altrose (CHO / HO— / —OH / —OH / —OH / CH$_2$OH) **A**

→ H$_2$/Ni → Same alditol (CH$_2$OH / HO— / —OH / —OH / —OH / CH$_2$OH) ≡ (CH$_2$OH / HO— / HO— / —OH / CH$_2$OH)

← H$_2$/Ni ← D-Talose (CHO / HO— / HO— / HO— / —OH / CH$_2$OH) **B**

A → C$_6$H$_5$NHNH$_2$ →

CH=NNHC$_6$H$_5$
C=NNHC$_6$H$_5$
—OH
—OH
—OH
CH$_2$OH

⟷ different phenylosazones ⟷

B → C$_6$H$_5$NHNH$_2$ →

CH=NNHC$_6$H$_5$
C=NNHC$_6$H$_5$
HO—
HO—
—OH
CH$_2$OH

D-galactose (CHO / —OH / HO— / HO— / —OH / CH$_2$OH) **C**

→ C$_6$H$_5$NHNH$_2$ → Same phenylosazone (CH=NNHC$_6$H$_5$ / C=NNHC$_6$H$_5$ / HO— / HO— / —OH / CH$_2$OH) ← C$_6$H$_5$NHNH$_2$ ←

D-Talose (CHO / HO— / HO— / HO— / —OH / CH$_2$OH) **B**

C → H$_2$,Ni →

CH$_2$OH
—OH
HO—
HO—
—OH
CH$_2$OH

⟷ different alditols ⟷

B → H$_2$,Ni →

CH$_2$OH
HO—
HO—
HO—
—OH
CH$_2$OH

(Note: If we had designated D-talose as **A**, and D-altrose as **B**, then **C** is D-allose)

24.32

```
    CHO                    COOH                          CHO
  ┌── OH                 ┌── OH                        HO ──┐
HO ──┤         Br₂     HO ──┤          pyridine        HO ──┤
      │         ──→           │        ──────────→           │
  ┌── OH         H₂O    ┌── OH        (epimerization)      ┌── OH
  ┌── OH                 ┌── OH                           ┌── OH
   CH₂OH                  CH₂OH                           CH₂OH
```

D-glucose

```
                              O
                              ‖
                              C ──┐
                          HO ──┤    │
               -H₂O      HO ──┤    O      Na-Hg        HO ──┐
               ──────→        │              ──────→   HO ──┤
                                          pH  3-5           ┌── OH
                          ┌── OH                           ┌── OH
                           CH₂OH                            CH₂OH
```

CHO

D-mannose

24.33

The conformation of D-idopyranose with four equatorial –OH groups and an axial
–CH₂OH group is more stable than the one with four axial –OH groups and an equatorial
–CH₂OH group.

more stable	less stable
4 equatorial –OH groups	4 axial –OH groups
1 axial –CH₂OH	1 equatorial –CH₂OH

24.34

(a) The anhydro sugar is formed when the axial –CH₂OH group reacts with C-1 to
form a cyclic acetal.

β–D–altropyranose   anhydro sugar

Because the anhydro sugar is an acetal (i.e., an internal glycoside), it is a non-reducing
sugar.

Methylation followed by acid hydrolysis converts the anhydro sugar to 2,3,4-tri-O-methyl-D-altrose:

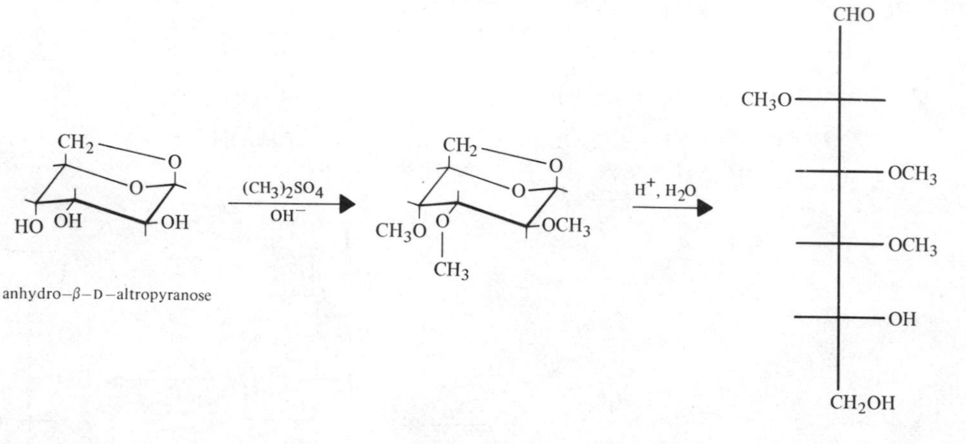

anhydro–β–D–altropyranose

$\xrightarrow{(CH_3)_2SO_4 / OH^-}$

$\xrightarrow{H^+,\ H_2O}$

2, 3, 4–tri–O–methyl–D–altrose

(b) Formation of an anhydro sugar requires that the monosaccharide adopt a chair conformation with the —CH$_2$OH group axial. With β-D-altropyranose this requires that two —OH groups be axial as well. With β-D-glucopyranose, however, it requires that all four —OH groups become axial, and thus that the molecule adopt a very unstable conformation:

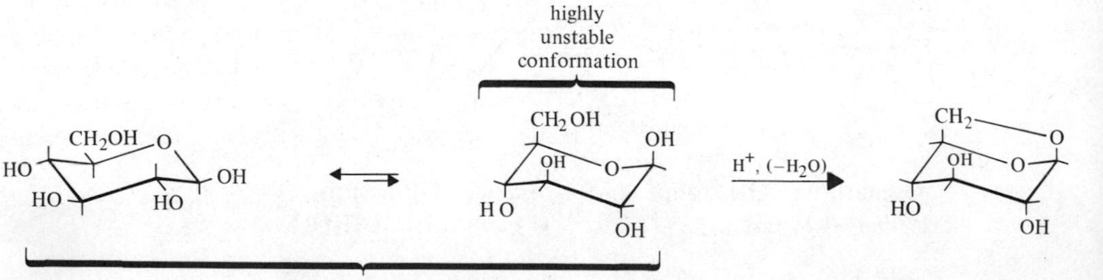

highly unstable conformation

β–D–glucopyranose

$\xrightarrow{H^+,\ (-H_2O)}$

anhydro–β–D–glucopyranose

24.35

The initial step in mutarotation—ring opening of the cyclic hemiacetal—requires that an acid donate a proton to the ring oxygen and that a base remove a proton from the anomeric hydroxyl. 2-Hydroxypyridine has, in close proximity, acidic and basic groups that can accomplish both of these tasks.

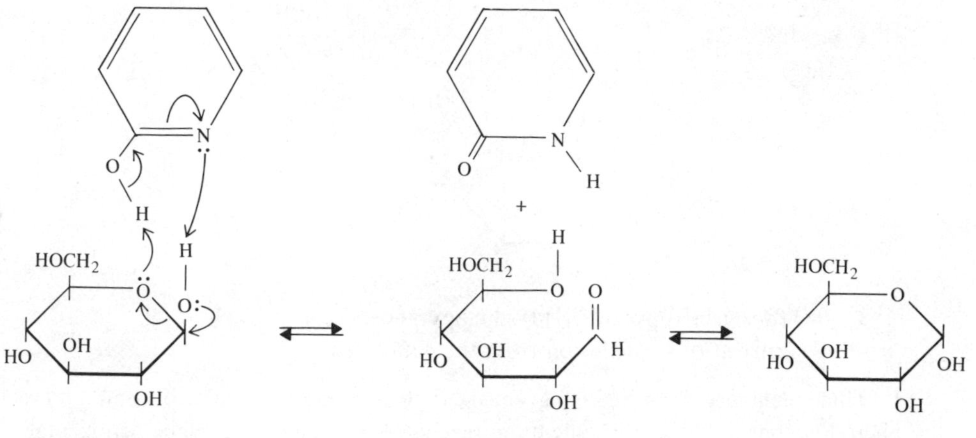

β–D–glucopyranose   aldehyde form   α–D–glucopyranose

24.36

1. The molecular formula and the results of acid hydrolysis show that lactose is a disaccharide composed of D-glucose and D-galactose. The fact that lactose is hydrolyzed by a *β-galactosidase* indicates that galactose is present as a glycoside and that the glycosidic linkage is *beta* to the galactose ring.

2. That lactose is a reducing sugar, forms a phenylosazone, and undergoes mutarotation indicates that one ring (presumably that of D-glucose) is present as a hemiacetal and thus is capable of existing to a limited extent as an aldehyde.

3. This experiment confirms that the D-glucose unit is present as a cyclic hemiacetal and that the D-galactose unit is present as a cyclic glycoside.

4. That 2,3,4,6-tetra-O-methyl-D-galactose is obtained in this experiment indicates (by virture of the free —OH at C-5) that the galactose ring of lactose is present as a pyranoside. That the methylated gluconolactone obtained from this experiment has a free —OH group at C-4 indicates that the C-4 oxygen of the glucose unit is connected in a glycosidic linkage to the galactose unit.

Now only the size of the glucose ring remains in question and the answer to this is provided by experiment 5.

5. That methylation of lactose and subsequent hydrolysis gives 2,3,6-tri-O-methyl-D-glucose—that it gives a methylated glucose derivative with a free —OH at C-4 and C-5—demonstrates that the glucose ring is present as a pyranose. (We know already that the oxygen at C-4 is connected in a glycosidic linkage to the galactose unit; thus a free —OH at C-5 indicates that the C-5 oxygen is a part of the hemiacetal group of the glucose unit and that the ring is six-membered.)

24.37

Melibiose has the following structure:

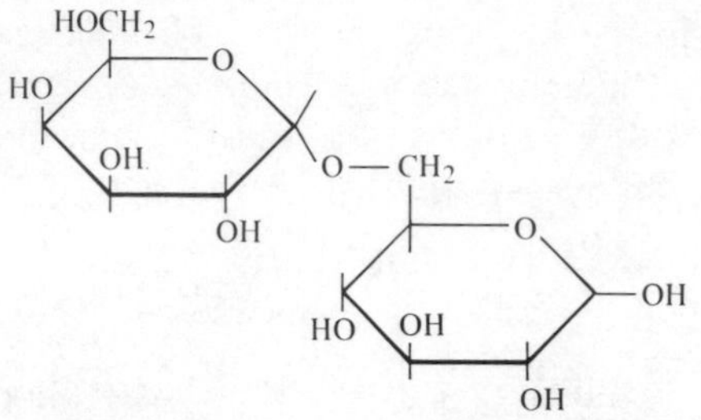

6-O-(α-D-galactopyranosyl)-D-glucopyranose
We arrive at this conclusion from the data given:

1. That melibiose is a reducing sugar, that it undergoes mutarotation and forms a phenylosazone indicates that one monosaccharide is present as a cyclic hemiacetal.

2. That acid hydrolysis gives D-galactose and D-glucose indicates that melibiose is a disaccharide composed of one D-galactose unit and one D-glucose unit. That melibiose is hydrolyzed by an α-galactosidase suggests that melibiose is an α-D-galactosyl-D-glucose.

3. Oxidation of melibiose to melibionic acid and subsequent hydrolysis to give D-galactose and D-gluconic acid confirms that the glucose unit is present as a cyclic hemiacetal and that the galactose unit is present as a glycoside. (Had the reverse been true, this experiment would have yielded D-glucose and D-galactonic acid.)

Methylation and hydrolysis of melibionic acid produces 2,3,4,6-tetra-O-methyl-D-galactose and 2,3,4,5-tetra-O-methyl-D-gluconic acid. Formation of the first product—a galactose derivative with a free —OH at C-5—demonstrates that the galactose ring is six-membered; formation of the second product—a gluconic acid derivative with a free —OH at C-6—demonstrates that the oxygen at C-6 of the glucose unit is joined in a glycosidic linkage to the galactose unit.

4. That methylation and hydrolysis of melibiose gives a glucose derivative (2,3,4-tri-O-methyl-D-glucose) with free —OH groups at C-5 and C-6 shows that the glucose ring is also six-membered. Melibiose is, therefore, 6-O-(α-D-galactopyranosyl)-D-glucopyranose.

24.38
Trehalose has the following structure:

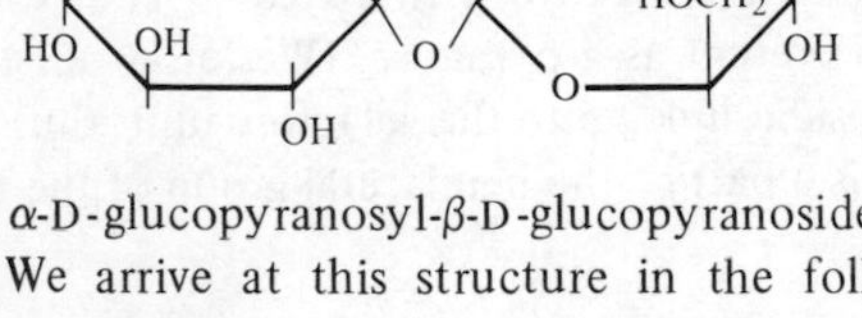

α-D-glucopyranosyl-β-D-glucopyranoside
We arrive at this structure in the following way:

1. Acid hydrolysis shows that trehalose is a disaccharide consisting only of D-glucose units.

2. Hydrolysis by α-glucosidases and not by β-glucosidases shows that the glycosidic linkages are *alpha*.

3. That trehalose is a non-reducing sugar, that it does not form phenylosazone, and that it does not react with bromine water indicate that no hemiacetal groups are present. This means that C-1 of one glucose unit and C-1 of the other must be joined in a glycosidic linkage. Fact 2 (above) indicates that this linkage is *alpha* to each ring.

4. That methylation of trehalose followed by hydrolysis yields only 2,3,4,6-tetra-O-methyl-D-glucose demonstrates that both rings are six-membered.

25

AMINO ACIDS AND PROTEINS

25.1

(a) $H_3\overset{+}{N}CH_2CH_2CH_2CH_2\underset{\underset{\displaystyle {}^+NH_3}{|}}{C}HCOOH$

(b) $H_2NCH_2CH_2CH_2CH_2\underset{\underset{\displaystyle NH_2}{|}}{C}HCOO^-$

(c) The α-amino group is less basic than the ϵ-amino group because of the proximity of the electron-withdrawing carboxylate group.

25.2

(a) $HOOCCH_2CH_2\underset{\underset{\displaystyle {}^+NH_3}{|}}{C}HCOOH$

(b) $^-OOCCH_2CH_2\underset{\underset{\displaystyle NH_2}{|}}{C}HCOO^-$

(c) $HOOCCH_2CH_2\underset{\underset{\displaystyle {}^+NH_3}{|}}{C}HCOO^-$ predominates at the isoelectric point rather than $^-OOCCH_2-$

$CH_2\underset{\underset{\displaystyle {}^+NH_3}{|}}{C}HCOOH$ because of the acid-strengthening inductive effect of the α-amino group.

(d) Since glutamic acid is a dicarboxylic acid, acid must be added (i.e., the pH must be made lower) to suppress the ionization of the second carboxyl group and thus achieve the isoelectric point. Glutamine, with only one carboxyl group, is similar to glycine or phenylalanine and has its isoelectric point at a higher pH.

25.3

(a) $C_6H_5CONHCH(CO_2C_2H_5)_2 \xrightarrow[\text{C}_6\text{H}_5\text{CH}_2\text{Br}]{\text{NaOCH}_2\text{CH}_3}$

$C_6H_5CONH\underset{\underset{\displaystyle C_6H_5}{\underset{\displaystyle |}{CH_2}}}{C}(CO_2C_2H_5)_2 \xrightarrow[\text{heat}]{\text{HBr}} \left[H_3\overset{+}{N}-\underset{\underset{\displaystyle C_6H_5}{\underset{\displaystyle |}{CH_2}}}{\overset{\overset{\displaystyle COOH}{|}}{C}}-COO^- \right] \xrightarrow{-CO_2} C_6H_5CH_2\underset{\underset{\displaystyle {}^+NH_3}{|}}{C}HCOO^-$

Phenylalanine

296

(b) $C_6H_5CONHCH(CO_2C_2H_5)_2$ $\xrightarrow[BrCH_2CO_2C_2H_5]{NaOCH_2CH_3}$

$C_6H_5CONHC(CO_2C_2H_5)_2$ $\xrightarrow{HBr}$

$\begin{array}{c} CH_2 \\ | \\ CO_2C_2H_5 \end{array}$ (attached)

$$\left[\begin{array}{c} COOH \\ | \\ H_3\overset{+}{N}-C-COO^- \\ | \\ CH_2 \\ | \\ COOH \end{array} \right]$$

$\xrightarrow{-CO_2}$ $HOOCCH_2CHCOO^-$

$\qquad\qquad\qquad$ $\overset{|}{\underset{+NH_3}{}}$

Aspartic acid

(c) $C_6H_5CONHCH(CO_2C_2H_5)_2$ $\xrightarrow[(CH_3)_2CHBr]{NaOCH_2CH_3}$

$C_6H_5CONHC(CO_2C_2H_5)_2$ $\xrightarrow[heat]{HBr}$

$\begin{array}{c} CHCH_3 \\ | \\ CH_3 \end{array}$ (attached)

$$\left[\begin{array}{c} COOH \\ | \\ H_3\overset{+}{N}-C-COO^- \\ | \\ CHCH_3 \\ | \\ CH_3 \end{array} \right]$$

$\xrightarrow{-CO_2}$ $CH_3CH\ CHCOO^-$

$\qquad\qquad\qquad$ $\begin{array}{cc} | & | \\ CH_3 & \overset{+}{N}H_3 \end{array}$

Valine

25.4

(a) NK + $BrCH(CO_2C_2H_5)_2$ $\longrightarrow$

$N-CH(CO_2C_2H_5)_2$ $\xrightarrow[(CH_3)_2CHCH_2Br]{NaOCH_2CH_3}$

$\begin{array}{c} CO_2C_2H_5 \\ | \\ N-C-CH_2CH(CH_3)_2 \\ | \\ CO_2C_2H_5 \end{array}$ $\xrightarrow[heat]{NaOH}$

$\begin{array}{c} COO^- \\ \\ C-NHCCH_2CH(CH_3)_2 \\ \| \qquad | \\ O \qquad COO^- \end{array}$ with COO^-

$\xrightarrow[Heat]{HCl}$ $(CH_3)_2CHCH_2CHCOO^-$ + CO_2 +

$\qquad\qquad\qquad\qquad$ $\overset{|}{\underset{+NH_3}{}}$

Leucine

$\begin{array}{c} COOH \\ \\ COOH \end{array}$

(b)

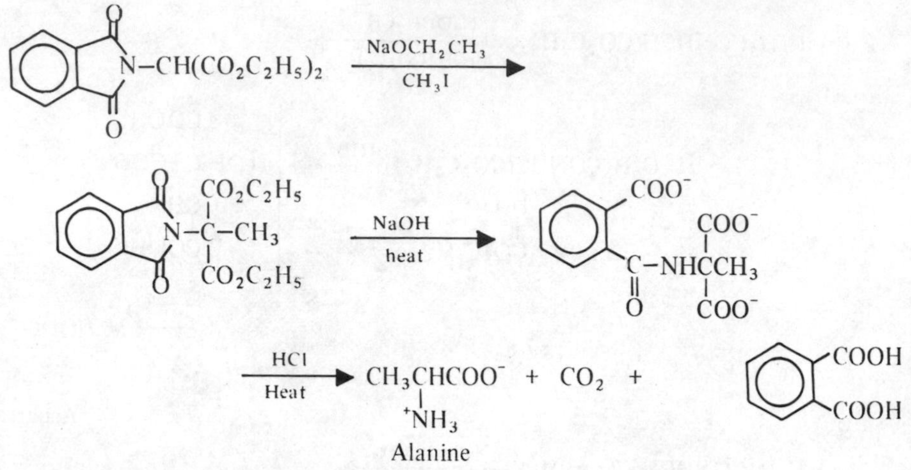

(phthalimide)N—CH(CO₂C₂H₅)₂ $\xrightarrow[\text{CH}_3\text{I}]{\text{NaOCH}_2\text{CH}_3}$

(phthalimide)N—C(CO₂C₂H₅)₂—CH₃ $\xrightarrow[\text{heat}]{\text{NaOH}}$

(benzene ring)—COO⁻ / C(=O)—NHC(CH₃)(COO⁻)(COO⁻)

$\xrightarrow[\text{Heat}]{\text{HCl}}$ CH₃CHCOO⁻ ($^{+}$NH₃) + CO₂ + (benzene ring)(COOH)(COOH)

Alanine

(c)

(phthalimide)NCH(CO₂C₂H₅)₂ $\xrightarrow[\text{C}_6\text{H}_5\text{CH}_2\text{Br}]{\text{NaOCH}_2\text{CH}_3}$

(phthalimide)N—C(CO₂C₂H₅)₂—CH₂C₆H₅ $\xrightarrow[\text{heat}]{\text{NaOH}}$

(benzene ring)—COO⁻ / C(=O)—NHC(CH₂C₆H₅)(COO⁻)(COO⁻)

$\xrightarrow[\text{heat}]{\text{HCl}}$ C₆H₅CH₂CHCOO⁻ ($^{+}$NH₃) + CO₂ + (benzene ring)(COOH)(COOH)

Phenylalanine

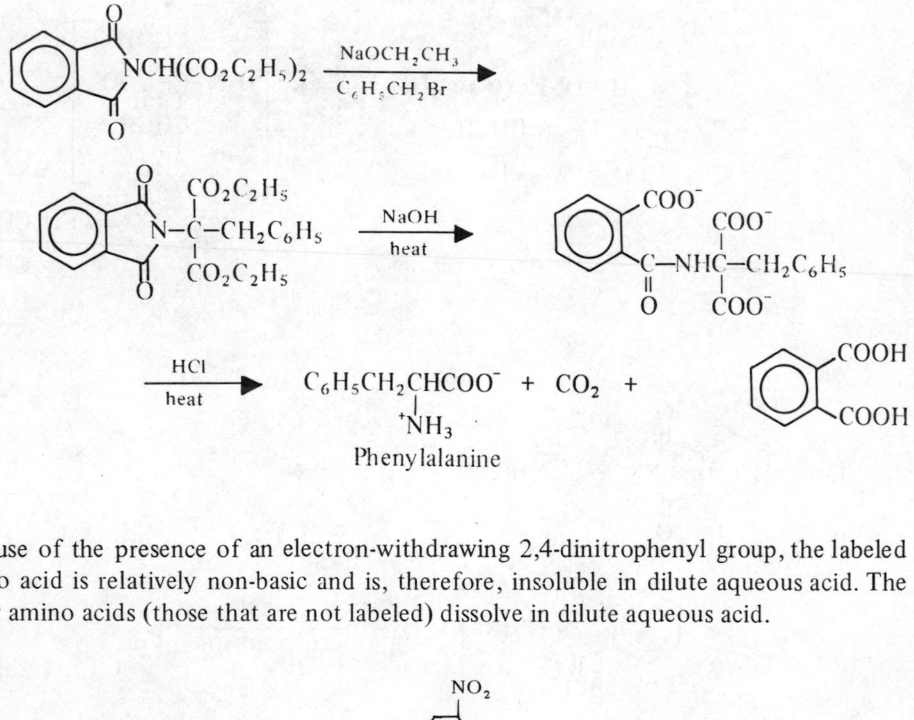

25.5

Because of the presence of an electron-withdrawing 2,4-dinitrophenyl group, the labeled amino acid is relatively non-basic and is, therefore, insoluble in dilute aqueous acid. The other amino acids (those that are not labeled) dissolve in dilute aqueous acid.

25.6

(a) H₃$^{+}$NCHCONHCHCONHCH₂COO⁻
(CHCH₃)(CH₃)
(CH₃)

Val·Ala·Gly

O_2N—(ring, NO₂)—F $\xrightarrow{\text{HCO}_3^-}$

O_2N—(ring, NO₂)—NHCHCONHCHCONHCH₂COO⁻
(CHCH₃)(CH₃)
(CH₃) $\xrightarrow[\text{heat}]{\text{H}_3\text{O}^+}$

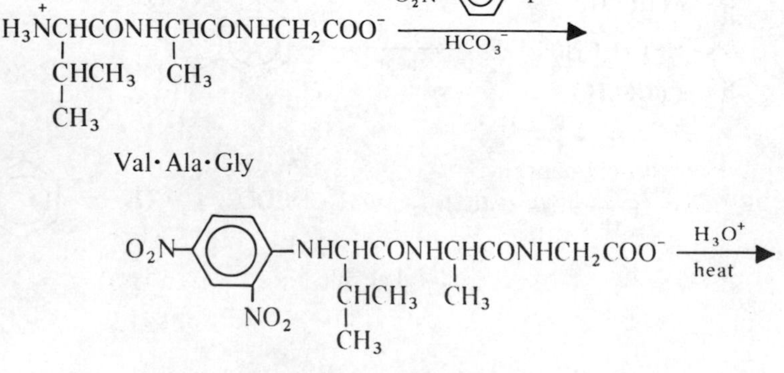

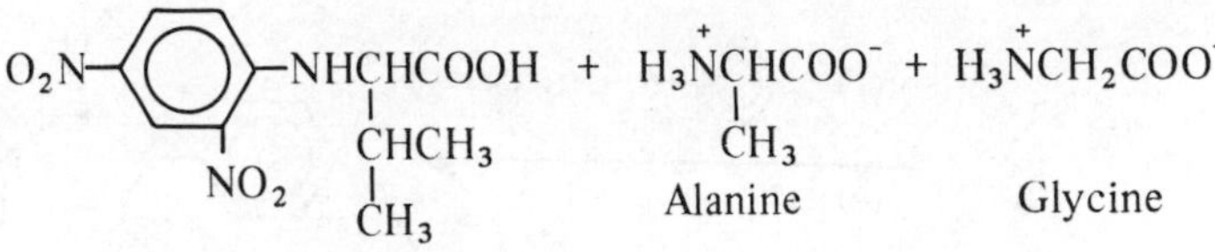

O_2N—⟨ring⟩—NHCHCOOH $+$ $H_3\overset{+}{N}CHCOO^-$ $+$ $H_3\overset{+}{N}CH_2COO^-$

(ring bears NO_2; main chain —CHCH$_3$ / CH$_3$) ; CH_3 ; Alanine ; Glycine

Labeled valine
(separate and identify)

(b) O_2N—⟨ring, NO_2⟩—NHCHCOOH $+$ O_2N—⟨ring, NO_2⟩—NHCH$_2$CH$_2$CH$_2$CH$_2$CHCOO$^-$

(left chain: CHCH$_3$ / CH$_3$) ; (right chain: $\overset{+}{N}H_3$)

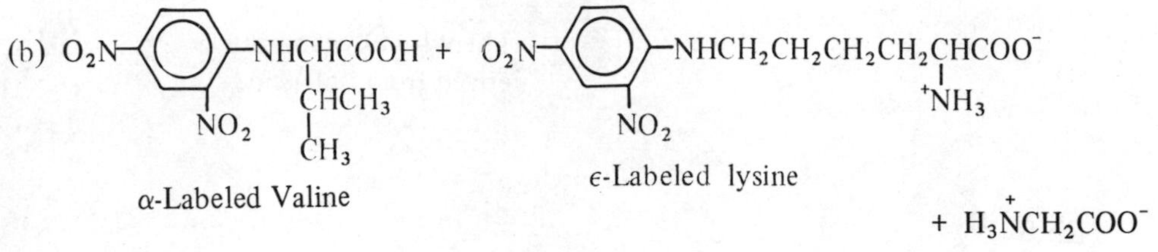

α-Labeled Valine ; ε-Labeled lysine

$+$ $H_3\overset{+}{N}CH_2COO^-$

Glycine

25.7

⟨ring⟩—N=C=S $+$ $H_2\ddot{N}$—CHCO—NHCHCO—NHCHCOO$^-$ $\xrightarrow{OH^-}$

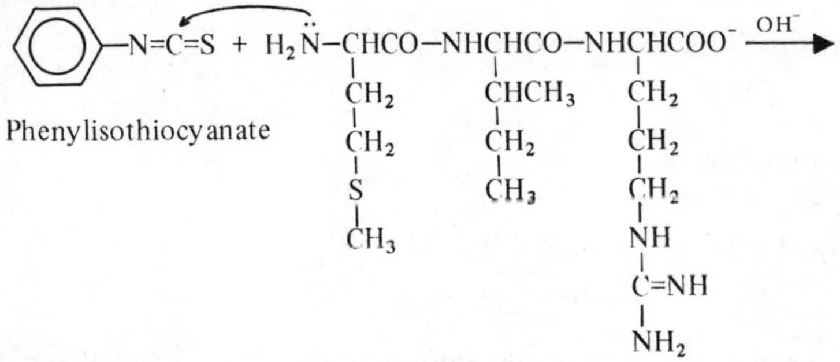

Phenylisothiocyanate

(chain 1: CH$_2$ / CH$_2$ / S / CH$_3$) (chain 2: CHCH$_3$ / CH$_2$ / CH$_3$) (chain 3: CH$_2$ / CH$_2$ / CH$_2$ / NH / C=NH / NH$_2$)

Met·Ile·Arg

⟨ring⟩—NH—$\overset{\overset{\text{S}}{\|}}{C}$—NH—CHCO—NHCHCO—NHCHCOO$^-$ $\xrightarrow{H^+}$

(chain 1: CH$_2$ / CH$_2$ / S / CH$_3$) (chain 2: CHCH$_3$ / CH$_2$ / CH$_3$) (chain 3: CH$_2$ / CH$_2$ / CH$_2$ / NH / C=NH / NH$_2$)

⟨ring⟩—N ring $\overset{\overset{\text{S}}{\|}}{C}$ NH / C—CH / O ... CH$_2$CH$_2$SCH$_3$

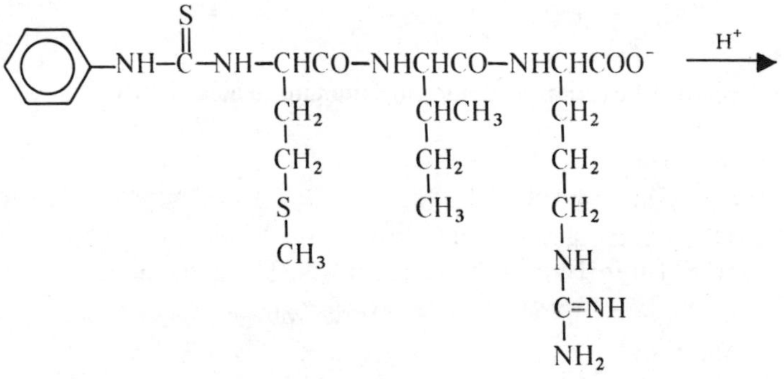

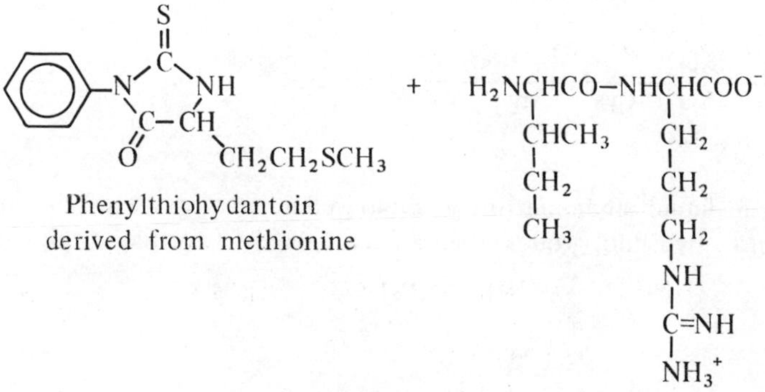

Phenylthiohydantoin
derived from methionine

$+$ H_2NCHCO—NHCHCOO$^-$

(chain: CHCH$_3$ / CH$_2$ / CH$_3$) (chain: CH$_2$ / CH$_2$ / CH$_2$ / NH / C=NH / $\overset{+}{N}H_3$)

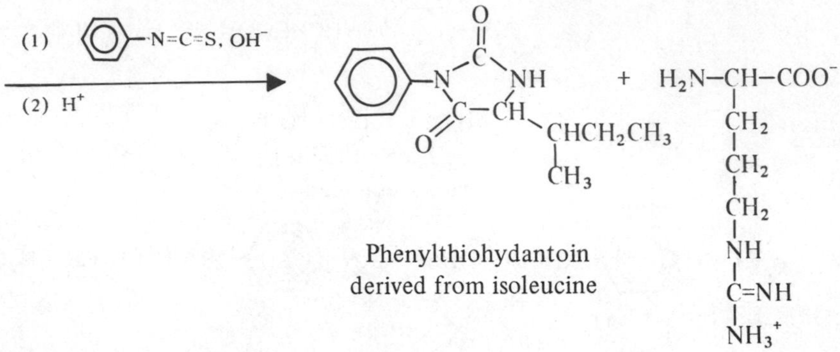

$$\text{(1)} \quad \bigcirc\!\!-N{=}C{=}S, \text{ OH}^{-} \quad \xrightarrow{\hspace{2cm}} \quad$$

$$\text{(2) H}^{+}$$

Phenylthiohydantoin
derived from isoleucine

$$+ \quad H_2N{-}CH{-}COO^{-}$$

25.8

(a) Two structures are possible with the sequence Glu·Cys·Gly. Glutamic acid may be linked to cysteine through its α-carboxyl group,

$$HOOCCH_2CH_2CHCO{-}NHCHCO{-}NHCH_2COO^{-}$$
$$\overset{|}{{}^{+}NH_3} \qquad \overset{|}{CH_2SH}$$

or through its γ-carboxyl group,

$$\overset{+}{H_3}NCHCH_2CH_2CO{-}NHCHCO{-}NHCH_2COO^{-}$$
$$\overset{|}{COO^{-}} \qquad\qquad \overset{|}{CH_2SH}$$

(b) This shows that the second structure above is correct, that in glutathione the γ-carboxyl group is linked to cysteine.

25.9

We look for points of overlap to determine the amino acid sequence in each case.

(a)
      Ser · Thr
            Thr · Hyp
Pro · Ser
Pro · Ser · Thr · Hyp

(b)
      Ala · Cys
          Cys · Arg
              Arg · Val
Leu · Ala
Leu · Ala · Cys · Arg · Val

25.10

Sodium in liquid ammonia brings about reductive cleavage of the disulfide linkage of oxytocin to two thiol groups, then air oxidizes the two thiol groups back to a disulfide linkage:

$$\begin{array}{c} CH_2 \\ | \\ S \\ | \\ S \\ | \\ CH_2 \\ | \end{array} \xrightarrow[NH_3]{Na} \begin{array}{c} CH_2 \\ | \\ SH \\ \\ SH \\ | \\ CH_2 \\ | \end{array} \xrightarrow{O_2} \begin{array}{c} CH_2 \\ | \\ S \\ | \\ S \\ | \\ CH_2 \\ | \end{array}$$

See also pages 623-624 in the text.

25.11

$$H_3\overset{+}{N}CH_2COO^- + (CH_3)_3CO\overset{O}{\overset{||}{C}}N_3 \xrightarrow[25°]{OH^-}$$

Glycine $\qquad$ tert-Butoxy-
$\qquad\qquad\qquad$ carbonyl azide

$$(CH_3)_3C-O\overset{O}{\overset{||}{C}}NHCH_2COOH \xrightarrow[(2)\ ClCO_2C_2H_5]{(1)\ (C_2H_5)_3N}$$

Boc-Gly

$$(CH_3)_3CO\overset{O}{\overset{||}{C}}NHCH_2\overset{O}{\overset{||}{C}}O\overset{O}{\overset{||}{C}}OC_2H_5 \xrightarrow[(-CO_2,\ -C_2H_5OH)]{}$$

Mixed anhydride

$$\begin{array}{c} H_3\overset{+}{N}CHCOO^- \\ | \\ CHCH_3 \\ | \\ CH_3 \\ \text{Valine} \end{array}$$

$$(CH_3)_3CO\overset{O}{\overset{||}{C}}NHCH_2\overset{O}{\overset{||}{C}}NHCHCOOH \xrightarrow[(2)\ ClCO_2C_2H_5]{(1)\ (C_2H_5)_3N}$$

Boc-Gly·Val $\qquad\quad$ CHCH_3
$\qquad\qquad\qquad\qquad\quad$ |
$\qquad\qquad\qquad\qquad\quad$ CH_3

$$(CH_3)_3CO\overset{O}{\overset{||}{C}}NHCH_2\overset{O}{\overset{||}{C}}NHCH\overset{O}{\overset{||}{C}}O\overset{O}{\overset{||}{C}}OC_2H_5 \xrightarrow{}$$

Mixed anhydride $\quad$ CHCH_3
$\qquad\qquad\qquad\qquad$ |
$\qquad\qquad\qquad\qquad$ CH_3

$$\begin{array}{c} H_3\overset{+}{N}CHCOO^- \\ | \\ CH_3 \\ \text{Alanine} \end{array}$$

$$(CH_3)_3CO\overset{O}{\overset{||}{C}}NHCH_2\overset{O}{\overset{||}{C}}NHCH\overset{O}{\overset{||}{C}}NHCHCOOH \xrightarrow[25°]{CF_3COOH \\ CH_3COOH}$$

Boc-Gly·Val·Ala $\quad$ CHCH_3 $\ $ CH_3
$\qquad\qquad\qquad\qquad\ $ |
$\qquad\qquad\qquad\qquad\ $ CH_3

(cont. on p. 302)

$$(CH_3)_2C=CH_2 + CO_2 + \overset{+}{H_3}NCH_2\overset{\overset{O}{\|}}{C}NHCH\overset{\overset{O}{\|}}{C}NHCHCOO^-$$

$$\underset{\underset{CH_3}{|}}{\underset{CHCH_3}{|}}\ \underset{CH_3}{|}$$

Gly·Val·Ala

25.12

(a) $2C_6H_5CH_2O\overset{\overset{O}{\|}}{C}Cl$ + $H_2NCH_2CH_2CH_2CH_2CHCOO^- \xrightarrow[25°]{OH^-}$

Benzyl chloro-
carbonate

Lysine

$\underset{NH_2}{|}$

$C_6H_5CH_2O\overset{\overset{O}{\|}}{C}NHCH_2CH_2CH_2CH_2CHCOOH \xrightarrow[(2)\ ClCOOC_2H_5]{(1)\ (C_2H_5)_3N}$

$\underset{NH}{|}$

$\underset{C_6H_5CH_2O\overset{}{C}=O}{|}$

$C_6H_5CH_2O\overset{\overset{O}{\|}}{C}NHCH_2CH_2CH_2CH_2CH\overset{\overset{O}{\|}}{C}O\overset{\overset{O}{\|}}{C}OC_2H_5 \xrightarrow[(-CO_2,\ -C_2H_5OH)]{\underset{CH_3\ \overset{+}{NH_3}}{\overset{CH_3CH_2CH-CHCOO^-}{|\ \ \ \ \ \ \ |}}}$

$\underset{NH}{|}$

$\underset{C_6H_5CH_2O\overset{}{C}=O}{|}$

$C_6H_5CH_2O\overset{\overset{O}{\|}}{C}NHCH_2CH_2CH_2CH_2CH\overset{\overset{O}{\|}}{C}NHCHCOO^- \xrightarrow[\underset{cold}{CH_3COOH}]{HBr}$

$\underset{\underset{C_6H_5CH_2O\overset{}{C}=O}{|}}{NH}\quad \underset{\underset{CH_2}{|}}{\overset{CHCH_3}{|}}$

$\underset{CH_3}{|}$

$2C_6H_5CH_2Br + 2CO_2 + \overset{+}{H_3}NCH_2CH_2CH_2CH_2CH\overset{\overset{O}{\|}}{C}NHCHCOO^-$

$\underset{NH_2}{|}\quad \underset{\underset{CH_2}{|}}{\overset{CHCH_3}{|}}$

$\underset{CH_3}{|}$

Lys·Ile

(b) $3C_6H_5CH_2O\overset{\overset{O}{\|}}{C}Cl$ + $H_2N\overset{\overset{NH}{\|}}{C}NHCH_2CH_2CH_2CHCOO^- \xrightarrow[25°]{OH^-}$

$\underset{NH_2}{|}$

$C_6H_5CH_2O\overset{\overset{O}{\|}}{C}NH\overset{\overset{NH}{\|}}{C}NCH_2CH_2CH_2CHCOOH \xrightarrow[(2)\ ClCOOC_2H_5]{(1)\ (C_2H_5)_3N}$

$\underset{\underset{C_6H_5CH_2O}{|}}{C=O}\qquad \underset{\underset{C=O}{|}}{NH}$

$\underset{C_6H_5CH_2O}{}$

$$C_6H_5CH_2O\overset{\overset{\text{O}}{\|}}{C}NH\overset{\overset{\text{NH}}{\|}}{C}NCH_2CH_2CH_2\overset{}{C}H\overset{\overset{\text{O}}{\|}}{C}O\overset{\overset{\text{O}}{\|}}{C}OC_2H_5 \xrightarrow[\;(-CO_2,\;-C_2H_5OH)\;]{\overset{\overset{+NH_3}{|}}{CH_3CHCOO^-}}$$

with C=O, C₆H₅CH₂O branch and NH, C=O, C₆H₅CH₂O branch

$$C_6H_5CH_2O\overset{\overset{\text{O}}{\|}}{C}NH\overset{\overset{\text{NH}}{\|}}{C}NCH_2CH_2CH_2\overset{}{C}H\overset{\overset{\text{O}}{\|}}{C}NHCHCOOH \xrightarrow[\substack{CH_3COOH \\ cold}]{HBr}$$

with C=O, C₆H₅CH₂O branch and NH, C=O, C₆H₅CH₂O branch and CH₃ branch

$$3C_6H_5CH_2Br + 3CO_2 + {}^+H_3N\overset{\overset{\text{NH}}{\|}}{C}NHCH_2CH_2CH_2CHCONHCHCOO^-$$

with NH₂ and CH₃ branches

Arg·Ala

25.13
The weakness of the benzyl-oxygen bond (p. 969) allows these groups to be removed by catalytic hydrogenolysis.

25.14
(a) An electrophic aromatic substitution reaction:

$$\text{(CH}_2\text{CH)}_n + CH_3OCH_2Cl \xrightarrow{BF_3} \text{(CH}_2\text{CH)}_n + CH_3OH$$

(with phenyl ring; product has CH₂Cl on para position)

(b) The linkage between the resin and the polypeptide is a benzylic ester. It is cleaved by HBr in CF_3COOH at room temperature because the carbocation that is formed initially is the relatively stable, benzylic cation.

25.15

$$\text{◯-CH}_2Cl + HO\overset{\overset{\text{O}}{\|}}{C}CHNH\overset{\overset{\text{O}}{\|}}{C}OC(CH_3)_3$$

with CH₃ branch

↓ base   **1 Add Boc·Ala**

$$\text{◯-CH}_2O\overset{\overset{\text{O}}{\|}}{C}CHNH\overset{\overset{\text{O}}{\|}}{C}OC(CH_3)_3$$

with CH₃ branch

2 Purify by washing

↓ CF_3COOH, CH_2Cl_2 **3 Remove protecting group**

(cont. on p. 304)

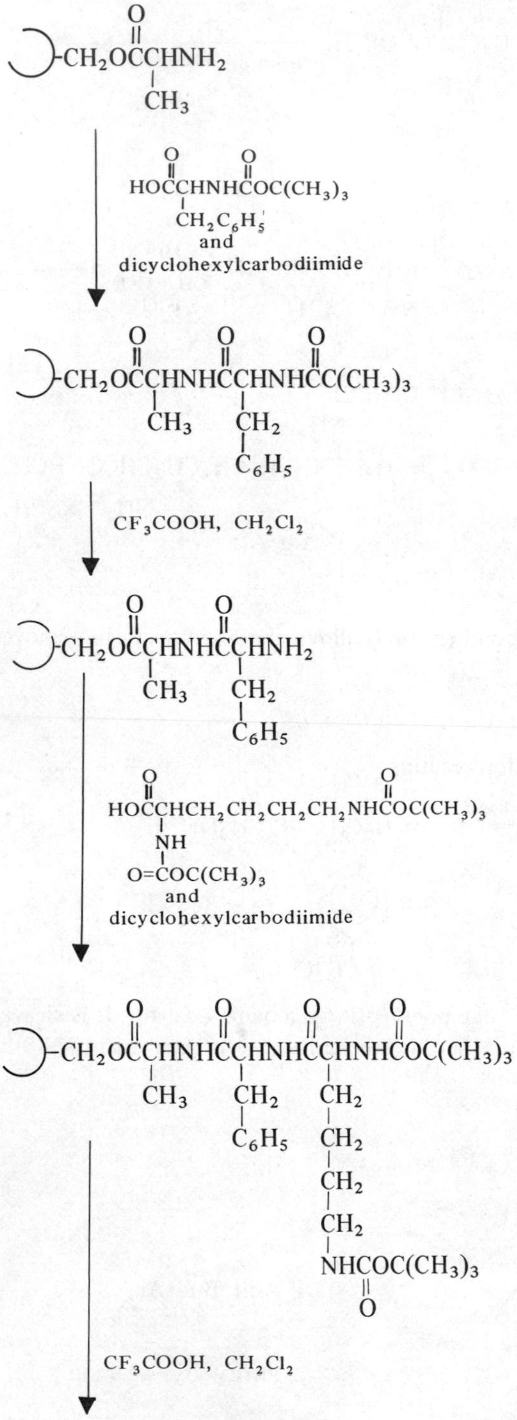

$$\bigcirc\!\!-CH_2OCCHNH_2$$

with O double bond on the OCC carbon, and CH_3 below.

4 Purify by washing

$$HOCCHNHCOC(CH_3)_3$$

with two O (double bonds), $CH_2C_6H_5$ below, and
and
dicyclohexylcarbodiimide

5 Add Boc·Phe

$$\bigcirc\!\!-CH_2OCCHNHCCHNHCC(CH_3)_3$$

with three O (double bonds), CH_3 and CH_2 below, and C_6H_5 below CH_2.

6 Purify by washing

$CF_3COOH,\ CH_2Cl_2$

7 Remove protecting group

$$\bigcirc\!\!-CH_2OCCHNHCCHNH_2$$

with two O (double bonds), CH_3 and CH_2 below, and C_6H_5 below CH_2.

8 Purify by washing

$$HOCCHCH_2CH_2CH_2CH_2NHCOC(CH_3)_3$$

with two O (double bonds), NH below, $O=COC(CH_3)_3$ below that
and
dicyclohexylcarbodiimide

9 Add Protected Lys

$$\bigcirc\!\!-CH_2OCCHNHCCHNHCCHNHCOC(CH_3)_3$$

with four O (double bonds), CH_3, CH_2, CH_2 below;
C_6H_5 below first CH_2; below second CH_2: CH_2, CH_2, CH_2, $NHCOC(CH_3)_3$ with O double bond.

10 Purify by washing

$CF_3COOH,\ CH_2Cl_2$

11 Remove protecting groups

$$\text{)}-CH_2O\overset{O}{\underset{\|}{C}}CHNH\overset{O}{\underset{\|}{C}}CHNH\overset{O}{\underset{\|}{C}}CHNH_2$$

with side chains:
CH₃ ; CH₂—C₆H₅ ; CH₂CH₂CH₂CH₂NH₂

12 Purify by washing

HBr, CF₃COOH

13 Detach tripeptide

$$\text{)}-CH_2Br \ + \ {}^-O\overset{O}{\underset{\|}{C}}CHNH\overset{O}{\underset{\|}{C}}CHNH\overset{O}{\underset{\|}{C}}CHNH_2$$

with side chains:
CH₃ ; CH₂—C₆H₅ ; CH₂CH₂CH₂CH₂NH₃⁺

14 Isolate product

Ala·Phe·Lys

25.16

(a) Isoleucine, threonine, hydroxyproline, and cystine.

(b)

$$
\begin{array}{c}
COO^- \\
H_3\overset{+}{N}\!-\!\!\!-\!H \\
CH_3\!-\!\!\!-\!H \\
CH_2 \\
CH_3
\end{array}
\quad \text{and} \quad
\begin{array}{c}
COO^- \\
H_3\overset{+}{N}\!-\!\!\!-\!H \\
H\!-\!\!\!-\!CH_3 \\
CH_2 \\
CH_3
\end{array}
$$

$$
\begin{array}{c}
COO^- \\
H_3\overset{+}{N}\!-\!\!\!-\!H \\
H\!-\!\!\!-\!OH \\
CH_3
\end{array}
\quad \text{and} \quad
\begin{array}{c}
COO^- \\
H_3\overset{+}{N}\!-\!\!\!-\!H \\
HO\!-\!\!\!-\!H \\
CH_3
\end{array}
$$

$$
\begin{array}{c}
COO^- \\
H_2\overset{+}{N}\!\!-\!\!\!-\!H \\
CH_2 \quad CH_2 \\
\diagdown H \diagup \\
OH
\end{array}
\quad \text{and} \quad
\begin{array}{c}
COO^- \\
H_2\overset{+}{N}\!\!-\!\!\!-\!H \\
CH_2 \quad CH_2 \\
\diagdown OH \diagup \\
H
\end{array}
$$

(With cystine, both chiral carbons are α-carbons, thus according to the problem, both must have the L-configuration, and no isomers of this type can be written.)

(c) Diastereomers

25.17
(a) Alanine

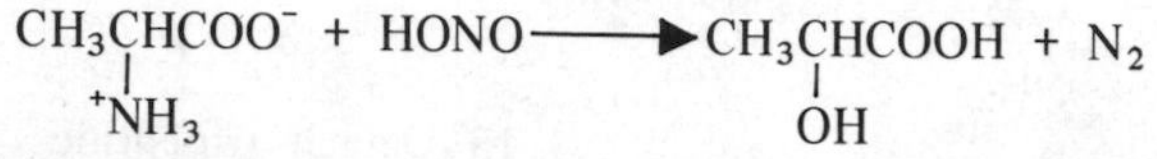

$$CH_3CHCOO^- + HONO \longrightarrow CH_3CHCOOH + N_2$$
$$\overset{|}{{}^+NH_3} \qquad\qquad\qquad \overset{|}{OH}$$

(b) Proline and hydroxyproline. All of the other amino acids have at least one primary amino group.

(c) HO —⟨benzene ring, Br above and Br below⟩— CH_2CHCOO^-
$$\overset{|}{NH_3{}^+}$$

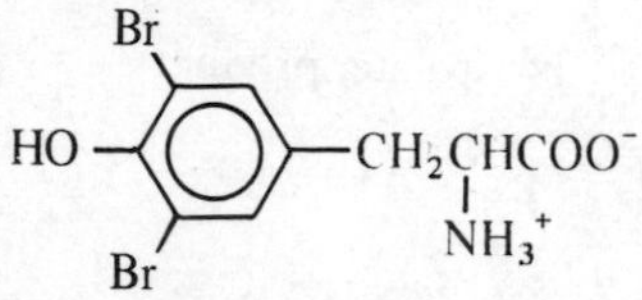

(d) ⟨benzene ring⟩— $CH_2CHCOOC_2H_5$
$$\overset{|}{NH_3{}^+}$$

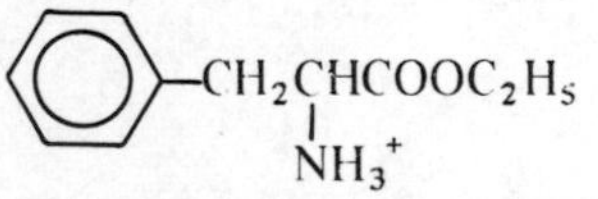

(e) CH_3CHCOO^-
$\overset{|}{NH}$
$\overset{|}{C=O}$
$\overset{|}{C_6H_5}$

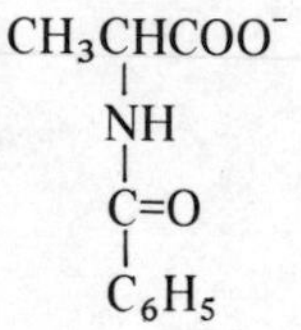

25.18
(a)

$$\overset{COO^-}{\underset{CH_2OH}{{}^+H_3N{-}{\mid}{-}H}} \xrightarrow[CH_3OH]{HCl} \overset{COOCH_3}{\underset{CH_2OH}{{}^+H_3N{-}{\mid}{-}H}} \quad Cl^-$$

(−)-Serine   A
         $(C_4H_{10}ClNO_3)$

$$\xrightarrow{PCl_5} \overset{COOCH_3}{\underset{CH_2Cl}{{}^+H_3N{-}{\mid}{-}H}} \quad Cl^- \xrightarrow[(2)\ OH^-]{(1)\ H_3O^+,\ H_2O,\ heat}$$

B
$(C_4H_9Cl_2NO_2)$

$$\overset{COO^-}{\underset{CH_2Cl}{{}^+H_3N{-}{\mid}{-}H}} \xrightarrow[dil.\ H_3O^+]{Na-Hg} \overset{COO^-}{\underset{CH_3}{{}^+H_3N{-}{\mid}{-}H}}$$

C   L-(+)-Alanine
$(C_3H_6ClNO_2)$

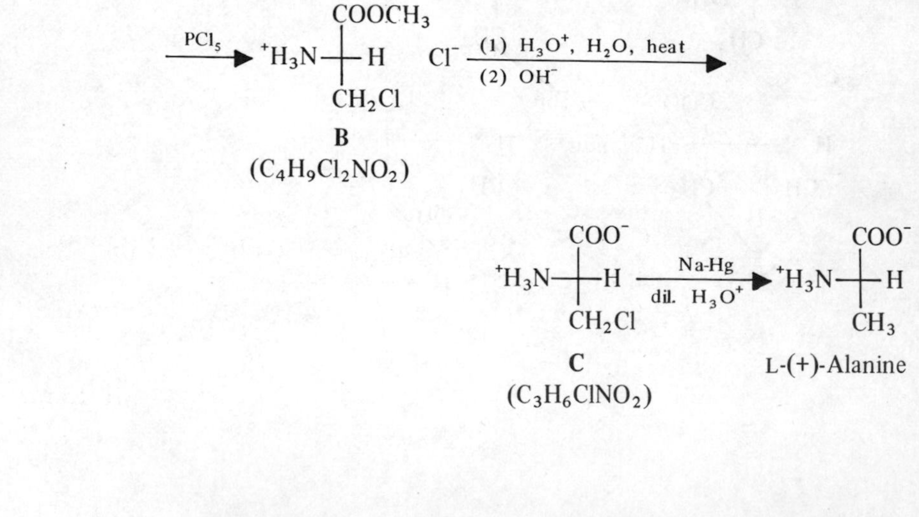

(b)

$$B \xrightarrow{OH^-} H_2N-\overset{\displaystyle COOCH_3}{\underset{\displaystyle CH_2Cl}{\overset{|}{\underset{|}{C}}}}H \xrightarrow{NaSH} H_2N-\overset{\displaystyle COOCH_3}{\underset{\displaystyle CH_2SH}{\overset{|}{\underset{|}{C}}}}H$$

$$\begin{array}{cc} \mathbf{D} & \mathbf{E} \\ (C_4H_8ClNO_2) & (C_4H_9NO_2S) \end{array}$$

$$\xrightarrow[\text{(2) } OH^-]{\text{(1) } H_3O^+,\ H_2O,\ \text{heat}} \ {}^+H_3N-\overset{\displaystyle COO^-}{\underset{\displaystyle CH_2SH}{\overset{|}{\underset{|}{C}}}}H$$

L-(+)-Cysteine

(c)

$$H_3\overset{+}{N}-\overset{\displaystyle COO^-}{\underset{\displaystyle CH_2\overset{O}{\overset{||}{C}}NH_2}{\overset{|}{\underset{|}{C}}}}H \xrightarrow{NaOBr,\ OH^-} H_2N-\overset{\displaystyle COO^-}{\underset{\displaystyle CH_2NH_2}{\overset{|}{\underset{|}{C}}}}H$$

$$\begin{array}{cc} \text{L-Asparagine} & \mathbf{F} \\ & (C_3H_7N_2O_2) \end{array}$$

$${}^+H_3N-\overset{\displaystyle COO^-}{\underset{\displaystyle CH_2Cl}{\overset{|}{\underset{|}{C}}}}H \xrightarrow{} NH_3$$

$$\mathbf{C}$$

(from part A)

25.19

(a) $CH_3\overset{O}{\overset{||}{C}}NHCH(CO_2C_2H_5)_2 + CH_2=CH-C\equiv N \xrightarrow[C_2H_5OH]{NaOC_2H_5}$

$$CH_3\overset{O}{\overset{||}{C}}NH-\overset{\displaystyle CO_2C_2H_5}{\underset{\displaystyle CO_2C_2H_5}{\overset{|}{\underset{|}{C}}}}-CH_2CH_2C\equiv N \xrightarrow[\text{reflux}]{\text{conc HCl}}$$

$$\mathbf{G}$$

$$HOOCCH_2CH_2\underset{\displaystyle \overset{|}{NH_3^+}}{CHCOO^-} + CH_3COOH + 2C_2H_5OH + NH_4^+ + CO_2$$

DL-Glutamic acid

(b)

$$CH_3\overset{O}{\overset{||}{C}}NH-\overset{\displaystyle CO_2C_2H_5}{\underset{\displaystyle CO_2C_2H_5}{\overset{|}{\underset{|}{C}}}}-CH_2CH_2C\equiv N \xrightarrow[68°,\ 1000\ psi]{H_2(Ni)}$$

$$\left[CH_3\overset{O}{\overset{||}{C}}NH-\overset{\displaystyle CO_2C_2H_5}{\underset{\displaystyle CO_2C_2H_5}{\overset{|}{\underset{|}{C}}}}-CH_2CH_2CH_2NH_2 \right] \xrightarrow{-C_2H_5OH}$$

$$\begin{array}{c} C_2H_5O_2C \diagdown \quad \diagup CH_2-CH_2 \\ C \diagdown \quad \diagup CH_2 \\ CH_3\overset{|}{\underset{||}{\overset{}{C}}}NH \diagup \quad C-NH \\ \overset{||}{O} \qquad \overset{||}{O} \end{array}$$

$$\mathbf{H}$$

(cont. on p. 308)

$$\xrightarrow[\text{reflux}]{\text{conc. HCl}} \overset{+}{H_3}NCH_2CH_2CH_2CHCOO^- + CH_3COOH + CO_2 + C_2H_5OH$$

$$Cl^- \qquad\qquad \underset{NH_3^+}{|}$$

DL-Ornithine hydrochloride

25.20

(a)
$$C_6H_5CH_2\overset{O}{\overset{||}{C}}H \xrightarrow[\text{HCN}]{NH_3} C_6H_5CH_2\underset{\underset{NH_2}{|}}{C}HC\equiv N \xrightarrow{H_3O^+} C_6H_5CH_2\underset{\underset{NH_3^+}{|}}{C}HCOO^-$$

Phenyl acetaldehyde DL-Phenylalanine

(b)
$$CH_3SH + CH_2=CH-\overset{O}{\overset{||}{C}}H \xrightarrow{\text{base}} CH_3SCH_2CH_2\overset{O}{\overset{||}{C}}H$$

$$\xrightarrow[\text{HCN}]{NH_3} CH_3SCH_2CH_2\underset{\underset{NH_2}{|}}{C}HC\equiv N \xrightarrow{H_3O^+} CH_3SCH_2CH_2\underset{\underset{NH_3^+}{|}}{C}HCOO^-$$

DL-Methionine

25.21

$$C_6H_5CH_2\underset{\underset{NH_3^+}{|}}{C}HCOO^- + HOOCCH_2CH_2\underset{\underset{O}{||}}{C}COOH \xrightleftharpoons{\text{transaminase}}$$

Phenylalanine α-Ketoglutaric acid

$$C_6H_5CH_2\underset{\underset{O}{||}}{C}COOH + HOOCCH_2CH_2\underset{\underset{NH_3^+}{|}}{C}HCOO^-$$

Phenylpyruvic acid Glutamic acid

Then:

$$HOOCCH_2CH_2\underset{\underset{NH_3^+}{|}}{C}HCOO^- + HOOCCH_2\underset{\underset{O}{||}}{C}COOH \xrightleftharpoons{\text{transaminase}}$$

Glutamic acid Oxaloacetic acid

$$HOOCCH_2CH_2\underset{\underset{O}{||}}{C}COOH + HOOCCH_2\underset{\underset{NH_3^+}{|}}{C}HCOO^-$$

α-Ketoglutaric acid Aspartic acid

This amounts to:

Phenylalanine + α-Ketoglutaric acid $\rightleftharpoons$ Phenylpyruvic acid + Glutamic acid
Glutamic acid + Oxaloacetic acid $\rightleftharpoons$ α-Ketoglutaric acid + Aspartic acid

Net: Phenylalanine + Oxaloacetic acid $\rightleftharpoons$ Phenylpyruvic acid + Aspartic acid

25.22
We look for points of overlap:

```
                              Phe · Ser
                 Pro · Gly · Phe
           Pro · Pro                  Ser · Pro · Phe
     Arg · Pro                              Phe · Arg
     Arg · Pro · Pro · Gly · Phe · Ser · Pro · Phe · Arg
```

Bradykinin

25.23

1. This shows that valine is the N-terminal amino acid and that valine is attached to leucine. (Lysine labeled at the ε-amino group is to be expected if lysine is not the N-terminal amino acid and if it is linked in the polypeptide through its α-amino group.)

2. This shows that alanine is the C-terminal amino acid and that it is linked to glutamic acid.

At this point, then, we have the following information about the structure of the heptapeptide.

Val · Leu (Ala, Lys, Phe) Glu · Ala

the sequence here is
unknown

3. (a) This shows that the dipeptide, **A**, is

Leu · Lys

(b) The carboxypeptidase reaction shows that the C-terminal amino acid of the tripeptide, **B**, is glutamic acid; the DNP labeling experiment shows that the N-terminal amino acid is phenylalanine. Thus the tripeptide **B** is:

Phe · Ala · Glu

Putting these pieces together in the only way possible, we arrive at the following amino acid sequence for the heptapeptide.

```
Val · Leu
      Leu · Lys
                  Phe · Ala · Glu
                              Glu · Ala
Val · Leu · Lys · Phe · Ala · Glu · Ala
```

25.24

At pH 2-3 the γ-carboxyl groups of polyglutamic acid are uncharged (they are present as −COOH groups). At pH 5 the γ-carboxyl groups ionize and become negatively charged (they become γ-COO⁻ groups). The repulsive forces between these negatively charged groups cause an unwinding of the α-helix and the formation of a random coil.

26 SPECIAL TOPICS V NUCLEIC ACIDS: PROTEIN SYNTHESIS

26.1
Adenine:

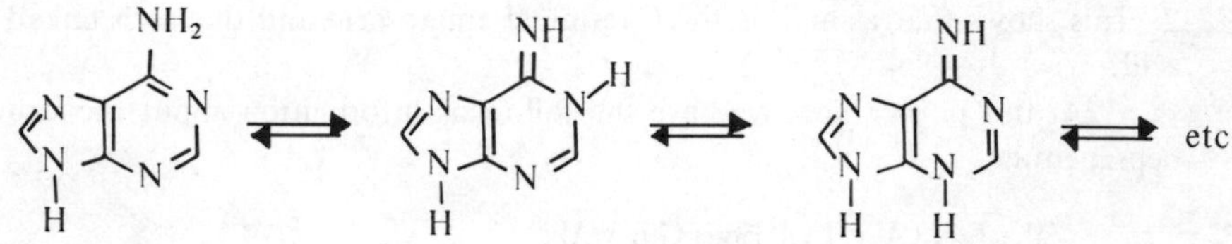

Guanine:

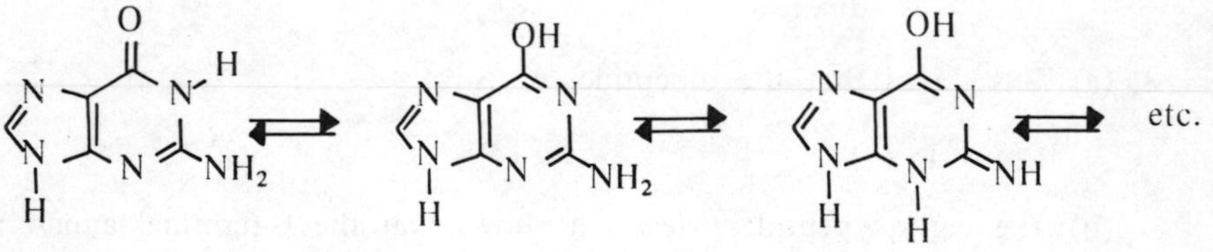

Cytosine:

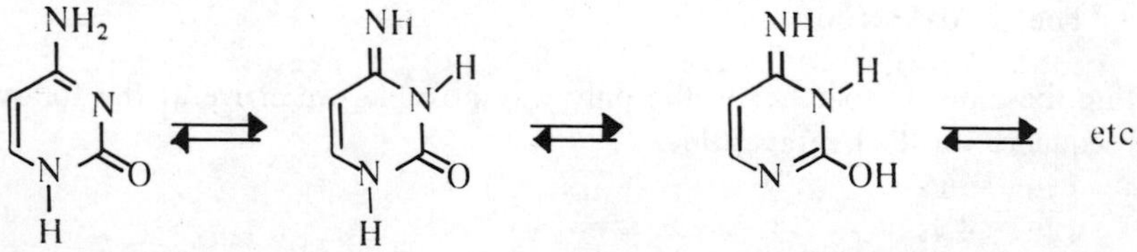

Thymine (R = CH₃) or Uracil (R = H):

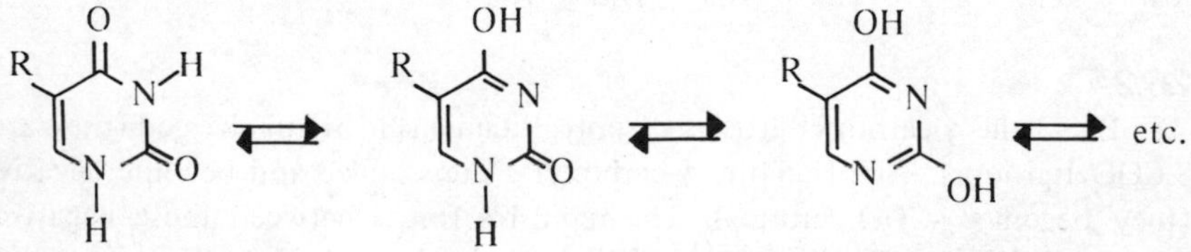

26.2
(a) The nucleosides have an N-glycosidic linkage that (like an O-glycosidic linkage) is rapidly hydrolyzed by aqueous acid but one that is stable in aqueous base.

310

(b)

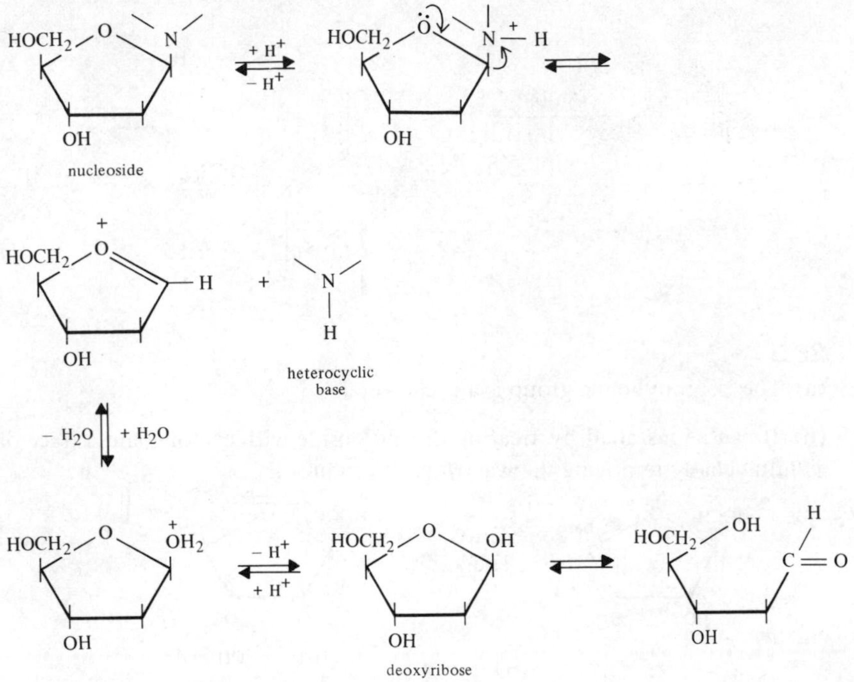

nucleoside

heterocyclic
base

deoxyribose

26.3
The reaction appears to take place through an S_N2 mechanism. Attack occurs preferentially at the primary 5′-carbon rather than at the secondary 3′-carbon.

26.4

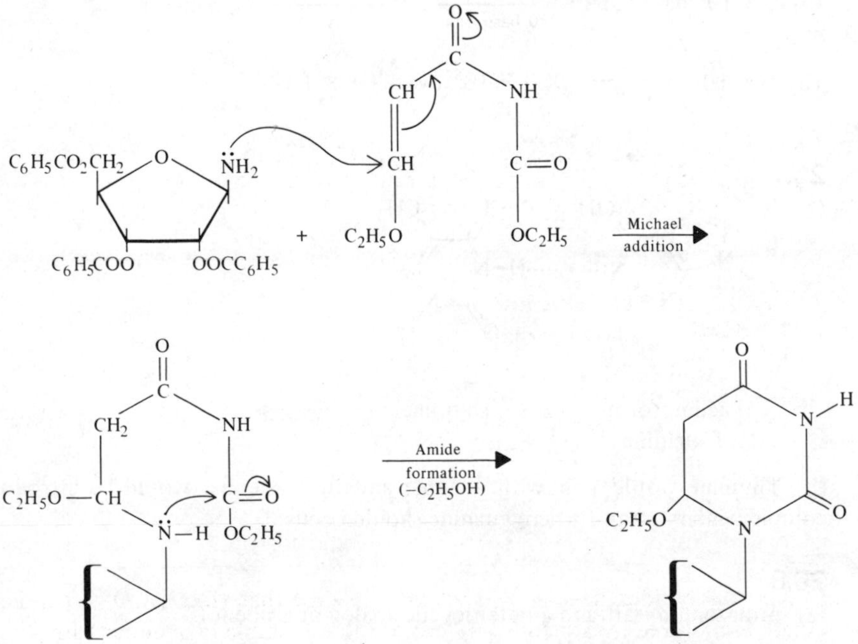

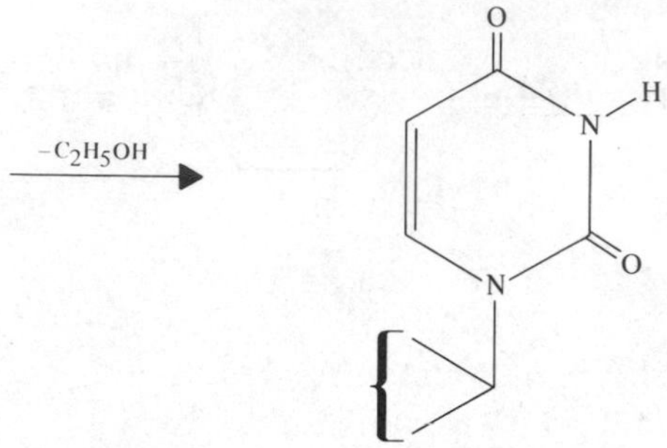

$-C_2H_5OH \longrightarrow$

26.5

(a) The isopropylidene group is a cyclic ketal.

(b) It can be installed by treating the nucleoside with acetone and a trace of acid and by simultaneously removing the water that is produced.

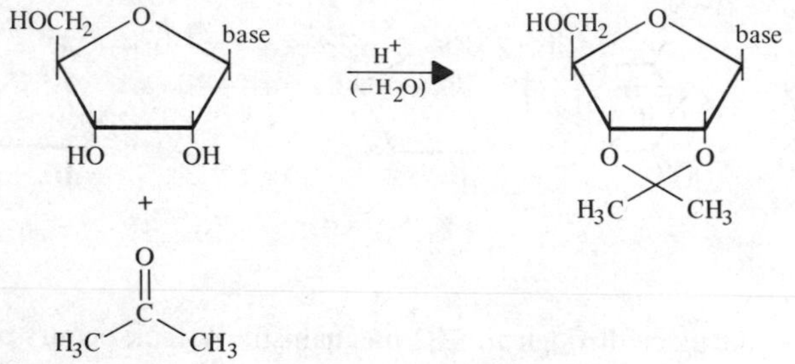

26.6

(a) 6×10^9 base pairs $\times \dfrac{34\text{Å}}{10 \text{ base pairs}} \times \dfrac{10^{-10} \text{ meters}}{\text{Å}} \cong 2 \text{ meters}$

(b) $6 \times 10^{-12} \dfrac{\text{g}}{\text{ovum}} \times 3 \times 10^9 \text{ ova} = 1.8 \times 10^{-2} \text{g}$

26.7

(a)

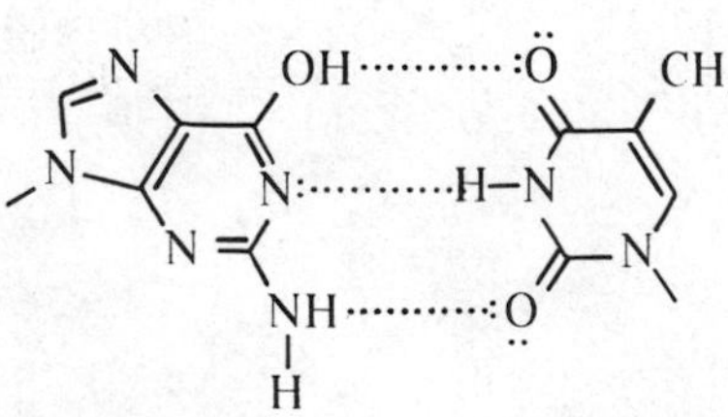

Lactim form   Thymine
of guanine

(b) Thymine would pair with adenine and thus adenine would be introduced into the complementary strand where guanine should occur.

26.8

(a) A diazonium salt and a heterocyclic analog of a phenol.

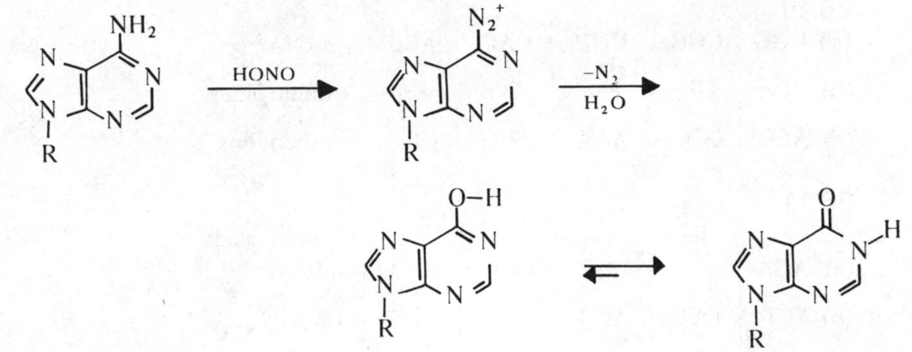
HONO
$-N_2$
H_2O

Hypoxanthine
nucleotide

(b)

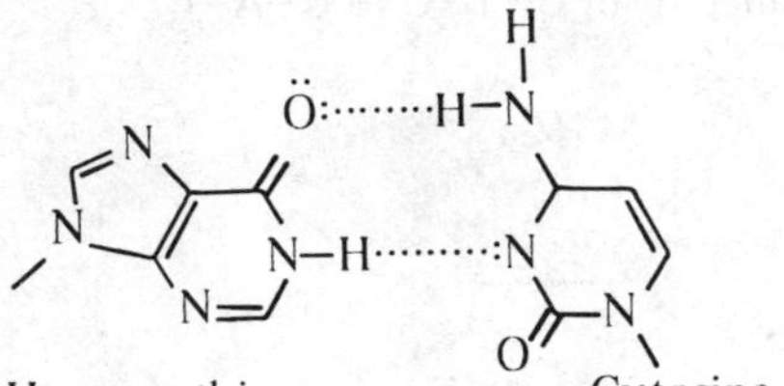

Hypoxanthine Cytosine

(c) Original double strand

First replication

Second replication

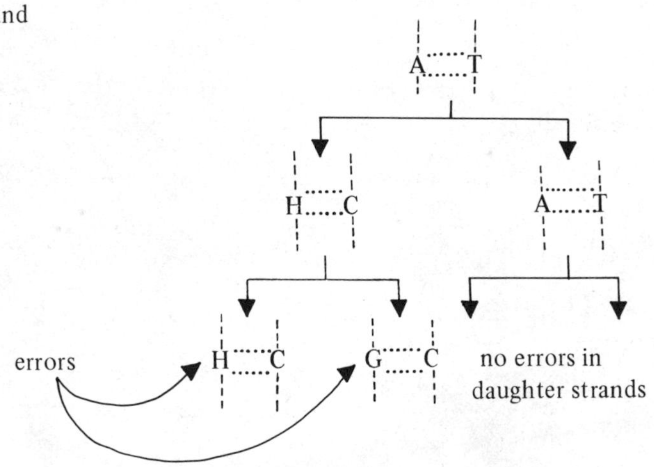
errors
no errors in
daughter strands

26.9

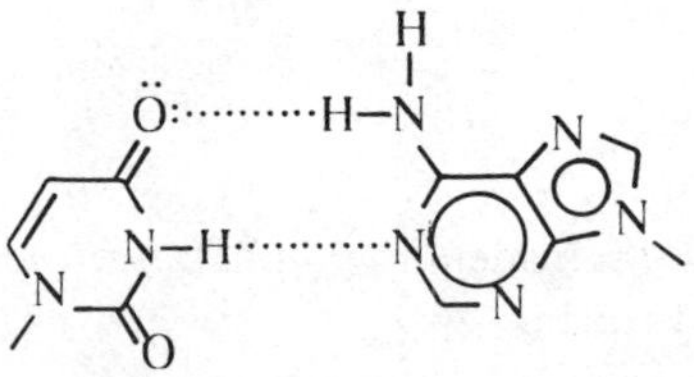

Uracil
(in mRNA)

Adenine
(in DNA)

26.10

(a) UGG ¦ GGG ¦ UUU ¦ UAC ¦ AGC   mRNA

(b) Tyr ¦ Gly ¦ Phe ¦ Tyr ¦ Ser   Amino acids

(c) ACC ¦ CCC ¦ AAA ¦ AUG ¦ UCG   Anticodons

26.11

    Arg · Ile · Cys · Tyr · Val   Amino acids

(a) AGA ¦ AUA ¦ UGC ¦ UGG ¦ GUA ¦ mRNA

(b) TCT ¦ TAT ¦ ACG ¦ ACC ¦ CAT ¦ DNA

(c) UCU ¦ UAU ¦ ACG ¦ ACC ¦ CAU ¦ anticodons

26.12

A change from C—T—T to C—A—T or a change from C—T—C to C—A—C.